Creating the International

Springer
London
Berlin
Heidelberg
New York
Barcelona
Hong Kong
Milan
Paris
Santa Clara
Singapore
Tokyo

3/03

David M. Harland and John E. Catchpole

Creating the
International Space
Station

Springer

Published in association with
Praxis Publishing
Chichester, UK

PRAXIS

David M. Harland
Kelvinbridge
Glasgow
UK

John E. Catchpole
Basingstoke
Hampshire
UK

SPRINGER–PRAXIS BOOKS IN ASTRONOMY AND SPACE SCIENCES
SUBJECT *ADVISORY EDITOR*: John Mason B.Sc., Ph.D.

ISBN 1-85233-202-6 Springer-Verlag Berlin Heidelberg New York

British Library Cataloguing-in-Publication Data
Harland, David M.
   Creating the International Space Station. –
   (Springer-Praxis books in astronomy and space sciences)
   1. Space stations – International cooperation
   I. Title   II. Catchpole, John, 1957–
   629.4′42

   ISBN 1-85233-202-6

Library of Congress Cataloging-in-Publication Data
Harland, David M.
   Creating the International Space Station/David M. Harland, John E. Catchpole.
      p. cm. – (Springer-Praxis books in astronomy and space sciences; v.4138)
   Includes index.
   ISBN 1-85233-202-6 (alk. paper)
   1. International Space Station. I. Catchpole, John, 1957–   II. Title.   III. Series.

TL797.H37 2002
629.44′–dc21

2001057665

Printed by MPG Books Ltd, Bodmin, Cornwall, UK

Project Copy Editor: Alex Whyte
Cover design: Jim Wilkie
Typesetting: BookEns Ltd, Royston, Herts., UK

Printed on acid-free paper supplied by Precision Publishing Papers Ltd, UK

*"We can lick gravity, but
sometimes the paperwork is overwhelming."*

Wernher von Braun

"The Earth is the cradle of the mind,
but you cannot live in the cradle forever."
Konstantin Tsiolkovsky

"So many worlds, so much to do ..."
Alfred, Lord Tennyson

"Science, like life, feeds on its own decay."
William James

"The dream of yesterday is the hope of today,
and the reality of tomorrow."
Robert Goddard

"Tonight I am challenging NASA to build a space station
and to do it within ten years ..."
Ronald Reagan

"Now is the time to send out the fur trappers and 49ers,
the pioneers and Conestoga wagons;
we need to build the trading posts, string the telegraph wires,
and lay down the railroad tracks..."
Dan Goldin

"Leadership abandoned is leadership lost."
Richard Kohrs, Director, Space Station Freedom

"It's easy to stand back and throw rocks,
but we're going to deliver the space station on time."
Dan Goldin

Table of Contents

List of illustrations

List of Tables

Authors' preface

Since it became evident that rockets would soon be able to launch people into space, one theme has dominated thinking, the concept of a 'space station'. In 1952, in the articles in the popular New York magazine *Collier's* in which artist Chesley Bonestell illustrated a detailed design proposed by Wernher von Braun, the idea was given form – as a giant 'wheel' in space, slowly rotating to generate 'artificial' gravity for its several thousand occupants.

Chesley Bonestell's classic 1952 depiction of a space station as a giant 'wheel' in space rotating to generate 'artificial' gravity for its thousands of inhabitants. Half a century later, however, the International Space Station is a very different architecture.

Although at the height of the Cold War the United States and the Soviet Union became distracted in the 1960s by a race to the Moon, both were laying the basis for space stations. In the 1970s, after the United States had reached the Moon and used its Apollo spacecraft technology to operate the Skylab station, it switched its effort to a re-usable Space Shuttle which would provide routine access to low orbit. During this protracted development, it was popularly expected, thanks largely to the efforts of Princeton physicist Gerard K. O'Neill, that the Shuttle would assemble structures so large that they would constitute 'colonies' in orbit, and that by the end of the century these would house thousands of people – entire families, with the children having been born in space. Meanwhile, having so spectacularly lost the race to the Moon that it was able to claim that it had never even tried, the Soviet Union focused on the development of a succession of ever more capable stations. In 1984, with the Shuttle finally in service, President Ronald Reagan ordered NASA to build a station within a decade ... and therein lies a story.

David M Harland
Kelvinbridge, Glasgow
and
John E Catchpole
Basingstoke

December 2001

Acknowledgements

We would like to thank, in no particular order, Roger Launius, Chief of NASA's History Office; Neville Kidger; Mike Gentry of the Johnson Space Center's Media Services; David Portree; Jim Oberg; David Woods; Kipp Teague; Ken Glover; Mike Hawes, Deputy Associate Administrator for Space Development (Space Station); Bill Readdy, Deputy Associate Administrator for the Office of Space Flight; Kathy Clark, ISS Chief Scientist at NASA HQ; Ed Hengeveld; Bob Dempsey of Command and Data Handling at NASA Johnson Space Center; Scott Benson of the Power and Propulsion Office at NASA Glenn Research Center; Dave Woolard of Pictures on the Wall; Patrick Moore; Brian Harvey; Stuart Clark; Dietrich Haeseler; Vladimir Semenov of Videocosmos; David Rickman; and the late Dan Gauthier. And, as usual Clive Horwood of Praxis is to be thanked for his boundless enthusiasm.

1

Apollo Applications

AMERICA ENTERS THE SPACE AGE

Within weeks of being formed on 1 October 1958, the National Aeronautics and Space Administration had established 'Project Mercury' to put the first human being into space. A few months later, the US Congress issued a report entitled 'The Next Ten Years In Space', in which it said that as soon as it was shown that a human being could be safely transported in space, the next logical step would be to develop a space station.

In February 1959, NASA began to lobby for a space station and missions to the Moon. Later that year, it formed the Research Steering Committee on Manned Space Flight, which ranked a station ahead of a lunar mission. Given a 20-year programme of methodical development, Wernher von Braun's 1952 visionary strategy of using a space station as a stepping stone for the Moon might have been pursued, but when President John F. Kennedy laid down the gauntlet in May 1961 for a lunar landing within that decade this 'logical' plan was short-circuited. Nevertheless, at the George C. Marshall Space Flight Center in Huntsville, Alabama, where von Braun was director, the accepted wisdom was that a flight to the Moon would require the assembly of the lunar spacecraft in Earth orbit, so it was argued that a small station would assist with the assembly of this composite vehicle. With the decision on 11 July 1962 to use a 'lunar orbit rendezvous' strategy, this assembly point in low orbit was rendered obsolete.

By mid-1963, in recognition of the fact that the case for a space station was ultimately independent of the Apollo lunar programme, NASA set out to identify goals that would *require* the development of a space station. Defining the station's role was important, because this would determine which part of the agency would 'receive' the programme – the Washington HQ defined the programmes and oversaw them, but they were managed by (and jobs went to) the field centres, which were in competition with one another. Apollo had been split nicely between the 'big three', with Huntsville managing the development of the rocketry, the Manned Spacecraft Center[1] in Houston, Texas, building and operating the

---

[1] In March 1973 the Manned Spacecraft Center was renamed the Johnson Space Center.

spacecraft and the Kennedy Space Center[2] in Florida dealing with launch operations.

As the Apollo lunar programme would involve flights lasting up to 14 days, the primary objective of the space station planners was to ensure that astronauts could live and work in space for much longer periods. The rule of thumb for extending mission endurance was to double the previous record, so they set their proving flight at a 28-day mission. From this, it was possible to define the systems that the station would need to support a crew. Once the systems had been 'certified', they would be incorporated into spacecraft that would venture into deep space to establish lunar bases and, ultimately, fly to Mars. The goal of this initial space station was, therefore, to set the scene for the long-duration and far-reaching missions that would follow Apollo's sprint to the Moon. This basic space station would also, of course, open the way to a more sophisticated orbital facility that would be able to undertake a variety of scientific investigations and not just focus on how people and machinery withstood the space environment.

However, in August 1963, Joe Shea, the Apollo programme manager in Houston, questioned whether it was wise to pursue a two-stage strategy and thought it would be better to aim initially for a multi-role station. Although configurations that could be served by either Gemini or Apollo spacecraft had been studied, the use of Gemini would enable the station to be developed sooner, albeit as a more limited facility. Apollo, on the other hand, offered the prospect of installing scientific apparatus in an otherwise vacant bay in the service module, and although this would obviate the need for a host station, such a mission would be limited to the Apollo spacecraft's two-week endurance, as required to satisfy lunar mission requirements. The 'Extended Apollo' study by Houston in 1963 evaluated three options: (1) an Apollo link up with a module possessing its own power and environmental systems; (2) a module that was dependent upon the Apollo spacecraft for support; and (3) a two-man version of the Apollo spacecraft fitted with solar panels instead of fuel cells, a mixed-gas atmosphere instead of pure oxygen, and a regenerative carbon dioxide filtration system for month-long missions. In each case, however, the scientific theme was to be human adaptation to weightlessness. As each of these missions could be launched using the Saturn IB rocket, they would not have to await the development of the mighty Saturn V 'moon rocket'. The 'dependent' module would be able to be launched together with its crew, because it would be carried in the adapter atop the Saturn IB's second stage, the S-IV-B, and be extracted using the transposition manoeuvre that was to be used to retrieve a lunar module *en route* to the Moon. The 'independent' module, being more substantive, would be launched separately and the crew would rendezvous with it. The study's recommendation was for a dependent module outfitted with biomedical apparatus to investigate the effects of exposure to weightlessness during a series of flights that would extend the endurance record to 120 days. Of course, orbital rendezvous technique would have to be developed as a preliminary to any scheme in which the station would be launched separately from its crew.

---

[2] Upon the death of President Kennedy, NASA's launch facility at Cape Canaveral in Florida was renamed Cape Kennedy, and hosted the newly created Kennedy Space Center, and in 1973, upon the completion of the Apollo programme, the cape reverted to its old name.

The Houston study had focused on options for a 'small' station; other teams had looked at larger configurations. In late 1962 the Langley Research Center in Virginia had conducted a 'medium' study that considered utilising Gemini spacecraft as 'taxis' to 'rotate' a succession of crews over a one-year period through a laboratory module that could be launched on either a Saturn IB or a Titan III. The 'large' station (appropriately called 'Olympus') was studied by Houston by way of parallel contracts awarded to both Douglas Aircraft and Lockheed. It was assumed that the Saturn V would be available, that the station would be capable of five years of habitability, and that the crew would eventually rise to 24 people, therefore it was implicit that this study had also to make recommendations on the types of spacecraft that would be required to replenish such an orbital facility. The results were published in 1964. Lockheed proposed a Y-shaped station that would rotate to provide a degree of 'artificial' gravity to minimise any problem that might arise from prolonged exposure to weightlessness. Douglas chose a zero-gravity solution in the form of a large cylinder incorporating a 'hangar' for servicing visiting spacecraft.

By 1964, therefore, while NASA had no shortage of options for space station development, it lacked the financial commitment to such a programme. Although timescale was not an issue in the absence of a commitment, time was nevertheless a factor: the sooner the groundwork was accomplished the more rapid would be the pace once the space station was approved.

THE 'SPENT STAGE' CONCEPT

'Extended Apollo' remained in 'advanced study' status until 6 August 1965, when George Mueller, a TRW vice-president, was recruited by NASA administrator James Webb as associate administrator for manned spaceflight to head up the newly established Apollo Applications Program Office in Washington. At this time, the plan's status was upgraded to 'project definition' in order to initiate preparations to fly the '28-day mission' as soon as it became clear that Apollo would be able to make a lunar landing within Kennedy's deadline.

Mueller's dilemma was complex: once the development of the Apollo spacecraft and its Saturn V launch vehicle was complete, and the flight items had been manufactured, the agency's industrial base would be redundant. As he saw it, he had to ensure that Apollo Applications provided continuity. However, NASA's budget had already peaked. He was particularly concerned for Huntsville, the largest field centre, because there was no prospect of its being assigned to design a new rocket to supersede the Saturn V. From Mueller's point of view, therefore, it was crucial that whatever Apollo Applications did, it would have to generate work for von Braun's rocket group. The studies undertaken by Houston and Langley – both of which were involved in spacecraft design – had focused on exploiting Gemini and Apollo spacecraft, and when Huntsville became involved it dusted off a plan to convert a rocket's 'spent stage' into a small space station.

When Douglas Aircraft – which was manufacturing the S-IV upper stage for an early form of the Saturn rocket – switched to the more advanced S-IV-B that was to be compatible with both the Saturn IB and Saturn V, it proposed modifying a surplus S-IV stage to serve as an 'enclosure' within which astronauts could assess 'spacewalking' tasks in safety. In fact, Douglas worked closely with von Braun's team and the study, which was formally reported in November 1962, was the result of discussions dating back to 1960. However, this was the first time that the study had assessed the adaptation of a specific rocket stage. From Douglas's point of view, the plan offered a way of exploiting previous development work and, indeed, the value of the S-IV to Huntsville was that it was already available. In the Douglas study, the S-IV would be launched as a 'live' stage which would place itself into orbit. A Gemini spacecraft would rendezvous and link up with it, utilising the docking system that was already in development for the much smaller Agena stage for use as a rendezvous and docking target in the mainstream Gemini programme. The docking system would be mounted on an interface module on the front of the S-IV. This module would also serve as an airlock allowing the crew to access the hydrogen tank after it had been purged, provide the oxygen to repressurise the tank once the vent valves had been sealed, carry the apparatus for the experimental programme, and deploy a solar panel to provide power. In an effort to broaden the utility of such a vehicle, Huntsville hired North American Aviation to assess likely missions, and its report was delivered in April 1965. In addition to refuelling spacecraft intended for deep space missions (an activity which reinstated the service platform that had been integral to the 'Earth orbit rendezvous' strategy for Apollo), the report suggested clustering stages equipped for specialised tasks around a central hub in order to create a 'modular' space station whose configuration could be modified to suit a dynamic programme of research. With the creation of the Apollo Applications Program Office a few months later, Huntsville updated its S-IV plan to use the larger S-IV-B, called it the 'wet workshop', and submitted it as one of three options to Mueller. In the 'minimalist' option, the stage was to be used as an unpressurised volume in which to evaluate an astronaut's ability to do productive work in weightlessness. In relation to this option, it is important to realise that spacecraft had previously been so cramped that, with the exception of Ed White's brief excursion from Gemini 4 in June 1965, NASA had had no experience of astronauts working in a 'free float' state, and an unpressurised enclosure offered an opportunity to evaluate tools and procedures envisaged for more advanced missions. The 'intermediate' option suggested that the tank be pressurised with a mixed-gas atmosphere, enabling the astronauts to work in a 'shirt-sleeve' environment, but utilising only the experimental apparatus that could be delivered by the Apollo. The interface module would serve as an airlock, allowing the astronauts to conduct external operations. Because it would depend upon the Apollo for attitude control, the utilisation of this facility would be limited by that vehicle's endurance. In the 'top of the line' option, the stage would be responsible for attitude control, thus permitting a longer mission, and a suite of experiments would be delivered in the interface module, which would deploy a solar panel for power. With typical foresight, von Braun had ordered the access cover at the top of the S-IV-B's hydrogen tank to have a quick-release hatch "to assist with checkout" (and, of course, the fact that this was wide enough for a space-suited astronaut to pass through was purely fortuitous). On 1 December 1965,

Mueller announced that Apollo Applications had selected Huntsville to adapt an S-IV-B as an 'orbital workshop' for the 'intermediate' option.

Once unleashed, Huntsville made rapid advancement with the project's definition by exploiting existing systems to minimise costs. Since the crew was to be launched together with the stage, it was decided to use the Apollo spacecraft (the Gemini spacecraft was configured for a Titan II) and fit the interface module's airlock with an Apollo docking system as this incorporated a pressurised tunnel. However, in view of the fact that the spent stage would have to be controlled in orbit by the Apollo spacecraft, on 11 February Robert Gilruth, the Manned Spacecraft Center's director, argued for the project to be transferred to Houston. Recognising that Houston had a close relationship with McDonnell – the manufacturer of the Gemini spacecraft whose environmental systems were to be utilised in the airlock module – von Braun transferred the development of the interface module to Houston, with the proviso that it would be delivered to Huntsville for vehicle integration. This, however, would prove to be only the first of a series of skirmishes for overall control of the project.

The use of a spent stage to make a start on studying the biomedical aspects of prolonged weightlessness was just one aspect of what Mueller hoped would develop into a multi-faceted programme whose scope would make the initial Apollo missions seem tame in comparison. Once the Saturn V became available, he intended to develop a larger follow-up workshop with scientific missions in low orbit, geostationary orbit and, later, polar lunar orbit as a preliminary to a series of advanced lunar missions that would establish bases for exploration. In the event, nothing came of the plans for geostationary and lunar missions, although some of the instruments that were developed to make scientific observations eventually proved successful.

In March 1966, Gilruth was critical of Apollo Applications, saying that it had defined its goals by what *could* be done using existing technology. It was counterproductive, he said, to set out to 'apply' Apollo technology as this would retard possible developments in spacecraft design. It would be better, he said, to undertake tasks that *should* be done in order to assist future programmes, and in the process design and develop *new* technologies and vehicles. If the ultimate goal was interplanetary exploration, then the space station ought to address the development of systems designed for such journeys. Of course, Gilruth was viewing the future from Houston's perspective, realising that after Apollo *it* would need a contract to develop a new spacecraft. A 'laboratory in the sky' would keep the astronauts and flight controllers occupied, but Houston's engineering base would be idle. Not surprisingly, however, Mueller exploited the independence of his office and lobbied for funding to launch his workshop in mid-1968, to gain experience of long-duration spaceflight. The fact that this mission would be conducted in parallel with the impending Apollo lunar landing was not seen as a problem.

AIR FORCE PLANS

NASA was not alone in seeking Congressional funding for a space station. In early 1962, the Air Force began to investigate whether NASA's Gemini spacecraft could

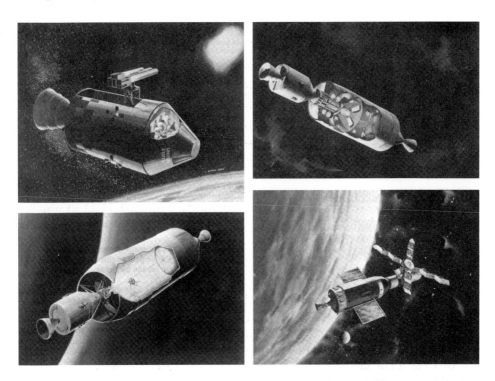

One early Apollo Applications concept was to mount a battery of instruments in a vacant bay of the Apollo spacecraft's service module (upper left). The first plan for using a 'spent stage' was simply to mount an airlock on its nose to accept an Apollo so that astronauts could rehearse EVA procedures in the safety of the unpressurised tank (lower left). Next, it was suggested that the tank be pressurised to provide a shirt-sleeved environment and outfitted to serve as an orbital laboratory (upper right). The incorporation of a Multiple Docking Adapter provided the basis for the cluster concept so that specialised modules (such as a solar telescope) could be added in space (lower right).

service an orbital reconnaissance platform. In fact, the Air Force already had a 'space plane' under development, but this 'DynaSoar' was regarded as a long-term venture and the Air Force was eager to stake its claim to space. When Robert McNamara, Kennedy's Secretary of Defense, authorised the Manned Orbiting Laboratory (MOL) on 10 December 1963 he also cancelled this ambitious spacecraft, so the decision to pursue the MOL represented a significant trade-off for the military. The Titan III rocket that was under development for DynaSoar would launch the MOL with its Gemini crew. Because the Titan III and Saturn IB rockets were similar in lifting capacity, the MOL was comparable to NASA's proposal for a 'small/independent' module that was capable of supporting a two-man crew for a month. Although they would obviously provide data on human adaptation to weightlessness, the crew's primary task would be to operate a state-of-the-art KeyHole-10 reconnaissance camera.

Project definition continued through 1964. On 25 August 1965 President Lyndon Johnson authorised development and awarded Douglas Aircraft (manufacturer of the S-IV and S-IV-B for Huntsville) the contract to build the MOL. The initial test flights were scheduled to start in mid-1968, so if all of these projects proceeded according to plan the final few years of the decade would see a tremendous amount of activity in space.

SOLAR OBSERVATORY

As soon as the Apollo Applications Program Office was formed, Meuller received a proposal from the Office of Space Science and Applications (OSSA) to install a battery of telescopes in an otherwise vacant bay of the Apollo spacecraft's service module. In order to overcome the need to abandon the instruments at the end of the flight (when the service module would be jettisoned and left to burn up) he decided that this 'Apollo Telescope Mount' (ATM) should be installed in a separate vehicle that a succession of crews would be able to utilise. The obvious carrier was the lunar module which Grumman was developing and, of course, the company was eager to find ways of expanding the role of its vehicle. It was decided to modify the descent stage to host the telescopes and to place the controls in the redesigned ascent stage, which the astronauts would access through the docking tunnel. A design requirement was that the ATM should be capable of being launched on a Saturn IB.

Gilruth argued that the lunar module had been designed as a short-duration vehicle for a very specialised function and was unsuitable for conversion for very long missions. Of course he insisted that if any modifications *were* to be made, then the work ought to be managed by *his* centre. On 19 August 1966 Mueller decided to operate the ATM in conjunction with the orbital workshop. It would be tethered to the workshop and would draw power and consumables from it through an umbilical. This provision, it was argued, would eliminate the requirement to adapt the lunar module for sustained independent operation. The astronauts would transfer to it by spacewalking from the workshop's airlock. However, when a 'tether dynamics' experiment on a Gemini/Agena mission ran into trouble in October, Mueller had second thoughts and decided to operate the ATM in a docked configuration. However, the workshop had only one docking port, for the Apollo. In order to retain the present airlock, it was decided to add a cylindrical Multiple Docking Adapter (MDA) equipped with a ring of supplementary peripheral ports. To minimise its development cost, the MDA was to form a passive structural element.

The incorporation of the MDA radically altered the perception of the workshop's mission. It was no longer a spent stage with a shirt-sleeve environment for a succession of crews making ever-lengthier flights to assess their ability to work in weightlessness; it had become the 'core' for a 'cluster' of scientific modules. In addition to facilitating a rescue in the event of a spacecraft's malfunction, the MDA introduced the possibility of crew 'handovers' and resupply of consumables and equipment by Apollo spacecraft with logistics modules extracted from the launch vehicle's adapter. Overnight, the mission concept had expanded far beyond the 'intermediate' option. In November, acknowledging that the workshop cluster had crossed the subtle threshhold from a short-term focused facility to a long-term multi-role platform, Mueller announced that it would become the centerpiece of Apollo Application's low-orbit programme. On the other hand, the addition of the MDA had pushed the workshop's mass painfully close to the Saturn IB's lifting capacity, so a weight-saving effort was initiated to ensure that an acceptable mass was not exceeded. Despite the lack of funding, the target date for the launch of the 'core' remained mid-1968. The original biomedical study would be undertaken by the first crew prior to the ATM's arrival in early 1969, after which a succession of crews would study the Sun for two years over the period of maximum 'spot' activity. Thus the workshop's role of studying biomedical and human factors to support the design of future space missions was expanded to include astronomical research. Although this initially increased OSSA's role, Mueller – at Gilruth's insistence – assigned Houston responsibility for all the experiments on human adaptation to weightlessness and all the apparatus that was to be integrated into the Apollo spacecraft. Huntsville was to integrate the other experiments into the workshop. However, apart from the ATM there were no other experiments, so OSSA, eager not to waste this opportunity, invited open proposals. Unfortunately, there was no pool of experimenters to tap and even NASA's own scientists had found it difficult to have experiments flown aboard manned spacecraft (partly because crews were committed to testing rendezvous and docking and to assessing their ability to work on spacewalks, and partly because Houston was reluctant to impose an additional burden of experiments). In fact, OSSA had been biased towards the view that everything could be done by automated spacecraft: astronauts were considered at best to be irrelevant, and at worst injurious to scientific experiments in space. Mueller's goal was to demonstrate that there were scientific experiments that not only *could* be performed by astronauts, but were only viable *because* astronauts were present to perform them. Accordingly, the proposals that were submitted were ranked by the extent to which they *required* a human operator.

The ATM was an excellent illustration of the role envisaged by Apollo Applications for its astronauts. The prospect of astronauts operating telescopes is anathema to today's space science community because electronic CCD cameras are taken for granted. In the 1960s, OSSA relied upon small satellites with automated telescopes and television scanners which transmitted their output to the ground. The resolution of these rudimentary cameras was inferior to that of film, but it was difficult to return film to Earth. The Department of Defense's reconnaissance satellites were fitted with film-return capsules, but this system had proved costly to operate. In the case of the ATM, not only could film be used (successive crews could take it and return it to Earth) but astronauts trained to use the telescopes would be able to take the initiative in making solar observations in a way that an automated system would not. At that time no TDRS geostationary relays were available to enable the scientists to control their telescopes in real time; observations were programmed, recorded, and dumped when within range of a ground station. Mueller called the ATM "the most comprehensive array of instruments that has ever been assembled for observing the Sun". Adding the ATM to the 'cluster' not only gave the 'intermediate' stage a supplementary theme, but by housing a succession of crews for extended periods (certainly longer than would have been possible with the ATM operating independently) the ATM's utility was greatly increased. Mueller hoped that the ATM, and other cameras for lunar and terrestrial mapping, would establish that astronauts were a necessary and integral part of a balanced space science programme. In retrospect, it is remarkable that the symbiosis bestowed by the clustering of a habitat with dedicated science modules had not been developed earlier.

It is instructive to compare this mid-1960s view of the function of a space station with that painted a decade earlier by von Braun in his articles in *Collier's*, when it had seemed obvious that space stations would be placed in geostationary orbit to serve as communications relays and the crews would spend their time as 'operators' at the 'switchboard' and maintain the transmitters by replacing burned-out vacuum tubes. However, the development of solid-state electronics in the interim had revolutionised the technology. By the mid-1960s, NASA had launched its first generation of geostationary relay satellites and, as a result, communications was deleted from the list of tasks to be performed by a space station. Nevertheless, even as satellites with ever more electronic sophistication diminished the 'man-in-the-loop' requirement, the automated systems proved to be extremely vulnerable. The first Orbiting Astronomical Observatory, for example, which was launched in April 1966, was crippled during start-up tests when its power system fused. To Mueller at Apollo Applications, this amply demonstrated that there was a role for astronaut intervention.

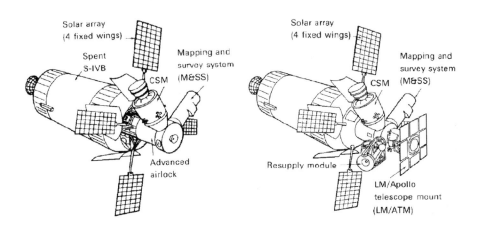

The operational flexibility of the cluster concept was appealing. In addition to the Apollo crew ferry, it facilitated a solar telescope, a terrestrial mapping system and a logistics module. Note however, that in this December 1966 exposition the details of the 'add on' modules had yet to be defined and the power requirements had yet to be addressed. (Courtesy Wade W. Wilkerson, McDonnell Aircraft Corporation.)

ORBITAL CLUSTER

On 23 January 1967, Mueller briefed Congress on the decision for the cluster concept, and three days later he explained his presentation to the Press. The following day, the Apollo 1 crew were killed when a fire swept through their spacecraft in a countdown rehearsal for a 14-day 'shakedown' flight scheduled in February. This accident not only threatened Apollo's chances of reaching the Moon by the end of the decade, but it also ruled out the possibility of starting Apollo Applications operations in 1968 – not that this had been a realistic prospect as development funding had yet to be assigned. With the Apollo lunar mission having priority for funding, the more grandiose Apollo Applications ideas were pared away. In fact, despite the progress made in defining the orbital workshop's mission, for some months the joke in NASA had been that Apollo Applications was at 'T minus two years and holding'. Mueller's dilemma was that if he allowed the development schedule to slip, funding would never be forthcoming. Nevertheless, on 30 March he acknowledged that the workshop could not be launched before the end of 1968. As NASA came to terms with the significantly lower budget for Fiscal Year 1968, the launch of the Apollo Applications workshop rapidly slipped in stages through 1969 and into 1970, by which time, if all went well, Kennedy's lunar landing challenge would have been met. Although the orbital cluster was to form the centrepiece of Apollo Applications, by September 1967 it had yielded its priority in the launch schedule to an independent mission. In January 1970, an Apollo was to ride a Saturn IB into orbit and retrieve a module from the S-IV-B. It had once been intended that this flight would test cameras developed for a 14-day 'mapping and survey' mission in lunar polar orbit, but this had been cancelled. Instead, this mission was to fly in an orbit inclined at 50 degrees to the equator to test OSSA's Earth-resources cameras. In such turbulent times, however, this plan could not survive and was cancelled in December 1967. Although Mueller's Earth-resources programme had stalled, it received a boost in March 1969 when one of the cameras was tested by Apollo 9, and the utility of multispectral imaging was subsequently established in the form of the Earth Resources Technology Satellite (later renamed 'Landsat').

Meanwhile, work on the orbital cluster had hit a snag. In mid-1967 a cooling system had been added to the ATM's instrument package and the solar panels had been enlarged to power it. Also, the vault that was to protect the stored film from radiation had been upgraded and was almost as large as an office desk, with the mass of a small car. As a result, the mass of the ATM had exceeded the Saturn IB's lifting capacity. Furthermore, the addition of a pair of large solar panels to power the workshop's experiments (which were now defined, if not yet funded) mean that this, too, was overweight. In contrast to the year before, when Apollo Applications had been criticised for not having scientific experiments, it now had so many that the crew would not have sufficient time to perform them all. Nevertheless, despite a continuing paucity of funding, Apollo Applications was making real progress as 1967 yielded to 1968.

Although the 11-day 'shakedown' mission of Apollo 7 in late October 1968 had reinvigorated the agency, it was Apollo 8 at Christmas which convinced the

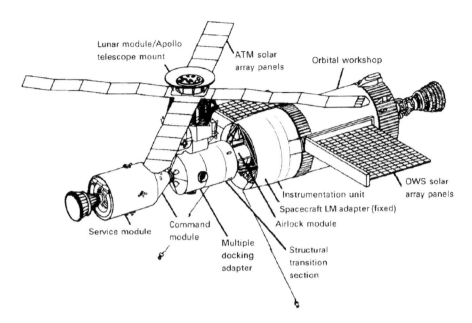

Lunar module/Apollo telescope mount

ATM solar array panels

Orbital workshop

OWS solar array panels

Service module

Command module

Multiple docking adapter

Instrumentation unit

Spacecraft LM adapter (fixed)

Airlock module

Structural transition section

By September 1968 the cluster concept had matured, the Apollo Telescope Mount had been defined and large solar panels had been added to provide power, although the Orbital Workshop was still to be a 'wet' conversion of a live stage. (Courtesy of the Office of Manned Space Flight, NASA HQ.)

American public that NASA *really* stood a chance of making a lunar landing in 1969. Webb, having nurtured that programme from its inception in 1961, retired a few days before Apollo 7's launch. When Tom Paine took over as administrator he expressed his support for Apollo Applications, but this optimism was dispelled by the election of Richard Nixon to the White House, because while he was content to allow the already funded lunar missions to fly, he was lukewarm to a series of ever more elaborate space stations. Nixon charged his Vice President, Spiro Agnew, with chairing a Space Task Group to recommend a 'post-Apollo' strategy, so it was clear that the future for the programme called 'Apollo Applications' was bleak.

Mueller's long-term plan had been to supersede the converted 'wet' workshop with a fully equipped 'dry' workshop launched on a Saturn V, and to proceed to a modular station that would be assembled over a 10-year period. In April 1969, however, he told Congress that it would be better to skip the 'wet' workshop. In truth, as soon as the spent stage workshop had embraced the cluster concept it would have been better to integrate it with the ATM and launch it using a Saturn V, but Webb had refused to sanction the transfer of any of these rockets from the lunar programme. With Apollo making rapid progress towards its goal, von Braun lobbied to transfer the now-overweight workshop to this larger rocket. Rather than simply launch the workshop in a 'dry' state, which would not take advantage of the tremendous lifting power of the Saturn V, he proposed that the ATM's instruments be mounted on a frame that would hold the package above the MDA for launch and then swing it over to one side once in orbit. At the end of June this recommendation was accepted, but Paine did not formally set aside a Saturn V until 22 July, by which time Apollo 11 was on its way back from its historic lunar landing. Gilruth in Houston enthusiastically supported this redesign because it obviated the need to modify a lunar module to operate telescopes and the complications of docking such a hybrid vehicle with the workshop. With the future of the orbital workshop secure – although with its launch postponed until 1972 to provide the time to update its configuration – Mueller resigned to take up a senior management post at General Dynamics.

Although Apollo Applications' orbital workshop survived Nixon's budgetary axe, the Air Force's MOL was cancelled during the adminstration's initial review. In this case, however, the cancellation was justified as reconnaissance satellites – just as was happening with communications satellites – would soon be capable of performing tasks autonomously which, just a few years earlier, had been thought to require a human operator.

In February 1970, NASA distanced its new prestige project from its lacklustre origins by renaming it 'Skylab' and closing the Apollo Applications Program Office. Meanwhile, the Space Task Group set up by Nixon had decided that Skylab would mark the end of the Apollo spacecraft, because if the agency was to have a future in space operations it had to develop a cheaper-to-operate transportation system that would provide 'routine' access to low orbit. The irony, of course, was that this 'reusable' spacecraft was similar in form to the space plane that von Braun had envisaged two decades earlier.

2

The world's first space station

THE CHELOMEI VEHICLES

While Sergei Korolev was developing the Soviet Union's programme to compete with NASA in the 'race' to land the first man on the Moon, Vladimir Chelomei, Korolev's long-time rival in rocketry, had secured support from the military to build a reconnaissance platform in low Earth orbit. As he already had a bigger rocket[1] than Korolev's Semyorka,[2] he designed a 20-tonne platform and a large spacecraft to service it.

Chelomei's reconnaissance platform, which he called Almaz (meaning diamond), was a single pressurised unit with a stepped-cylinder configuration. Its main section was a cylinder 4.15 metres in diameter, and the 3-metre-wide forward part had a conical re-entry spacecraft mounted at its front. The design allowed the crew of three to enter the capsule on the launch pad by way of a side hatch, and, in orbit, a hatch in the heat shield gave them access to a tunnel that linked with another hatch for entry to the module behind. The main compartment was to be equipped with ocean-surveillance radar, a high-resolution optical-imaging system and a scanner that transmitted imagery electronically when the spacecraft came within range of a ground station. A pair of large solar panels would generate sufficient power to run this equipment and the life-support system for the crew. The main compartment was to be split into two sections divided by a control panel, behind which, in the 3-metre-wide section, would to be the living quarters. The wider section would be dominated by the conical bulk of the long-focal-length folded-optics camera system with its massive film canisters. The space station was to be stocked to support a tour of duty lasting several months, with cosmonauts working shifts around the clock.

In effect, Almaz was to perform the same mission as the US Air Force's Manned Orbiting Laboratory. However, unlike its Amercian counterpart, it was not simply to be used once and then discarded. Chelomei had designed a cargo ship to enable

---

[1] Chelomei's UR-500 rocket subsequently became known as the 'Proton'.
[2] Korolev's rocket was affectionately referred to as 'Semyorka', meaning 'old number seven' because it was his seventh design.

This model displayed at the Memorial Museum of Cosmonautics in Moscow in 1994 to mark the 80th birthday of Vladimir Chelomei shows the configuration of the Almaz reconnaissance platform that he designed in the late 1960s. The main vehicle has two compartments, one large one containing the optical system and, towards the front, the living quarters for the crew, who would ride into orbit in the conical section and enter the station via a hatch in the heat shield. The long structure on the nose held the launch escape system, the propulsion systems for attitude control and the de-orbit manoeuvre and, adjacent to the descent capsule, the parachutes. (Courtesy of Dietrich Haeseler.)

Almaz to be resupplied, and because this was to ride on a Proton, it was almost as large as the Almaz itself. The cargo ship was to dock at a port at the rear of the station. Since the technology for automated rendezvous and docking had yet to be proved, this ferry would be flown by a crew and have a re-entry spacecraft of the same type as that on the Almaz. This offered the prospect of exchanging crews, ensuring that the work of the platform could be continuous. Because it served a combined crew transport and resupply function, the spacecraft was referred to as the TKS, an acronym for the Russian words describing its role. Once in orbit, the crew would pass through the hatch into the main compartment of the ferry. Because the docking system was on the far end, the ferry would be flown 'backwards', with the crew of the ferry observing through a small porthole adjacent to the docking unit. In effect, the two vehicles would dock tail-to-tail. Once the vehicles were joined, the docking assemblies would be disengaged and rotated aside to expose the hermetically sealed transfer tunnel. Once the cargoes had been transferred, the retiring crew would board the ferry and return to Earth. This process of replenishment and crew exchange would be continued until the mission was complete, at which time the final crew would power down the Almaz, retreat to its re-entry spacecraft and return to Earth.

The Almaz/TKS concept had tremendous potential. Unfortunately, Chelomei's organisation had no prior experience in developing systems and vehicles for human spaceflight, and the pace of development was so slow that the hoped-for initial launch in 1968 proved unobtainable. Meanwhile, Korolev's Soyuz spacecraft suffered a series of failures during flight trials, culminating in the loss of its first pilot, Vladimir Komarov, in April 1967. An ambitious test had been planned in which two such spacecraft would dock to allow two cosmonauts to transfer from one to the other in space, but the second launch was cancelled when a stuck solar panel halved

A model of the TKS ferry designed by Chelomei to resupply the Almaz reconnaissance platform. Like the Almaz, it was to have been launched with a crew. A cosmonaut (arrowed) would have observed through a porthole during the docking at the rear of the Almaz. (Courtesy of Dietrich Haeseler.)

Soyuz 1's power and a faulty attitude control system made the spacecraft difficult to stabilise. Although Komarov managed to orient the spacecraft manually for re-entry, he died when the parachute failed to deploy and the capsule smashed into the ground. Korolev had not lived to witness this disaster, he had died of complications during an intestinal operation a year earlier. Although in January 1969 Soyuz 4 and Soyuz 5 docked and two cosmonauts spacewalked from one to the other, it was apparent by this time that America would win the race to the Moon and so the Soviets decided to concentrate on the development of the space station. With Chelomei's re-entry spacecraft still to conduct its initial trials, and the Proton rocket yet to demonstrate the reliability needed to trust it with a crew, the rival design bureaux were told to cooperate and service the station using Soyuz spacecraft launched by the trusted Semyorka rocket. Korolev's bureau, which was now headed by Vasili Mishin, his former deputy, duly adapted the Almaz platform to use proved systems. Chelomei had intended to mount a pair of engines, one on each side of the aft-mounted axial docking port, but because this engine block was new, it was decided to replace it with the propulsion module used by the Soyuz spacecraft. This action, however, prevented access through the rear hatch, but the deletion of the re-entry vehicle permitted the docking port to be relocated to the front, on the end of a short transfer compartment that would provide access to the station's forward hatch. This configuration was a 14.6-metre-long structure with a 2.2-metre-wide propulsion module at the rear of the Almaz's 4.15-metre-wide and 3-metre-wide stepped-cylinder, plus a 2.2-metre-wide forward docking adapter. Rather than risk unproved solar panel deployment frames, two pairs of Soyuz panels were to be used, one pair set on the propulsion module and the other on the narrow section of the stepped cylinder. The energy in solar insolation at the Earth's distance from the Sun is about 1 kW per square metre. Each pair of Soyuz panels had an area of 12 square metres. The transducer efficiency was just 10 per cent, so each panel pair produced a peak of only 1.2 kW (the output falling off sinusoidally with divergence from the ideal perpendicular angle of illumination) and much of this power was actually lost in the low-voltage (28 volts) direct-current distribution lines. Even with the contribution from a docked ferry, the 3.6-kW peak would be barely sufficient to run the

A TKS being prepared for launch. The cylindrical section on the right is the adapter for the upper stage of the Proton launch vehicle. (Courtesy of Vladimir Semenov of Videocosmos with thanks to David Rickman.)

A TKS in its aerodynamic shroud, and with the launch escape system undergoing assembly. (Courtesy of Vladimir Semenov of Videocosmos with thanks to David Rickman.)

A Proton launch vehicle configured for a TKS being set up on the pad. (Courtesy of Vladimir Semenov of Videocosmos with thanks to David Rickman.)

environmental systems to support the crew. This configuration would be launched unmanned, and its crew would follow several days later in their Soyuz ferry. To ensure the minimum of delay in effecting the redesign, the fabrication work was done on the same assembly line at the Khrunichev factory in Moscow that had been established for Almaz.

Undeterred by the order to let his rival steal the limelight, Chelomei continued work on his reconnaissance platform and TKS ferry and, as events transpired, they went on to play a significant role in future space station programmes.[3]

LIVING ... AND DYING IN SPACE

On 19 April 1971, a Proton put Salyut 1 in orbit. Many years later it was revealed that it had been intended to be called 'Zarya' but this was the radio call sign of the Yevpatoria Control Centre in the Crimea, so the station was hastily renamed Salyut (salute) to celebrate the anniversary of Yuri Gagarin's pioneering first orbital flight a decade earlier.

Soyuz 10 was launched on 23 April with Vladimir Shatalov, Alexei Yeliseyev and Nikolai Rukavishnikov, who was a member of the team that designed the station. On its 22nd orbit, the ferry's final transfer brought it within a few kilometres of its target, and Shatalov, the cosmonaut with the most experience of rendezvous, took control for the final approach. The station had automatically oriented itself with its front facing the approaching spacecraft. When the 18-tonne station's bulk appeared in his periscope, he likened it to "a train entering a terminus". The mood was tense, but the docking was achieved without a hitch. The probe on Soyuz 10's nose slipped straight into the conical drogue and the capture latches engaged to establish the initial 'soft' docking. The probe then retracted to draw the 1.5-metre-wide annular collars together for a 'hard' docking. Unfortunately, the electrical connections within the collars failed to mate, and the crew were denied access to the 0.8-metre-wide hermetically sealed tunnel to the station. After only 5.5 hours, during which (as the official news agency Tass put it) Soyuz 10 "completed its planned experiments", it undocked and returned to Earth.

Although the orbital compartment of the Soyuz, which carried the docking system, could not be recovered to unambiguously identify the fault, the engineers double-checked the system on the next spacecraft, and this was dispatched on 6 June as Soyuz 11. Alexei Leonov, Valeri Kubasov and Pyotr Kolodin had trained for this mission, but when the preflight medical examination on 3 June revealed an inflamation in one of Kubasov's lungs, the crew were replaced by the back-up cosmonauts Georgi Dobrovolsky, Viktor Patsayev and Vladislav Volkov. This time the hatch opened readily, and the cosmonauts were able to enter the station. The automatic television camera showed the trio performing somersaults in the (by comparison to the ferry) voluminous compartment.

---

[3] The International Space Station's Zvezda and Zarya 'core modules' are directly descended from Chelomei's designs.

A technically accurate artist's illustration of a Soyuz spacecraft moving in to dock with the nose of Salyut 1.

The objective of the mission was to stay in space for a month, during which a range of biomedical experiments were to be performed to study the effects of prolonged exposure to weightlessness. The experiments included measuring bone density, studying the reaction of the vestibular system to different stimuli, measuring energy expenditure, measuring the capacity of the respiratory system to assess stamina, monitoring the electrical activity of the heart, measuring the flow of blood within the major blood vessels by taking seismocardiograms, measuring arterial pressure, measuring the force and rhythm of the heart, and collecting blood samples for subsequent analysis. In effect, they were acting as guinea pigs to prove that cosmonauts could survive a tour of duty aboard an orbital station and return safely to Earth. Looking back, it is difficult to appreciate just how truly pioneering this mission was.

As the days passed, the crew made television broadcasts which featured on the domestic nightly news, showing off the station and its apparatus. It was a busy time for the three men, working in shifts. On 18 June, by which time life had settled down to a routine, a small electrical fire in a cable was discovered. The cosmonauts were so alarmed that they urged controllers to let them return to Earth immediately. In weightlessness, however, a fire in an oxygen-nitrogen mix is rapidly suffocated because there is no convection to draw off the carbon dioxide and suck in oxygen. Although they powered up the Soyuz just in case, once it became evident that they were in no danger they reentered the station and resumed work. Unfortunately, this scare appears to have broken their pioneering spirit, and a week later it was decided to let them return to Earth early. On 29 June, therefore, they packed the film

A photograph looking 'forward' in Salyut 1's main compartment, over the vehicle's control panel and through the hatch to the docking compartment beyond. Note that at this stage in the programme's development it was thought appropriate to provide seats for the crew.

canisters and the results of their experiments into the cramped descent module and returned the station to its automatic flight regime. The transfer tunnel was overpressurised to check for a leaky hatch, then vented. Laughing and joking because they were happy to be on their way home, they undocked and withdrew. The automatic system oriented the spacecraft for the de-orbit manoeuvre, performed this burn, and jettisoned the orbital and service modules precisely on schedule. There was a premature loss of signal, but the recovery team that gathered around the capsule on the Kazakh steppe were unaware of this, so when they opened the hatch and peered inside they were appalled to find that the cosmonauts were dead in their couches.

An inquiry and examination of the descent module eventually determined that a valve that was designed to equalise the internal and external pressures shortly before landing had been shaken open by the shock from the explosive bolts that had jettisoned the orbital and service modules. Analysis of the telemetry revealed that the leaking air had acted like a thruster and had set the descent module adrift, but the automatic attitude control system had stabilised it in time for re-entry. Soyuz cosmonauts did not wear pressure suits, so they first succumbed to unconsciousness resulting from asphyxiation as the pressure fell, and then expired from embolism of the blood in vacuum. They were exposed to vacuum for 12 minutes before the capsule was repressurised by the open valve upon entering the lower atmosphere. During the two weeks leading up to the official explanation, there was speculation that the men had been killed by the 3-*g* deceleration forces imposed by re-entry. The pessimists insisted that the deaths proved that there really *was* a limit to the time that the human organism could spend in weightlessness. However, the fact that they had died *prior* to re-entry indicated that deceleration forces had played no part in their demise. It was a technical failure. After due ceremonies, the bodies of Dobrovolsky, Patsayev and Volkov were interred in the Kremlin Wall, alongside their colleagues Yuri Gagarin, Vladimir Komarov and former chief spacecraft designer Sergei Korolev.

In the aftermath of this disaster it was decided that henceforth Soyuz crews would wear a lightweight pressure suit for launch and re-entry. To accommodate the

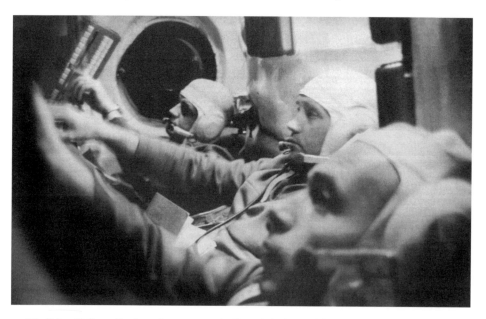

Vladislav Volkov (farthest from camera), Georgi Dobrovolsky and Viktor Patsayev, the ill-fated crew of Soyuz 11, photographed through a porthole in the Soyuz spacecraft trainer.

independent life-support system, the vehicle had to be limited to two cosmonauts. The fact that the Soyuz modifications would take a year or so to complete and verify precluded another visit to Salyut 1, so it was de-orbited over the Pacific Ocean in October. The first crew's inability to gain access had been frustrating and the death of their successors had been a tragic accident, but the station had proved itself, and there was no evidence to show that there was a limit to the time a crew could spend in space, so the long-term prospects looked promising.

3

Skylab

NASA'S LABORATORY IN THE SKY

After two years of severe budgetary cuts, Tom Paine resigned, leaving George Low, his deputy, as NASA's acting administrator. Paine had readily cancelled Apollo 20 to release a Saturn V to launch Skylab because that had been the only way to rescue the workshop from its weight problems, but the loss of Apollos 18 and 19 for purely financial reasons had been penny-pinching – after all, the rockets and spacecraft had already been manufactured. It was all the more frustrating as this left the agency with a pair of Saturn Vs it would never be able to use.

In Houston, the strategy was initially 'minimal' changes to the workshop's design to avoid any delay, but it was soon acknowledged that the relaxation of the weight constraint meant that the ATM could be affixed prior to launch, thereby eliminating some outstanding docking issues. Also, because the workshop would be able to be outfitted prior to launch, it would be productive as a science platform as soon as the commissioning crew took up residence.

In its original conception, the S-IV-B's hydrogen tank was to have served as an unpressurised enclosure in which to rehearse spacewalking tasks in safety, but it had later been decided to pressurise the tank, making it a shirt-sleeve workshop for conducting various experiments. In this early plan, the Apollo spacecraft was to remain powered up throughout the mission and the astronauts were to retire there at night. In August 1967, although it had been decided to build living accommodation into the tank, 'habitability' was not a term in the systems' engineering vernacular, so George Mueller had asked Huntsville to hire an industrial designer as a consultant. A two-layer grid had been laid across the lower end of the tank to serve as 'floors' for circular 'rooms'. The consultant's report in February 1968 was highly critical. The cluttered interior had to be simplified for ergonomic efficiency, the 'harsh' fluorescent lighting had to be replaced by 'warmer' lamps, which had to be located more logically to make the workshop less forbidding, and the bare metal had to be painted according to a psychologically suitable scheme. The problem with decorating the interior of the workshop was that no commercial paint vendor had certified its product as being able to withstand immersion in liquid hydrogen. If the paint flaked, it would jeopardise the smooth running of the live S-IV-B's engine. With the switch to a 'dry' workshop this issue became irrelevant. Never having built a spacecraft before, Huntsville asked for the astronauts' advice. Astronauts tended to see the concern over the decor as frivolous. In their view, the purpose of flying was to meet the mission objectives and anything that was not strictly necessary for achieving this was a distraction. When the consultant recommended partitioning off a part of the lower level to form a 'wardroom' to serve as a kitchen-dinette, Huntsville readily agreed. However, when it was suggested that a large porthole be installed in the wardroom, the engineers argued that doing this would weaken the structure. As the matter was debated, the word was passed down the line from Washington that it would be "unthinkable" *not* to have a window. To engineering-oriented NASA, the orbital workshop was not simply a platform for conducting science, it was a vehicle for assessing systems and procedures for improving space flight, so the window, the flooring and the decor were regarded as experiments in themselves.

In Houston, Caldwell Johnson, a Langley engineer with a reputation for devising innovative systems, was appointed in June 1968 to lead the 'Habitability, Crew Quarters' experiment. He soon concluded that the issue of habitability could not be conducted in the normal sense of devising an environment and then running a planned series of controlled tests designed to assess its suitability by varying key parameters. The astronauts would have to live and work productively in the test environment, so this would need to be as efficient as could be devised, despite the lack of practical experience of living in orbit. In an effort to apply engineering rigour to the analysis, he systematised the 'crew–spacecraft relationship' in terms of its key factors, namely:

- environment
- architecture
- mobility and restraint
- food and water
- clothing
- personal hygiene
- housekeeping
- internal communications
- off-duty activity

and set about devising tests, to be conducted in orbit, designed to identify the 'best' solution to each issue. The 'floor', for example, had a triangular grid and several forms of shoe-cleats were to be tested to determine which was the most effective as a 'restraint aid', and in which circumstances, so that on later flights this could be regarded as a proved system rather than an experiment. If all went well with these mundane space systems experiments, science would be able to be conducted as a bonus of being able to live and work effectively in space.

At the heart of the issue, of course, was the human body. The initial motivation for building a workshop was to assess how the body adapted to weightlessness over an extended period. So far, the prophecies of doom had been shown to be false. The longest Mercury flight occurred in May 1963, when Gordon Cooper spent 34 hours in space. With Pete Conrad on Gemini 5 in August 1965, he had launched with '8 days or bust' sewn onto the mission patch. In December of that year, Frank Borman and Jim Lovell had spent a fortnight aboard Gemini 7 to prove that the body could endure weightlessness for the time required for an Apollo to fly to the Moon and allow the astronauts to spend a few days on its surface. Although marvellously manoeuvrable, riding the Gemini spacecraft has been likened to sitting in the front seat of a Volkswagen, so this endurance mission had been something of an ordeal. It was found that, even on such a 'short' flight, the muscles and bones degraded. A key objective of Skylab's programme was therefore to determine the rate at which this took effect and, if it levelled off, to establish the body's space-adapted state. The worst case, of course, was where this degradation would *not* level off, because an irreversible trend would impose a limit on the time that astronauts could spend weightless, which in turn would mean that space stations would have to generate 'artificial' gravity, and this would greatly complicate the design. One insight into this issue was added by Apollo. It was noticed that the degradation of bone and muscle was less pronounced in the astronauts who had landed on the Moon than in those who had remained in lunar orbit. This held out the prospect that even if the body's reaction did not level off, it might be possible to slow the degradation by spending 'therapy' sessions in a centrifuge on a space station. In the longer term, of course, it was likely that spacecraft designed to fly to Mars could provide artificial gravity for most of the interplanetary cruise. However, the demise of Salyut's crew while returning to Earth had served only to increase the NASA medical community's concern. As they pointed out, although the cosmonauts had felt fine as they set off for home, their deaths meant that it had yet to be *demonstrated* that the body could survive the stress of returning after a 24-day flight; and from that perspective the record was still 14 days.

As soon as Mueller had decided to convert a spent stage and fly a 28-day mission, Houston's Medical Research and Operations Division had defined three experiments to track the degradations discovered during by the Gemini missions. For example, it had proved surprisingly difficult to perform work on spacewalks, so the medics had developed a methodical investigation of metabolic and cardiovascular reactions. The Metabolic Activities experiment was to follow up the problems involved in spacewalking, and determine whether physical work in space was really a more demanding activity than on Earth, or whether some other factor had induced a state of exhaustion. An instrumented exercise bicycle – velo-ergometer – would provide calibrated degrees of resistance against which the astronaut would work while his energy expenditure was estimated by measuring the ratio of the oxygen inhaled to the carbon dioxide exhaled. The Cardiovascular Function experiment was to investigate changes in the body's cardiovascular system as it atrophied when fluids migrated into the upper torso in the immediate aftermath of entering weightlessness. An astronaut was to slip his legs into a cylinder which would make a hermetic seal at his waist. Once the pressure in this Lower-Body Negative-Pressure chamber had been slightly reduced to draw the blood into his legs (just as if he was once again standing on Earth) a vector-cardiograph would record how his heart responded to the increased load. In addition to the specific actions required for these experiments, the Mineral Balance Experiments required the astronauts to maintain a log of their food consumption.

One of Johnson's most significant contributions was the 'food system' in the wardroom. On Apollo, the astronauts had lived on two types of 'space food'. Solids comprised bite-sized cubes of compressed and processed raw materials. Liquids were dehydrated and reconstituted by injecting water and kneading the bag until the mush was thoroughly mixed, by which time it had invariably gone cold. Astronauts could accept such hardship in return for a chance to walk on the Moon, but it was decided that a crew facing several months "boring holes in the sky" aboard Skylab would need something better. The food system (which was itself an experiment, of course) was built into the wardroom table. The table had heated receptacles for tins to provide an alternative to the rehydrated foods. In addition, a freezer was installed to store freeze-dried produce, fresh fruit and vegetables. Furthermore, astronauts would eat with utensils. It was the switch to a Saturn V that made these enhancements to habitability practicable – life aboard a converted spent stage workshop would certainly have been considerably harder – but the more varied diet would significantly complicate sampling bodily wastes for the Bioassay Experiment. The Apollo crews had urinated into a central collection tank and had defecated into individual 'faecal containment bags'. For Skylab, a proper toilet was developed. Strangely, considering the thought that had gone into the layout of the 'rooms' to provide a comforting (to the habitability specialists, at least) sense of 'vertical', the toilet was on the 'wall' of the cubicle reserved for this function. This, however, did not affect its utility, as the prime design requirement was that the toilet be capable of taking the samples needed for the Bioassay Experiment. After tests, it was decided that urine should be frozen and that the toilet should heat the faeces in vacuum conditions to dry it. To certify the toilet for use in space, it was carried on the KC-135 trainer and (over an exhausting two days of flying trajectories designed to induce weightlessness for periods of 30 seconds or so) a team of volunteers managed to secure nine good 'data points'. This was undoubtedly very welcome news to the Air Force; the basic design of the toilet had been derived from its cancelled MOL project, so that effort had not been wasted.[1] To verify (as best as could be achieved in the Earth's gravity) that the biomedical apparatus worked as intended, three yet-to-fly astronauts spent 56 days living in an appropriately fitted simulator. Although this intensive trial revealed several problems, the modified flight hardware performed perfectly.

Previously, it had not been feasible to perform biomedical studies on astronauts *in space*, because the spacecraft had been too cramped and the crew's focus had been otherwise directed, but Skylab's 6.6-metre-wide compartment had a volume of 350 cubic metres (including the airlock module and multiple docking adapter) which was fully *150 times* that of an Apollo Command Module.

Although the original motivation for the orbital workshop had been to conduct biomedical studies, the introduction of the ATM added a major scientific theme. It incorporated six instruments:

---

[1] Apparently, when it was proposed that the toilet be mounted on the wall of the Apollo 14 spacecraft in 1971 in order to verify its utility in space, Alan Shepard, the mission commander, firmly vetoed the idea.

- White Light Coronagraph
- X-ray Spectrographic Telescope
- Ultraviolet Scanning Spectroheliometer
- X-ray Telescope
- Extreme-Ultraviolet Spectroheliograph
- Ultraviolet Spectrograph.

The White Light Coronagraph's occulting disk would enable it to observe the corona (the tenuous but extremely hot ionised gas which streams out from the Sun, which was normally able to be observed only during a solar eclipse) on an ongoing basis. The two X-ray telescopes would be able to study the corona without an occulting disk. The other instruments would view the solar disk at wavelengths absorbed by the Earth's atmosphere, and so denied to terrestrial observers.

At the end of 1965, when the decision was taken to build the ATM, Washington's Office of Space Science and Applications was processing the results from its first two automated Orbiting Solar Observatories and looked forward to launching the ATM in 1968 in order to observe the Sun through the period of maximum spot activity. In fact, as the instruments had initially been proposed for the next generation of 'Advanced OSOs', the electronic detectors were replaced with film cameras. The unprecedented spatial resolution of the pictures would enable the composition, density and temperature of the features on the solar disk to be resolved, and the physical and magnetic state of the material in the corona to be calculated. With more than a tonne of instruments, the ATM would be a much more capable observatory than the 120-kilogram payloads of its predecessors. Despite the advanced state of their specification in 1965, the fabrication and integration of the instruments took longer than expected. In part, however, this delay derived from the decision to enable the Ultraviolet Scanning Spectroheliometer (which had retained its electronic detector) to be operated in 'real time' from the ground during periods when the workshop was unoccupied. The photographs from the other instruments would not become available until the film was retrieved by the astronauts and returned to Earth, so mounting the instruments on the workshop involved trading off real-time operations for a considerable improvement in the quality of the data that would have been transmitted from an electronic detector. Delaying Skylab to the early 1970s meant that the level of spot activity would be in decline by the time that this unprecedentedly capable solar observatory became available. Nevertheless, the equipment was so advanced that there was every prospect of making significant discoveries.

Skylab had to serve as a stable platform that ensured the ATM faced the Sun. When deployed, the ATM would be canted out at right angles to the workshop. Its solar panels had been mounted in a cross-pattern to provide their maximum power when observations were underway. Skylab's attitude control computer was more sophisticated than previous systems. It could adopt and hold solar-inertial attitude completely automatically. Although nitrogen gas thrusters would make substantial manoeuvres, three 66-kilogram, half-metre-wide electrically driven gyroscopes were mounted in the ATM's framework. Any two of these could be used to stabilise the vehicle, but the preference was to use all three. The big wheels stored and selectively shed momentum in order to impart the 'turning moment' required to rotate the spacecraft about a specific axis. The gyros were referred to as CMGs because they controlled the spacecraft's momentum. The ATM used attitude and 'rate' sensors to direct the CMGs to adopt, and maintain, an attitude to within 0.1 degree of arc. Even so, for the telescopes to achieve a pointing accuracy of 2.5 seconds of arc the entire instrument assembly was gimballed within the ATM's frame.[2] In order to enable the astronaut operating the ATM to aim the instruments at a specific feature, two small sighting telescopes were added and their output was displayed on a TV monitor on the ATM's control panel, which was in the multiple docking adapter. These sighting scopes had filters which only passed light at a specific wavelength in the hydrogen spectrum (a wavelength at which the structure of the solar disk is clearly distinguished) and once the operator zoomed in and placed a crosshair to designate the target for the main instruments, a camera could be mounted on the monitor screen to photograph the feature in order to provide context to assist in interpreting the detailed results. Because the ATM would not be tended on a full-time basis, it was fitted with a sensor to monitor the Sun's total X-ray emission and to alert the crew if this rose above a certain threshhold, so that they could assess the situation and decide whether to start making observations. If ever there was a space-based telescope facility that would demonstrate just what a human operator could achieve, it was the ATM. In addition to displaying the status of each instrument, the control panel presented Skylab's attitude and position in orbit. After much testing, all the controls had been set within arm's reach of a conveniently seated operator. It was a masterpiece of ergonomic design.

The solar scientists developed a list of research themes, in which the priorities were: the 'chromospheric network' and its coronal extension; the so-called 'active' regions, including sunspots; flare activity; and prominences and filaments. It was hoped that the astronauts would be able to accumulate sufficient data to document the development and evolution of these highly dynamic phenomena. When this was explained to Houston's crew activities planners they were appalled. Their task was to determine what the crew would do on a day-by-day basis during the mission, with the aim of ensuring that the various activities integrated in such as way as to maximise the chance of achieving the mission's overall objectives. The scientists were adamant that it was impracticable to schedule observing tasks to specific days; the ATM's schedule would have to be driven by the state of the Sun. Even worse, from the planners' perspective, was the scientists' demand that the ATM take precedence over other scheduled tasks whenever the Sun presented an opportunity for them to 'tick a box' on their list of targets. Given the investment in the ATM, and the logic of the scientists' case, the planners had little choice but to acquiesce.

The ATM, however, was not the only observational package. When Mueller

---

[2] Much is made these days about how it would be impossible to perform telescopic observations from the International Space Station because the crew's activities would disturb the pointing, but evidently this was not a problem on Skylab.

adopted the cluster concept, it was envisaged that one of the science modules would carry a camera system to undertake a survey of the Earth. The Office of Space Science and Applications had been frustrated by the cancellation of the Apollo Applications flight scheduled in 1970 to test this 'Mapping And Survey System' (a pressurised module which would be retrieved from the S-IV-B), but Apollo 9 had tested one of the multispectral cameras in March 1969. The decision to switch to the Saturn V rendered the cluster concept redundant because the workshop was to be fully equipped prior to launch, and in the redesign the camera package was initially ignored. However, NASA was pleasantly surprised by the rapidly developing public awareness of the Earth's fragile environment (a realisation stimulated in part by the pictures taken by the Apollo astronauts showing the Earth rising over the Moon's limb) and by the matching mood in Congress that 'natural resources' was definitely something worth pursuing, so when money was offered to assess the possibilities the agency was responsive. In fact, the Office of Space Science and Applications already had an Earth Resources Technology Satellite under development for launch in the early 1970s but the opportunity to expand Skylab's observational mission in a way that was relevant to the national economy was irresistible. Because the sensor suite was already under development, it was impracticable to issue the usual request for proposals from scientists who would define their scientific objectives and list the equipment they would require to be built. In this case, therefore, NASA made an announcement of the instruments that it intended to mount on Skylab, namely:

- Multispectral Camera
- Terrain Camera
- Infrared Spectrometer
- Multispectral Scanner
- Microwave Radiometer
- L-Band Radiometer

and invited proposals for observational programmes utilising these instruments. In effect, the agency declared that it had a 'facility' that it would operate on behalf of 'users' who would receive data that matched their requirements. It was a major departure from Houston's procedure but it offered the advantage that the scientists could be kept at arm's length, and (weather permitting) observations could be planned because the ground track was predictable, which eased the task of planning the crew's activities.

The Multispectral Camera (an element of which had been tested by Apollo 9) comprised six different cameras using a variety of film types to document each site in visible and near-infrared wavelengths. All the other instruments were state-of-the-art sensors not previously used in space. The Infrared Spectrometer was sensitive in two broad bands spanning most of the infrared spectrum but it sensed a narrow spot only a few hundred metres across. However, it could be adjusted to observe specific targets. In contrast, the Multispectral Scanner split this spectral range into 10 narrow bands and scanned a 75-kilometre-wide swath centred on the station's ground track. The Microwave Radiometer extended the spectral coverage into the far-infrared, and the L-Band Radiometer could sense surface temperature. The function of the Terrain

Camera was to document the spacecraft's ground track in natural-light in order to provide context to assist in the interpretation of the other data. The Earth-resources package was placed on the opposite side of the multiple docking adapter from the ATM. For the early 1970s, this was a comprehensive suite of environmental monitoring instruments.

Introducing this second observational package to Skylab annoyed the solar scientists, however, because it was not practicable to undertake both types of observation simultaneously. To observe the Earth, Skylab would have to set an 'orbit rate', in effect rotating once per orbit, so that its sensors remained pointing directly downwards, during which time solar observations would not be possible. Furthermore, the ATM had been made responsible for controlling Skylab's attitude, so it had to be reprogrammed to be able to adopt and maintain this nadir-inertial attitude.

Another complication of adding the Earth-resources observations to Skylab's programme was that it could not be conducted on the planned orbit. Given the opportunity, NASA will launch due east from Cape Canaveral at latitude 28.5 degrees, as this is the most efficient azimuth and makes maximum use of a rocket's lifting power. However, Florida is the most southerly State and a spacecraft in such an orbit is unable to observe the United States by nadir viewing. It was therefore decided to increase Skylab's orbital inclination to 50 degrees, which would take it as far north as southern Canada. It was also decided to employ an orbit that repeated its ground track every five days, which would assist in planning the terrestrial observations. Just as the plan to use the solar telescopes was driven by the state of the Sun, the plan to use the terrestrial sensors was subject to the weather. For each 'pass' over the US on which the flight plan required Earth studies, one of several viable sites was to be selected, depending upon which looked as if it would be least affected by cloud coverage. Those sites that were clouded would become candidates five days later, and while working within the constraints of each project – there was no point in undertaking a harvest survey in winter, for example – the planners simply whittled down the list until all the sites had been covered. On the other hand, it was hoped to be able to follow a complete crop cycle to assess the feasibility of monitoring agricultural productivity from space.

By May 1973, when Skylab was finally ready to launch, its mission had expanded out of all proportion to its origins as an unpressurised enclosure in which to rehearse spacewalking tasks. In fact, it had so much apparatus aboard that the astronauts had to specialise. The mission commander had responsibility for the Apollo spacecraft, the pilot would specialise in the workshop's systems and the Earth-resources instruments, and the scientist-astronaut would have expertise with the ATM and the biomedical programme.

SKYLAB IN TROUBLE

At 1330 local time on 14 May 1973, the controllers in the firing room at the Kennedy Space Center dispatched their final Saturn V, and control was passed to Houston as

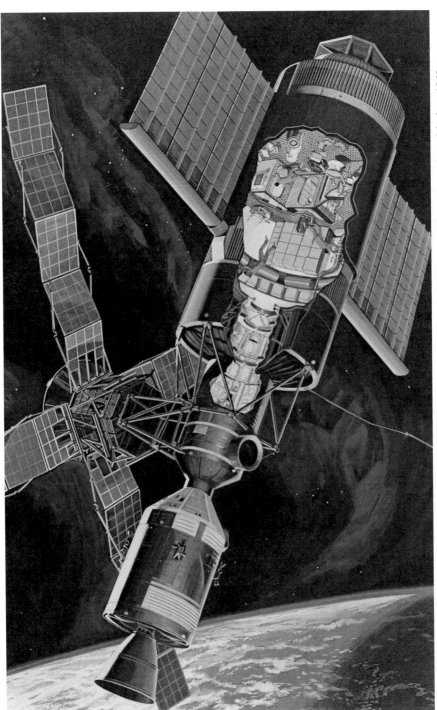

A technically accurate artist's cutaway view of Skylab in the pre-integrated 'dry' configuration selected in late 1969.

The final Saturn V lifts off on 14 May 1973 carrying Skylab in its distinctive aerodynamic shroud.

it cleared the tower. As if to mark this transition from Apollo to Skylab, in February the Manned Spacecraft Center had been renamed the Johnson Space Center in honour of Lyndon Johnson.

On the rear of the S-IV-B, where the J-2 engine would have been if it was a live stage, there was a large square plate. This was the workshop's radiator, and the first act in the activation sequence was to jettison its cover. A minute or so later, the aerodynamic shroud split into four segments which peeled away from the top of the stage, exposing the ATM on its frame directly above the MDA. The articulated frame pivoted the massive ATM out to the side and clamped it into position. This sent a sigh of relief through the control room, as the deployment of the ATM had been the 'show stopper'. The mission would have been over if the ATM had failed to deploy. The ATM unfolded its cruciform of solar panels and, moments later, the spacecraft flew south of the Apollo tracking network's coverage in its steeply inclined orbit.

While the controllers waited for Skylab to fly into range of Carnarvon on the western coast of Australia, they pondered the significance of an erratic signal suggesting that the micrometeoroid shield had deployed prematurely. To protect the wall of the hydrogen tank against penetration by micrometeoroids (the tank would form the habitable part of the workshop), a metal 'cover' had been stretched around the workshop and fastened by a seam of small trunnion bolts. Once safely in orbit, linear pyrotechnics were to fire to release 'fold out panels' so that a series of spring-loaded torsion rods could elevate the body of the cover a few centimetres from the surface of the workshop. The idea was that micrometeoroids would lose so much energy punching through the shield that they would be unlikely to do significant damage to the wall of the pressurised structure. It was an ingenious design which the Huntsville engineers had exploited to simplify the workshop's thermal regulation system, so it actually served two functions. The erratic signal was what the flight controllers termed a 'funny', by which they meant that they did not understand what had prompted it. As the fault that caused the premature deployment of the micrometeoroid shield might have affected other systems, the controllers were eager to see whether the two solar panels on the workshop had successfully deployed as the vehicle ran through its automatic sequence.

When Carnarvon acquired communications, the controllers were astonished by the telemetry: although one signal indicated that the panels had released, the flat power meters implied that the transducer arrays had not deployed – either that, or the panels had somehow become detached from the vehicle! Clearly, Skylab was in trouble because, as a pair, the main panels were to have provided 5 kW of power. The ATM was producing the specified 4.5 kW. Without a crew and with none of the scientific instruments running the vehicle could survive on 3 kW, so the situation was not immediately threatening.

In fact, even as the controllers attempted to make sense of the confusing solar panel telemetry, they discovered that the temperature inside the workshop was rising rapidly. Skylab had *two* serious interrelated problems. It was now realised that far from having deployed prematurely the micrometeoroid shield was missing, and with the workshop's skin exposed to the Sun the tank inside was heating up. An analysis of the telemetry during the ascent revealed that the first sign of trouble occurred 63 seconds after launch, just as the vehicle went supersonic and was subjected to its greatest aerodynamic stress. The conformal shield had evidently been ripped off by the wind shear. But had this taken the solar panels with it or had the debris simply fouled their deployment mechanisms? The second significant anomaly occurred immediately after the S-II released Skylab, at which time the S-IV-B's attitude control system had countered a lateral impulse. As would later be determined, it was at this point that one of the solar panels had ripped off. The shield's premature departure had snapped the latches holding the booms in position, but the airstream had forced them tight against the vehicle. It was only when they were hit by the blast as the S-II fired its retro-rockets to make a clean separation that one of the panels was forced to open beyond its deployment limit and was ripped off at the hinge. The boom on the other side of the vehicle, although flapping loose, had been luckier.

If the thermal problem could be overcome and the workshop appeared to be still habitable, it should be practicable to dispatch the first crew for a 28-day mission to address the biomedical issues, even with a drastically reduced science programme. If a visual inspection showed that the solar panels were still on the spacecraft, then it might be feasible for the astronauts to clear the obstruction and so complete the automated deployment. As long as the workshop was not overly stressed by the heat, the situation was not beyond redemption. It had been planned to launch the crew the following day, but this was postponed for five days (one cycle of Skylab's repeating orbit) to give the engineers time to stabilise the situation. With each daylight pass, however, Skylab's workshop absorbed more energy, and the temperature rapidly rose to 60 °C.

When Skylab was in orbital darkness it drew power from chemical batteries that were recharged during the daylight pass. The controllers' immediate problem was that if they rotated Skylab so that the workshop's bare skin would not be exposed to the Sun, then the ATM's panels would not be able to recharge the batteries. But something *had* to be done because the surface had become so hot that Huntsville was concerned that it might suffer structural failure. Skylab was reoriented to aim its axial docking port at the Sun, so that the workshop's skin would radiate some of the heat it had absorbed, and when the batteries were placed in jeopardy Skylab was returned to solar inertial attitude to allow the batteries to be recharged. However, alternating back and forth between these two attitudes jeopardised the station's long-term future because the manoeuvring used up irreplaceable propellant. As a compromise which balanced power against thermal stress, Skylab was set with its axis at 45 degrees to the Sun. However, the efficiency of the transducers fell off sinusoidally with the angle of illumination, and the station was only just able to maintain a viable power supply as the workshop gradually cooled to 42° C, which was hot, but not dangerously hot. The control situation was complicated by the fact that the 'rate gyros' utilised by the ATM to sense its rotational state were also overheating and their spurious readings were prompting the computer to perform manoeuvres to correct a non-existent drift. It was in fact *inducing* drift, so Houston's controllers had to infer the vehicle's true orientation by interpolating from the heating and power levels before they could intervene. If the station ran out of propellant, it would become a drifting hulk. On 17 May, the launch of the crew was slipped another five days, by which time the station had used almost a quarter of its propellant. Meanwhile, the flight controllers discovered another serious problem. Having oriented Skylab at 45 degrees to the Sun to allow the panels on the ATM to provide power and, at the same time, control the temperature in the workshop, the temperature in the airlock, which was permanently shaded by the ATM, had dramatically fallen. Despite directing power to its heaters, by 21 May the airlock's system were near-freezing. If the water froze and the resulting expansion cracked a pipe, the water-based heat-exchanger – which was to cool the spacesuits during spacewalks – would be damaged. There was no option but to reorient Skylab to illuminate the airlock and warm it. During this time, of course, the ATM's panels were unable to keep the batteries charged, with the result that the workshop's temperature rose once again and the necessary rectifying manoeuvres consumed even more propellant. It was evident that the most important task was to find some means of protecting the surface of the workshop so that Skylab could adopt and maintain solar-inertial attitude.

A 'sunshade' could only be installed by astronauts on the scene, and it would have to be sufficiently lightweight and compact to be carried inside the Apollo spacecraft. Numerous ideas were offered, but only two appeared to be practicable. One idea was to affix a V-shaped frame to the structure of the ATM and deploy it to unfold a foil shield across the sunny side of the workshop. It was successful in water-tank trials, but a spacewalk from the Apollo spacecraft, flying in formation, would probably involve two astronauts making their way over to Skylab to work in the close confines of the ATM's framework with the risk that their umbilicals may snag on its solar panels. When Pete Conrad, the commissioning crew's commander, saw a demonstration of the alternative proposal, he gave it his enthusiastic backing. Jack Kinzler of Houston's Technical Services Division had devised a 'parasol' to be deployed from *within* the workshop by passing a bundle of telescoping rods with a foil shield wrapped around them through the scientific airlock mounted in the wall on the workshop. Once the assembly had been inserted into the airlock, which was only 20 centimetres wide, springs would extend the main spar and as soon as this cleared the airlock four other spars would swing out and form a 42-square-metre foil over the workshop's exposed skin. It would not need to be a close fit as it had only to shade the workshop's curved surface from the Sun. Although a trial indicated that the foil would degrade through solar exposure, it would provide the time required to develop a permanent solution for the next crew to implement.

Meanwhile, the flight controllers had some good news. A close study of Skylab's telemetry had strengthened the suspicion that *one* of the main solar panels might still be in place, although fouled by the remnant of the micrometeoroid shield. Perhaps the astronauts might be able to release it. The Huntsville engineers modified a cable-cutter and a universal gripping tool to fit onto the end of a 3-metre-long handle. This would enable an astronaut stationed in the hatch of the Apollo spacecraft flying in formation alongside Skylab to cut away the obstruction and extract it from the deployment mechanism, in the hope that the solar panel boom would automatically extend once it was free. This was NASA as its best: a spacecraft in danger, all field centres pulling together and racing the clock. On 25 May, Conrad, the absolute epitome of this 'can do' spirit, led Paul Weitz and Joe Kerwin out to their waiting rocket. If they were able to salvage Skylab they would probably be able to fly the planned 28-day mission, but if the station could not be saved they would be back on Earth within 24 hours.

Rather than refurbish Pad 34 for the Saturn IBs that were to launch the Skylab crews, one of the two Saturn V facilities on Pad 39 had been modified; after all, now that Skylab had been dispatched there would be no more Saturn Vs. The multi-tiered swing-arms on the fixed service structure had been modified to accommodate the smaller rocket, which had been mounted on a 39-metre-tall pedestal (inevitably dubbed the 'milkstool') to enable the 'white room' at the end of the topmost arm to mate with the Apollo spacecraft just as if it was on a Saturn V. If nothing else, this *ad hoc* configuration emphasised the enormous size of the rocket that America was forgoing in its quest for cheaper access to low orbit.

At ignition, the flame from the eight H-1 engines blasted down through the pedestal's truss. Clamps restrained the rocket as its engines ran up to full thrust, and continued to do so while the controllers confirmed the vehicle's status. Nothing like this had ever been attempted and there was concern that if the engines had to be shut off the rocket might collapse as its structure 'relaxed', but there were no anomalies and the clamps released.

THE REPAIRMEN

"Houston, Skylab 2," Conrad called as the rocket cleared the tower and control passed from the firing room to Houston, "we fix anything!"

The initial orbit was to be lower than Skylab's, so that the Apollo spacecraft could catch up and rendezvous during its fifth revolution of the Earth. Most of the orbital manoeuvres had to be performed without reference to Houston, in part as a result of the high inclination but primarily because of severe VHF interference, but Conrad, having flown both Gemini and Apollo missions, had considerable experience in space rendezvous.

"Tallyho the Skylab!" Conrad yelled jubilantly 8 hours into the mission, as the station, now only a few kilometres away, emerged from the Earth's shadow. It was evident that at least one of the main solar panels was indeed still in place. "I can already see the partially deployed solar panel." As the separation reduced, he confirmed the engineers' analysis just before passing out of communication once again. "Be advised, the meteoroid shield is pushed up under the panel." As they crossed the coast of California 15 minutes later, and the radio link was regained, he elaborated: "As you suspected, solar wing-two is gone – completely off the bird. Solar wing-one is, in fact, partially deployed." The surviving boom had rotated about 15 degrees before snagging on a thin strap of metal. He flew the Apollo around to gain a better look at the material that was fouling the hinge, and saw that it was not as bad as had been feared. "We ought to be able to get it out," he ventured, referring to the plan to use long-handled tools to remove the debris that was fouling the boom. He also reported that the micrometeoroid shield had been so completely stripped away that the scientific airlock was clear; if it had been blocked

On the way to rendezvous with Skylab, a Saturn IB lifts off from the 'milkstool' which enabled it to use the facilities scaled for the much larger Saturn V.

this would have interfered with the deployment of the parasol. Conrad was therefore in an optimistic mood when he flew around to soft-dock with Skylab's axial port. "Boy, I've had some big things on my nose in space before," he mused as he stared at the enormous structure looming in the optical reticle in his rendezvous window. "It sure beats the Agena or the LM."

Once they had eaten and prepared their tools, Conrad undocked and withdrew again to attempt to free the boom.

"Fly safe," urged Dick Truly, Houston's CapCom, a few seconds before the spacecraft flew out of communication.

Once Conrad had manoeuvred the Apollo alongside the fouled solar panel, Weitz swung open the large side hatch and stationed himself with his upper body projecting outside. Kerwin held his legs to provide a degree of stability while Weitz tried to free the boom's hinge using the long-handled gripper, but the tool was awkward to use and although the strap that was fouling the boom was narrow it was tightly stretched. After half an hour of making almost no progress, the astronauts relaxed their radio discipline, and when the link was re-established the controllers, hearing some frustrated language, knew that things had not gone well. Upon realising that they were back in communication, Conrad informed Houston that he was calling a halt. "We ain't gonna do it with the tools we've got."

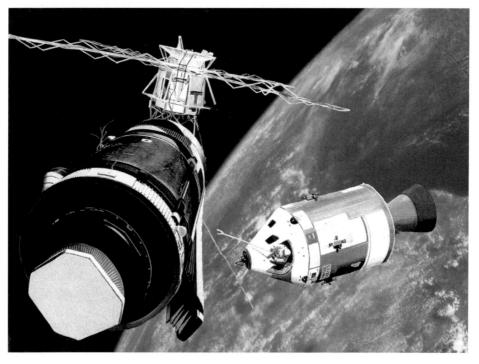

A rendition of Paul Weitz performing the 'stand up' EVA in the hatch of the Apollo spacecraft, using a long-handled tool to attempt to clear the debris that was fouling Skylab's remaining solar panel. (Courtesy of Ed Hengeveld.)

His mood somewhat dampened, Conrad flew around once more and prepared to dock. This time, however, when the Apollo's probe hit the tip of the conical drogue the capture latches failed to engage. The radio link came alive as they overflew a tracking ship, and Conrad told Houston that three attempts to dock had failed. "I'm standing by for any suggestions." It was recommended that they work through the established troubleshooting procedures. When communications were re-established, he told the controllers that he had decided to disable the probe's retraction system and repeat a procedure that had been devised for a similar problem on Apollo 14 *en route* to the Moon. Once the probe had aligned the two vehicles, the spacecraft's thrusters would be fired to ease the main docking collars into contact and hold them together until the main latches had locked. "If we ain't docked after that," he warned Houston, "you guys have run out of ideas." It took several hours to prepare, but it worked at the first attempt. "We got the hard dock!" he said ebulliently upon re-establishing communications, thereby prompting a hearty cheer from the flight controllers. Four hours late, Conrad, Kerwin and Weitz stripped off their suits, had their supper, and retired for the night. The plan had always been for the crew to sleep in the Apollo on the first night and enter Skylab the next day. Having 'slept in' to recover from their 22-hour day, the astronauts spent the morning working on the probe in an effort to identify the fault with its latches then, around midday, Kerwin and Weitz cracked the hatch and ventured into Skylab's MDA. After checking out its systems, they passed through the airlock compartment to make a brief inspection of the orbital workshop beyond.

"The OWS appears to be in good shape," Weitz announced. "It's a little warm." In fact, this was masterful understatement; it was 50 °C. "It feels..." he searched for an analogy for the dry heat, "like the desert."

With that, they gathered in the airlock for lunch, choosing this area because it was cool and had sufficient windows to give each man his own view. Conrad, the only crew member who had been in space before, felt at home in his weightless state and after lunch, enjoying Skylab's scale, he gleefully propelled himself through the axial hatch in the domical 'roof' and did somersaults in the workshop's vast upper compartment.

Most of the afternoon was devoted to moving items that had been set in one place to balance the vehicle during launch but had to be relocated for use in flight. "The mobility is super," Conrad said during one of the all-too-brief communication sessions. "Everything that we've been supposed to unfold or move has been easier than we could have hoped for." This news vindicated those in Huntsville who had been criticised for having planned to reconfigure a 'wet' workshop. However, the temperature in the OWS obliged the astronauts to retreat periodically to the airlock to cool off. The main item on that day's agenda was the deployment of the parasol. This had been scheduled for a series of orbits offering favourable communications in the hope that it could be televised by a camera that Kerwin put in one of the Apollo's forward-facing windows, but their progress was slower than anticipated. "We're progressing slow but sure," Conrad assured, two and a half hours into the operation.

The parasol was contained in a 1.3-metre-long, 20-centimetre-square canister taken from a scientific experiment that was to have been performed in the airlock, therefore attaching it to the airlock was straightforward. The canister contained a central rod with four ribs compressed into five sections. The ribs were connected to a tightly folded sheet of aluminised mylar with a bright-orange nylon cover. The procedure involved a time-consuming task of attaching a series of extensions to the central rod to ease the structure out into space. When they were ready, Kerwin, in the Apollo, pointed out that they were now in orbital darkness, so Conrad waited until sunrise to enable Kerwin to provide a commentary on their progress. By then, however, they were out of communication, so television was impracticable. "We had a clean deployment," Conrad announced upon regaining communications. "But it's not laid out in the way it's supposed to be." The sheet was badly wrinkled, but soon began to straighten out in the heat of the Sun.

While the crew slept, the controllers put Skylab into solar-inertial attitude and telemetry indicated that the parasol was an effective shield. In fact, the computer model predicted that the temperature in the workshop would take three or four days to drop to a nearly nominal 26 °C. It would be hot, but habitable. Mueller had hoped that Apollo Applications would show the value of the human presence in space, and so far Skylab had proved their utility as repairmen.

Most of the next day was devoted to completing the unstowing and testing of scientific apparatus, but Weitz took time out during a favourable communications pass to describe to Rusty Schweickart, the commander of the back-up crew, precisely how the strap was snagging the solar panel boom so that Schweickart could test a procedure in the neutral buoyancy tank designed to release it. The temperature in the workshop was falling, but by nightfall it was still uncomfortably hot, so rather than spend another night in the Apollo the astronauts rigged their sleeping bags in the

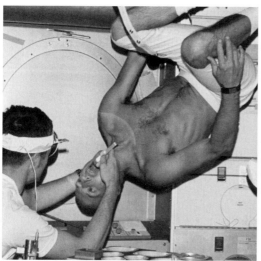

Pete Conrad, commander of the first Skylab crew, on the exercise bicycle (right) and, floating upside down for convenience, is subjected to a dental inspection by physician Joe Kerwin (left).

cooler MDA. The original flight plan had allowed two days to activate Skylab before starting on the science programme. Although the activation was running a day late, and the power crisis would limit the scientific work, it was remarkable in the circumstances that the 'laboratory in the sky' had any future at all.

The Gemini spacecraft had been so cramped that its crews had resisted any medical experiments. In part, this reflected the traditional antipathy that pilots had towards flight surgeons. The fact that a flight surgeon had authority to ground a pilot was not something to be taken lightly in performing a 'scientific' experiment. In the case of Skylab, however, such experiments were the primary reason for this first mission. As the geologists had discovered during the later Apollo missions, the best way to gain the astronauts' support was to convince them that the scientific work was a significant mission objective and not simply a distraction from the mission's true objectives. Once the astronauts realised that they would be assessed by how well they conducted such work, they would give it serious attention. Appropriately, the scientist on this first crew, Joe Kerwin, was a physician.

The biomedical tests began on 31 May when a blood sample, which was taken from each man, was spun in a small centrifuge to separate its components and frozen for return to Earth for detailed analysis. Previous missions had revealed that the red blood cell count diminished during spaceflight and that these normally disk-shaped cells were distorted. It was hoped that sampling on a daily basis would enable the way in which the cells adapted to weightlessness to be monitored in detail. Another daily test was body-mass measurement: to 'weigh' the body in weightlessness, the crewman was strapped into a frame on a spring-mount and the manner in which the frame oscillated after application of a calibrated impulse provided a measure of his mass. As a main benefit, of course, Skylab was large enough to accommodate bulky items of apparatus. Once Weitz was settled in the Lower-Body Negative Pressure chamber, Kerwin performed the cardiovascular tests. The next test used a rotating 'chair' to study the vestibular system's adaptation to weightlessness. Finally, Weitz strapped on the gas analyser's face mask to measure his metabolism while riding the velo-ergometer. This test, however, did not go according to plan. He was expected to use waist- and shoulder-straps to 'sit' on the seat, but they so inhibited his action that he had to stop. This upset the experimenters' calibrated exercise programme, complete with 'controls' that were meant to assist the interpretation of the data. When Conrad was tested, he persevered and completed the programme. Upon reflection, however, Kerwin pointed out that as the heat could be influencing the data they should take things easy until the temperature in the workshop had levelled off. To overcome the balky harness, the astronauts experimented with new ways to ride the 'bicycle'. Weitz poked his fingers through the grid of the ceiling (really the floor of the upper compartment) to help him to maintain his position as he pedalled. The ever-inventive Conrad 'stood' on the ceiling and 'pedalled' the bike by hand! As the human body adapts to weightlessness, the unloaded spine's vertebrae open up. As the mission progressed, Conrad was delighted to observe that he had become temporarily taller than his wife. Of course, he would shrink as soon as he returned to Earth.

The observational programme was started in earnest on 29 May, when the ATM

Owen Garriott at the control panel for Skylab's Apollo Telescope Mount.

received its 'first light', and Kerwin spent most of the day studying the Sun. As the temperature in the workshop had stabilised, the crew transferred their sleeping bags into their individual cabins. The next day, after a run on the ATM, Skylab was reoriented to nadir-inertial attitude and Weitz secured the first remote-sensing data with a long pass from the Pacific off Washington State down across Central America to Brazil. However, although Skylab returned to face the ATM's solar panels to the Sun prior to flying into the Earth's shadow, five of its storage batteries dropped offline in orbital darkness. Clearly, taking Earth-resources data would have to be limited while the station was short of power.

In fact, Schweickart's trials were sufficiently encouraging for NASA to permit the astronauts to venture out and try to sever the strap that was restraining the solar panel boom. If the Gemini spacewalks had proved anything, it was that a freely floating astronaut cannot do useful work. Having seen Dick Gordon flail about while attempting to work on an Agena rocket stage, Conrad realised that it would not be straightforward to even reach the boom, let alone cut the strap. EVA was part of the Skylab mission, but it was to replace the ATM's film packs. All the handholds and foot restraints were in the ATM's support structure. How could they make their way across the workshop's open skin to the hinge of the solar panel boom?

The one-week-in-space milestone on 1 June was supposed to be a rest day, but Conrad decided to use it to catch up on loose ends. Afterwards, they made a television broadcast. Conrad had bet a colleague that he would be able to 'walk' around the inner surface of the workshop's cylindrical wall. He started on hands and knees in a slow crawl and then transitioned to an upright gait as he build up speed. His antics gave a whole new meaning to the term 'spacewalking'. A prankster, he had emerged from the Waste Management Compartment after an hour and a half, chuckling that he could now boast of having circled the world while sitting on the toilet.

While Skylab countinued ATM runs, Schweickart's team worked on the EVA procedure. This was received by teleprinter for the crew's consideration on 5 June. The next day, in a televised rehearsal in the workshop, Conrad and Kerwin suggested a few refinements, which were approved, and so on 7 June, the day they exceeded Gemini 7's 14-day endurance record, Weitz moved into the Apollo spacecraft to enable him to act in an emergency, and Conrad and Kerwin donned their suits in the airlock. Although he thought the plan was feasible, Conrad was not particularly confident of the outcome.

Skylab's external hatch was a left-over from Gemini. It opened outwards into the framework which supported the ATM's structure. Radio communications were lost while the airlock was being depressurised and by the time they were regained Conrad and Kerwin, stationed among the struts, had already assembled five rods to form an 8-metre-long handle for the cutter. At that point Skylab passed into darkness, so they decided to await the Sun before attempting the next stage of the operation. With the link often out for an hour at a time, they were effectively on their own and would have to resolve difficulties without ground support. Back in sunlight, Conrad and Kerwin manoeuvred around to the side of the Fixed Airlock Shroud that formed the rim of Skylab's cylindrical body. From there they could peer along the shell at the boom's hinge. In Schweickart's plan, Kerwin was to swing the cutter and engage the strap so that Conrad, utilising the handle as a hand rail, could make his way across the exposed surface of the vehicle to the hinge. With no footholds on this part of the shroud, Kerwin had to loop one arm around the mount of an antenna. It was difficult to manipulate the long-handled cutter using only one hand. Once it looked as if the cutter had engaged the strap, Kerwin let go of the antenna mount in order to tie his end of the handle in place, but as soon as it was free, his body reacted against the action, floated out of position and the cutter worked loose. After half an hour of this, he tied himself to the antenna mount, which allowed both his hands to be free. This enabled him to emplace the cutter at the first attempt – which was just in time, as Skylab entered the Earth's shadow a moment later. At sunrise, Conrad translated hand-over-hand along the cutter's handle. At the hinge, he affixed a hooked tether to a vent one-third of the way along the boom's length. The other end of the tether was fixed to the rim of the Fixed Airlock Shroud. After inspecting the jaws of the cutter to verify that the strap was correctly positioned, Conrad made his way back beyond the hinge to wait while Kerwin fired the cutter. However, no matter how hard Kerwin yanked on the rope connected to the cutter's trigger, nothing happened. No sooner had Conrad started in for a second look than his action vibrated the handle and the cutter fired. With the strap severed, the boom rotated, but only for another 20 degrees. Schweickart had foreseen this, and had devised a procedure to snap the bracket restraining the boom, and it was with this contingency in mind that Conrad had affixed the tether. Once Conrad had rejoined Kerwin at the rim of the Fixed

Airlock Shroud they tugged on the tether, but the boom refused to open further. Determined not to be beaten after having severed the strap, Conrad moved back along the handle to the hinge, adopted a crouching position between the tether and the skin of the workshop and 'stood up' with the tether strung over his shoulder for increased leverage; but even this inspired improvisation did not snap the bracket until Kerwin, still tied to the antenna mount, tugged on the end of the rope, whereupon the bracket snapped, the boom rotated perpendicularly and locked, and Conrad flailed around wildly on his 16-metre-long umbilical until Kerwin pulled him to safety.

Having plenty of time available, Conrad decided they should do some opportunistic maintenance on the ATM. The cover of one of the X-ray telescopes was ominously sticky, so Kerwin climbed up and pinned the cover open. Also, as the film cassette on the Ultraviolet Spectroheliograph had been feeding erratically, he replaced it. With the Earth passing below, the view from the top of the ATM looking down over the bulk of the workshop was astounding.

"We've got the wing out, and locked," Conrad announced upon re-establishment of communications as they were heading back to the airlock, prompting a round of applause in the control room. When he pointed out that the three transducer arrays seemed to have stopped short of full deployment, Houston ventured that as the frozen hydraulics warmed in the Sun they would unfold, as they eventually did. Thanks to a great deal of ingenuity, both on the ground and in space, Skylab's future was now assured. It was a happy crew that gathered in the wardroom for supper that night.

Although the loss of the other solar panel meant that Skylab was 2.5 kW below nominal, Huntsville had built in sufficient margin for the 7 kW that was available in solar-inertial mode to pursue the planned science programme, so the astronauts settled down to a routine fairly similar to that of the original flight plan.

When the ATM was conceived, it had been intended to launch it in time to study the Sun through the peak of its 11-year sunspot cycle, but the switch to the 'dry' configuration had resulted in its being flown as the Sun neared its minimum level of activity, and the scientists were worried that the entire mission might be completed without observing a solar flare. The X-ray sensor that was supposed to alert the astronauts to heightened solar activity – and the possible onset of a flare – turned out to be rather too sensitive because it was triggered by the enhanced radiation of the South Atlantic Anomaly, a broad region in which the inner Van Allen belt dips down slightly towards the Earth, through which Skylab passed frequently. On 15 June, Weitz decided to spend his 'day off' at the ATM. After a few hours of snapping interesting points of detail for the record, he spotted the sign of a flare on the rise. Within seconds he had adjusted the instruments and was able to follow the activity through its peak and well into its decline before Skylab flew into the Earth's shadow. The ATM scientists were delighted. Solar scientists who had steadfastly maintained that there was no role for humans on an observatory in orbit began to wonder whether they might have been too hasty in their dismissal of Skylab.

Another major milestone was passed on 18 June, when Conrad, Kerwin and Weitz matched the 24-day record set by the ill-fated Salyut crew. The next day,

Conrad and Weitz went out to collect the film from the ATM. First, however, Houston asked Conrad to attempt to 'fix' a battery recharger that had failed earlier. It was believed that its relay had siezed, and he was requested to tap its cover with a hammer in the hope that the shock would release it.

"Boy," Weitz laughed, "is he hitting it!"

"Did anything happen?" Conrad asked after delivering several savage blows.

"Thank you very much gentlemen, you've done it again!"

On the lunar surface four years earlier, Conrad had released a jammed nuclear fuel element from its container by hammering it. "Never come to the Moon without a hammer," he had quipped. Now it was clear that his motto had wider application.

With this chore attended to, Conrad moved up to the ATM and pulled the film packs for return to Earth. While there, he moved so that he could visually inspect the Apollo. It had been in space for twice as long as its predecessors and the Sun was peeling its paint, but it was otherwise in good condition. Three days later, they boarded their spacecraft with their scientific results and undocked.

"Bye bye, Skylab," Conrad yelled to the vehicle that had been their home for 28 days. As commander, he had the satisfaction of knowing that he was leaving it in considerably better shape than he had found it and, against all the odds, he had completed his assigned mission which, when all was said and done, *was what mattered.*

For the first time (and despite the mishap which had befallen the Salyut crew) an Apollo crew made atmospheric re-entry without wearing full pressure suits. In another departure – in order to avoid the stress of struggling out of the capsule into the raft and being plucked by a bosun's chair into a helicopter – it had been decided that they would remain in the capsule until it was safely on the flight deck of the aircraft carrier. It had been expected that they would have to be carried, but after a few moments of acclimatisation to recover from the dizziness as the blood drained into their legs, they were able to walk. This answered the basic biomedical question: the human body could indeed survive a month in space. This pioneering crew had not merely survived exposure to weightlessness, they had *worked* their passage. In every sense, therefore, the first Skylab mission had been a triumph.

Having invested years of their careers in its instruments, Skylab's many scientific investigators had been shocked to hear of the damage that the vehicle had suffered during launch. The deployment of the parasol to overcome the thermal stresses had prompted optimism, but it was not until the astronauts released the solar panel boom that the scientific programmes were secure. The scientists' frustration was all the greater because it was evident that their instruments were producing high-quality data. As Robert Parker, an astronaut serving as Skylab's 'programme scientist' put it, the experimenters were 'starved' for data. His function was to mediate requests for observing time. Each experimenter had a severe dose of 'tunnel vision', considering that Skylab was a platform for his particular instrument and that time spent on anything else was 'wasted'. Even astronaut Karl Henize, an astronomer with a stellar photography experiment on Skylab, was frustrated. The biomedical experiments had priority on this first mission, so scheduling the competing activities was a major headache. Nevertheless, once Skylab's power crisis was resolved, Conrad, Kerwin and Weitz were exhuberant in their enthusiasm with the result that, overall, despite losing time, they were able to complete almost 100 per cent of the biomedical tests and 80 per cent of the solar work, but only 60 per cent of the terrestrial work because Skylab could not depart from solar-inertial attitude for very long in the early days. Of course, this is relative to the ideal baseline mission. When the output was tallied up, it included 29,000 pictures of the Sun, 10,000 pictures of the Earth and some 14 kilometres of sensor data stored on magnetic tape.

SECOND CREW

Skylab had been designed to function for at least 8 months, this being the time needed to follow the initial 28-day mission with two 56-day missions and allow sufficient time for the results of one mission to be assessed before starting the next. Although it would have been more adventurous for the final crew to relieve their predecessors in orbit, such a handover was not an objective of the programme. However, the 'rate gyros' which enabled the ATM to control Skylab's orientation were proving unreliable, and there was some concern that the station might not survive its planned duration. To provide a degree of redundancy, it was decided to send up another package of gyros with the next crew. In order not to waste the station's limited lifetime, it was decided to bring forward the second crew's launch by three weeks. One result of the attitude control problem was that little ATM work was possible while Skylab was untended, so the solar scientists were eager to have it reoccupied. The Saturn IB on its 'milkstool' was ready to go; it had been rolled out to Pad 39B towards the end of the first mission, just in case the prolonged exposure had damaged Conrad's Apollo spacecraft and it had proved necessary to mount a rescue mission.

As Director of Flight Crew Operations, Deke Slayton chose who would fly each mission. The experimenters hoped he would assign *two* scientist-astronauts but he insisted that each crew have a second pilot in case the commander was incapacitated. In this case, Jack Lousma flew as the second pilot and engineer and the scientist was physicist Owen Garriott. On 28 July, Al Bean, who had walked on the Moon with Conrad, led his crew out to the pad. Although this was his first mission, Lousma dozed off during a lull in the countdown and the medical sensors tracked his heart rate down to a mere 38 beats per minute.

The launch window was very brief on a rendezvous mission. For an efficient rendezvous, the launch had to be timed so that the chaser would be in the plane of its partner's orbit at the moment of orbital insertion. As usual, the Saturn IB performed flawlessly. However, no sooner had he begun the rendezvous than Bean reported that "sparklers" were flying in formation with them. On investigation, the telemetry revealed that there was a leak in one of the four RCS thruster quads and tiny globs of fuel were drifting alongside the spacecraft. An Apollo could operate with only two thruster units and with the third providing a level of redundancy, so they deactivated the faulty thruster. As had their predecessors, they drew alongside Skylab during their fifth orbit. Once a TV camera had been set in one of the windows, Bean performed a fly-around inspection. As the thruster plumes impinged upon it, the parasol was seen to 'flap', almost with sufficient force to rip it off. This time, the docking was straightforward. It had been hoped to reactivate Skylab within two days to provide time to prepare for the spacewalk on which they were to deploy the replacement sunshield. However, this optimistic schedule fell apart within hours when all three men succumbed to 'space sickness'.

One in three of the astronauts on the Apollo missions had suffered brief periods of nausea and several had vomited, but they had all soon recovered. In this case, however, the Skylab crew not only found the experience debilitating, but it persisted overnight and through the next day, during which time so little was being achieved that Bean ordered everyone into their bunks in an attempt to allow their vestibular systems to settle. The flight surgeons had been sceptical of the early rumours of cosmonauts suffering sickness, because the Mercury and Gemini crews had shown no such symptoms. They were particularly surprised when Frank Borman, who had survived a fortnight in Gemini 7 without undue effect, vomited as Apollo 8 was *en route* to the Moon. They had concluded that this 'stomach awareness', as they termed it, was a form of motion sickness that manifested itself only in a vehicle that was large enough for its occupants to move freely. In an attempt to investigate this vestibular phenomenon, Conrad's crew had subjected themselves to the diabolical rotating chair designed to induce motion sickness. Given that the first crew had shown no ill effects, the medics regarded the sickness that afflicted Bean's crew as a welcome opportunity to investigate this aspect of adaptation to weightlessness. By 1 August, fully recovered, the astronauts belatedly began the formal biomedical programme. Once this data had been digested, the flight surgeon consented to the spacewalk being rescheduled three days hence. The astronauts were awakened the next morning by a master alarm. For no obvious reason, a second thruster unit on the Apollo had developed a leak and Houston was concerned that this might imply a generic fault that sooner or later would cripple the spacecraft. If this had been an independent flight the mission would have been terminated, but as Skylab was a safe habitat the flight controllers pursued the standard procedure of not unnecessarily closing off options and took time to study the situation further. They discovered that it was not a 'show stopper' and that the two leaks were unrelated. However, while this issue was being worked, it was decided to postpone the spacewalk to 6 August.

With the ATM unavailable, Skylab was reoriented on 3 August to perform the first Earth-resources run. The schedule called for one track per day for the first few weeks, a hiatus during the middle of the flight while lighting over the United States was poor from Skylab's perspective, and then a resumption later in the mission. Although each run collected at most 35 minutes of data, it took an hour or so to prepare the apparatus. In fact, this experiment usually involved the entire crew. One man, most often Garriott, would mount the Terrain Camera in the scientific airlock facing the ground (the airlock on the other side of the workshop from that from which the parasol was projecting) while the other two worked in the MDA, with Bean aiming the Infrared Spectrometer and Lousma (whose speciality this was) working at the control panel. An attempt to photograph the Earth's horizon to record 'airglow' and the ozone layer for the Naval Research Laboratory the next day had to be abandoned when the apparatus refused to retract into the scientific airlock and had to be jettisoned to clear the airlock. In addition to two Earth-resources runs the following day, Lousma and Garriott prepared the EVA suits for the coming spacewalk.

Because the parasol's deployment mechanism had been 'proven', it had been suggested that the parasol ought simply to be replaced by an improved version in order to obviate the spacewalk, but an EVA was required in any case to load film into the ATM. Furthermore, replacing the parasol would require the workshop to be exposed during the time it took to erect the replacement, and Skylab would be in serious trouble if this deployment failed. If the deployment of Huntsville's A-frame sunshade was frustrated, at least the parasol would still be in place.

This time, because the spacewalkers would be working near the airlock, they would be able to utilise the mobility and stability aids mounted just outside the hatch, so the task was expected to be straighforward. While Garriott remained at the 'workstation' near the airlock and assembled the rods of the A-frame, Lousma made his way into the ATM's support structure to affix the bracket on which it would be mounted. Each of the rods required the connection of a dozen tubes, through which a rope had to be threaded to form a drawstring loop. The assembly progressed considerably more slowly than when they had rehearsed it under water, and Garriott's commentary on the differences served as a useful assessment of training in the neutral buoyancy tank. In fact, it took an hour to assemble the first rod. He finished the second rather more rapidly, having adapted his technique to match the constraints that applied in space. There was no rush, however, because the suits would sustain them for almost a full 8-hour 'working day'. Once the two rods had been mounted on the bracket they were splayed out in a V-shape over the workshop. All that required to be done then was to tie the sheet of the shield to the ropes and cycle the loops to deploy it. Although the sheet initially retained deep folds it flattened out as it was heated by the Sun and was soon stretched tightly over the top of the parasol. With two levels of protection, the temperature in the workshop gradually fell to its originally planned level. Being more resistant to the space environment, this new shield was expected to see Skylab through to the end of its operating life. As the deployment procedure had consumed four hours, which was substantially longer than predicted, Skylab was serving one of its originally envisaged roles of learning how to do productive external work. With the sunshade deployed, Garriott loaded film packs into the ATM's instruments and then affixed an experiment to one of its struts. NASA had made several studies of micrometeoroids (most notably by the Pegasus satellites) using electronic detectors which radioed their data to Earth. In this case, however, the detector was to be retrieved and returned for study so that the impacts could be analysed, and not simply counted. Before climbing off his perch, Garriott visually inspected the RCS thrusters of the Apollo for leaking propellant but saw nothing. Back at the airlock, he and Lousma checked the radiator for a coolant leak but saw nothing amiss, which was surprising because both the primary and secondary coolant loops were gradually losing pressure. Due to the multiplicity of assignments, the EVA lasted six and a half hours. Clearly, the spacewalking technique had greatly improved since the Gemini experiments.

Al Bean sets out to install the film in the Apollo Telescope Mount.

The next day, 7 August, Garriott started full-scale ATM work. Although the Sun was fairly inactive he inspected the features on view and, because some of the instruments transmitted imagery to the ground as well as presenting it on the ATM screen, he probed the experts for commentary on what he was seeing, in order to compare the ATM trainer with the real thing. Two days later, he had the satisfaction of documenting a 'medium' flare, and the following day a solar observatory in the Canary Islands noted another; Skylab's overly-sensitive 'alert' was not being used, so when the news was relayed and Bean started up the ATM, he was on-line to record the enormous eruption that developed a little later. Such 'coronal transients' were rare during solar minimum, so catching one in all its complexity was a welcome observation. Given the unusually active Sun, the ATM was operated every day from this point, and on 21 August Bean found a huge prominence looking "like a bubble" on the Sun's eastern limb. Although he had to break off for Skylab's shadow passes, over the next few hours he was able to monitor its expansion through the corona. It was as if the Sun had laid on this timely display especially for Skylab's benefit.

Although a great deal of effort had gone into the design of NASA's first space station, it was inevitable that some of its systems would prove to be less reliable than designed, but some had degraded surprisingly rapidly. Conrad had had trouble with the freezers and the airlock's coolant loops. During Bean's tenure, the pressure in both the primary and the back-up coolant loops in the airlock slowly fell, even though there was no sign of a leak. When the dehumidifier developed a leak, Bean spent a day fixing it. Maintenance was becoming a significant overhead, but the manner in which a system failed, and how it was repaired, was valuable engineering knowledge that would hopefully be put to profitable use on a future space station. For the astronauts, however, chasing leaks was not as satisfying as seeing explosions on the Sun.

Throughout this fortnight of ATM work, the initial phase of the Earth-resources studies were concluded. However, the rapidly worsening condition of the attitude control system's rate gyros led to a loss of nadir-inertial lock on 11 August, which prompted the decision to install the package of gyros that had been ferried up. Although the hardware was easily mounted in the MDA – in order to be near the ATM – it could not be activated until a spacewalk ran the cabling to the control system on the ATM, so this task was added to the list of chores for the ATM's next service call. In the meantime, Skylab would have to limp along with the few remaining gyros of the original set. The engineers had come to the conclusion that the gyros had failed because bubbles of gas in the floating chambers had caused them to overheat. Mounting the new units inside the station would obviate this. The root of the problem was that the ATM and its attitude control system had been designed as a converted lunar module that was expected to operate independently, with its own cabin, but when it had been integrated into the 'dry' workshop it had been reconfigured as an unpressurised shell.

While Garriott was observing the Sun on 13 August, Bean and Lousma conducted a key technology experiment. Ten years earlier, the Air Force had hoped to fly 'Blue Gemini' missions. It had developed a 'rocket backpack' so that an astronaut would be able to fly over to inspect 'hostile' satellites. When NASA agreed to assess this,

Gene Cernan was given the task on Gemini 9 but he had found moving purposefully while floating freely to be more difficult than expected and was unable to test the backpack. However, the manoeuvring unit had been loaded aboard Skylab, allowing it to be tested in the cavernous workshop, to certify it for use by astronauts working outside the Shuttle. Bean did not even need to don his spacesuit, he simply strapped himself into it. He assessed its stability and ability to move purposefully and reported that it required some additional work. Although it had excellent attitude control, its translational and rotational actions did not match his expectations as a spacecraft pilot.

The 24 August spacewalk by Lousma and Garriott to service the ATM and wire up the new package of gyros was completed without incident. They were out for four and a half hours but the pace was relaxed: working on the ATM had become part of the routine. The next day, as they took their predecessors' endurance record, Skylab was now in reasonable shape to sustain them for the full mission. NASA's strategy had been for successive endurance missions to double the record, but an 'increment' of a month appeared rather adventurous, so once Bean's crew exceeded the record they would be granted extensions of a week at a time, or would be recalled if their biomedical tests showed undue deterioration. Nevertheless, although the original plan had called for a 56-day mission, this was extended by 3 days in order to optimise the recovery.

In an effort to catch up from their initial 'space sickness', Bean had adopted the motto "one hundred per cent, and more if possible" and they had soon increased the assigned 8-hour 'work day' to 12 hours. To Houston's amazement, Bean asked that they be set more tasks because "we've got the ability, and the time and energy". Houston had five teams of controllers sharing the load of such a long mission. "I know y'all do down there," Bean taunted them, laying down the challenge of an

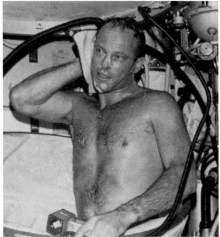

Owen Garriott in the Lower-Body Negative Pressure chamber (left) and Jack Lousma wipes off after a shower (right).

accelerated pace. The experimenters were delighted. They had 'ticked off' nearly all of the miscellany of 'corollary' experiments by the midway point. As they pressed on, almost one and a half times as many hours were spent on solar and terrestrial observations as were called for in the original plan. Bean proposed that perhaps the mission should continue to be extended on a week-by-week basis as long as the crew was willing, but Houston refused. The final week was spent preparing to depart. On 22 September, Bean and Garriott retrieved both the film from the ATM and the micrometeoroid experiment that they had previously deployed.

Upon their return to Earth on 25 September, Bean's crew were found to be in *better* shape than their predecessors. Skylab was certainly achieving its primary objective of establishing that the human body could adapt to the space environment, and later rapidly readapt upon returning to gravity. Although most of the changes that had caused such concern before Skylab stabilised in an adapted state, atrophy of the bone seemed to be a 'runaway' process that would pose a serious challenge to futher extension of the endurance limit.

WHAT NEXT?

NASA now faced a dilemma. On the one hand the concern over the degradation of Skylab's systems argued for dispatching the final crew as soon as possible. On the other hand, a comet had been discovered just before Skylab had been launched and this would be passing through the inner Solar System at the turn of the year and offered an ideal opportunity to expand Skylab's scientific yield. If the final mission was brought forward, the crew would be home before the comet reached perihelion, but the astronomers argued that it would be better to *delay* this final mission by a month to enable the comet to be followed as it moved away from the Sun.

The leaking thrusters in Bean's Apollo spacecraft had raised the possibility that his crew would need to be rescued, so the third Saturn IB had been prepared well in advance. When it had seemed that Skylab's attitude control problems might mean that leaving the station untended would risk its loss, a crew 'handover' had been considered. Two options presented themselves. The new Apollo could dock at the MDA's radial port the day before the retiring crew left, but the asymmetric configuration would have further complicated attitude control. The alternative was to have the new crew draw up alongside, wait until Bean's crew undocked and then move in and dock at the axial port. The risk in both scenarios, of course, was that if Bean's Apollo turned out to be unmanoeuvrable he would have to redock until a rescue spacecraft with a two-man crew and three spare couches could be prepared. On 16 August, when Bill Schneider, Skylab's Programme Director, decided that Bean's crew should fit the augmentation package, he added that the final mission would *not* be advanced but would launch on schedule. However, it soon became evident that the launch would have to slip two days to 11 November in order to accommodate the rate at which the plane of Skylab's orbit was precessing.

On 2 October Kenneth Kleinknecht, Skylab's Programme Manager, cited the outstanding performance of Bean's crew as proof that an astronaut "was able to do more than we thought he could do". In all likelihood, this final mission would be the agency's last opportunity for man-in-the-loop space experiments for at least a decade. Considering the performance of Bean's crew as the benchmark, the activities planners set out to schedule the final mission at a similar pace. The experimenters were also eager to expand their programmes in order to follow up issues raised by their initial results. A week later, after an inventory showed that Skylab had plenty of air and water and that the constraint was food, it was decided that sufficient food could be ferried up to undertake an 84-day mission. As previously, the extensions beyond the 56-day baseline would be incremental. The astronomers were delighted because the extra month would enable the comet to be monitored as it withdrew from the Sun. The biomedical team was delighted, too, because this extension would yield new data. With all the extra experiments, the comet and the high level of expectation, it was evident that this final mission would be very demanding.

An inspection on 6 November discovered hair-line cracks in the tail fins of the Saturn IB. It had been on the pad since 14 August, when it had seemed possible that it would be needed to rescue Bean's crew. The first stage was ten years old, and the weather had prompted stress corrosion. It was a common problem and the Cape's engineers had spare fins in storage. This time, however, their task was complicated by the need to exchange the fins with the rocket perched high on the 'milkstool' because there was no time to return it to the assembly building, repair it and roll it out again within one of Skylab's 5-day cycles. The delay was welcomed by the astronomers because it meant that they would be able to track their comet that little bit farther as it withdrew from the Sun. After much hype that it might be a 'Great Comet' of the type that had been common in the late nineteenth century, it turned out that it would not be well positioned for terrestrial viewers, so Skylab would be uniquely able to study it near perihelion.

ERROR OF JUDGEMENT

Because the entire Skylab programme was to be completed over an 8-month period, Deke Slayton had named the three crews well in advance. For the final crew he had relaxed his rule that the commander had to have flight experience, and selected three 'rookies'. Gerry Carr was in command, Bill Pogue was the second pilot and engineer, and Edward Gibson, another physicist, was the scientist. Skylab had therefore been something of a 'dead end' for the back-up crews, particularly for Conrad's back-up, Rusty Schweickart.

The Saturn IB's performance was typically flawless and the rendezvous was on time, but it took three attempts to engage the docking probe's capture latches. In an effort to preclude the 'space sickness' which had afflicted Bean's crew upon entering Skylab, Carr's crew was to spend the first night in the Apollo spacecraft in the hope that their vestibular systems would be acclimatised to weightlessness prior to having to cope with the voluminous workshop. Even so, no sooner had they completed the tricky docking manoeuvre than Pogue vomited. Carr decided that since they were out of communication with Houston they should neglect to mention "the barf". Unfortunately for this subterfuge, the tape which recorded spacecraft telemetry while out of communication and dumped it upon flying into range, also recorded the crew's conversations. The next day, using an open circuit, Al Shepard reprimanded Carr for his "error of judgement" and Carr accepted that it had been "a dumb decision", if only because it directly impacted on the biomedical studies.

The long-standing procedure was for vomitus to be frozen and returned to Earth for analysis. Lest this sound bizarre, recall that the Skylab crews were obliged to document *everything* they ate and to collect samples of *all* wastes to document their adaptation chemically. NASA's interest in 'space sickness' was not just scientific. If Shuttle crews were struck down as badly as Bean's crew had been, this would constrain flight operations as it would be unwise to schedule a spacewalk on the first few days because vomiting in a helmet could be fatal. Given the importance of studying this aspect of space adaptation, Carr's rebuke was fully justified.

A POOR START

The following morning, the newcomers opened the hatches and entered their new home only to find that the previous residents had left a team of caretakers. One empty spacesuit was astride the velo-ergometer, one was strapped on the toilet and the other had been inserted into the pressure chamber.

Even though they were not sick Carr, Gibson and Pogue soon fell behind in the activation of the station, in part because some items were not where they expected them to be but mainly because it took time to learn to function in such a capacious vehicle. As a result, what had been assigned two days spilled over into a third day. Station activation was not a matter of just switching systems on; the various operating modes of each system had to be verified. Anything that did not perform as intended had to be investigated and rectified. Skylab had fared well while untended, but there were some outstanding repairs from the previous occupation, the most crucial of which was the airlock's coolant system. When the primary loop had lost pressure the secondary had taken over, but this too had gradually deteriorated, and Pogue's task on the third day was to try to replenish the primary loop. When the system had been conceived, replenishment in space had not been 'designed in'. Fortunately, the pipe was accessible. A tank had been supplied, and once Pogue had stripped off the insulation and clamped the valve onto the pipe he turned a screw to pierce the pipe and injected 20 kilograms of 'coolanol' from a flask using a nitrogen-fed manifold. The potentially messy repair was completed without incident. If this effort had been frustrated, it would not have been possible to extend the mission beyond the turn of the year.

Several new biomedical tests had been introduced, and Pogue attempted one the next day. The astronauts were to photograph one another daily using an infrared 'thermal imaging' camera to highlight areas of heat loss and thus provide insight into the migration of body fluids. Another new test involved repeatedly measuring the girth of the body at 50 specific points in order to determine how it changed shape. In consenting to the 84-day flight, the flight surgeon had insisted upon an increase in the exercise regime to try to limit the physical deterioration. On Conrad's mission, each man had exercised for half an hour each day and Bean's crew had exercised for one hour each day, so the daily exercise for Carr's crew was raised to an hour and a half. In light of the fact that the velo-ergometer had not worked the leg muscles as expected, Bill Thornton, a physician-astronaut, had devised a new exercise utility in the form of a teflon sheet to be affixed to the floor grid so that an astronaut wearing a harness could realistically walk on this 'stationary treadmill'. This proved to be effective at stressing leg muscles which could otherwise rapidly atrophy and lead to 'chicken leg' syndrome.

Houston's ambitious schedule called for the science programme to ramp up rapidly, but Carr warned on 20 November, "We haven't had time to stow everything properly, and this place is really getting to be a mess." They had ferried up a tonne of supplies which had to be unloaded and stored. It was worth spending a little more time making Skylab ship-shape now, he said, for otherwise time would be wasted later in tracking down misplaced items; he then recommended that the schedulers delete some assigned tasks.

Pogue and Gibson went out to service the ATM on 22 November. Their first task was to document Skylab's physical state, but the camera failed after Pogue had taken only a few shots. Once Gibson had loaded film into the ATM, he pinned open the cover of one of its sighting telescopes. Late in the second crew's tour the Microwave Radiometer's dish antenna had stuck, and Pogue was to try to repair it. However, as the antenna was on the other side of the Fixed Airlock Structure from the ATM workstations, Gibson affixed a foot restraint to the trusswork and held Pogue's legs, leaving both his hands free to insert a pin to lock the antenna in pitch. Carr, from inside, then verified that it scanned in yaw to either side of the ground track. Fixing the antenna had been assigned such a high priority that if they ran out of time they planned to return the next day. In fact, they were out for six and a half hours. The astronaut repairmen had demonstrated their worth once again.

That night, the astronauts were awakened by the master alarm in response to the failure of one of Skylab's three CMGs. This particular unit had first given cause for concern on 3 November, when it had momentarily fluctuated. The rotational bearing was contained within a bath of oil, and it seemed that the lubricant had degraded allowing the bearing to rub and hence overheat. There was no option but to turn it off. Skylab could be controlled by the two remaining units but its changes of attitude would not be as efficient. The Huntsville engineers immediately began to run tests to determine whether the lubrication problem was generic, in which case Skylab was in serious trouble.

CATCHING UP

Meanwhile, the crew tried to catch up with their flight plan. Reflecting upon the first week, Carr reported that it was "very, very demoralising" to slip further and further behind. "The best word I can think of to describe it, is 'frantic'." He suggested that it

The scale of Skylab's Orbital Workshop is evident in this view from the hatch in the top of the tank, looking through the grid-floor to the lower level. Gerry Carr (right) and Ed Gibson (left) are in the opening between levels.

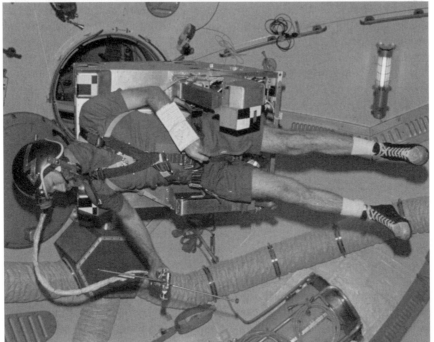

Gerry Carr tests the Astronaut Manoeuvring Unit (left) and demonstrates zero-g by balancing Bill Pogue on his finger (right).

was a matter of adapting to their environment. It "just takes a great deal of time" to learn to move about effectively. In his view, during the first few days their schedule had been "too busy". The training regime had been unrealistic. "When you get up here, it's a whole new world." Before launch, Carr had warned that on such a long mission it would be impracticable to sustain the extreme pace of their predecessors but it appeared that the planners had ignored him. In fact, the planners felt let down by their new charges.

The first week had indeed been frantic. On the other hand, the longer it took to activate the station and to complete the vital repair of the airlock's primary coolant loop, the longer it would have been before they could go out to service the ATM and repair the Earth-resources scanner. Now that this was done, they could start the programme of scientific observations. However, by the end of the second week the planners' continuing efforts to ensure that his crew was never idle had made Carr irritable. As always, the crew took its lead from its commander and a 'them and us' situation arose with neither side willing to speak plainly about their frustrations on the open radio circuit. In the early days of his tour, while struggling to rescue the stricken space station, Conrad had sharply criticised the overly optimisitic daily schedules. Bean had also made several pointed remarks, but in the opposite sense. Not having flown before, and having been reprimanded for concealing Pogue's bout of sickness, Carr was rather more circumspect. Carr's dilemma highlighted the benefit likely to result from handing over a fully operational station from one crew to the next. Quite apart from the time to be saved from not having the retiring crew mothball the station only for the newcomers to reactivate it, each specialist could brief his counterpart on how the systems differed from the simulators and the two crews could share the unloading of stores. Furthermore, the retiring commander would be able to advise his successor on how best to handle various situations. Although, the planners would have to have to allow time for the new crew to adapt, at least Carr's crew would not have become frustrated by falling behind straight away. However, ongoing operations had never been a programme objective. Mueller's original idea had been to use the initial workshop to certify systems for follow-on space stations but it was now evident that NASA would not be in a position to develop another station for a decade or more, and since any station that could be assembled by the Shuttle would necessarily be built from small modules it would have little in common with Skylab in terms of technology. In effect, therefore, Skylab had become an end in itself, which was why the planners were determined to maximise the data yield.

On the week starting 26 November, having concluded that the astronauts had acclimatised, the planners began subtly to ramp up the pace. By 5 December, Carr had had enough. No one on the ground "would be expected to work a 16-hour day for 84 days," he said, pointedly, "so I really don't see why we should even try to do it up here." The nub of the issue was that people simply did not wake up at the same time each day, run through a strictly timed series of assignments, fall asleep at an assigned time and get up the next day to do it all over again. Carr had decided to split the daily 90-minute exercise into two 45-minute sessions, but the schedulers made no allowance for winding down and freshening up following each session.

Who, among the ground staff, followed a strenuous workout by a delicate scientific experiment? Despite the differences between the astronauts and their colleagues on the ground, the review on the 28th day cleared the way for a basic 56-day mission.

THE CHRISTMAS VISITOR

When a second CMG showed intermittent fluctuations early in December, there was renewed concern over flying the 'extended' mission. In an attempt to maximise the efficiency of the unit's lubricant the flight controllers activated the internal heater to keep the oil warm, but the fault recurred every few days and it was clear that this unit was close to failing completely. If Skylab was reduced to using thrusters, manoeuvring would have to be limited to conserve propellant. An analysis found that Skylab had sufficient propellant for a week or two of manoeuvring and the Apollo had propellant to control the docked combination for a further 10 days, so although the loss of the CMG would not require an immediate evacuation, the observational programme would be seriously impacted. At its perihelion passage the comet would be so near the Sun's glare that terrestrial telescopes would not be able to see it. Neither would other astronomical spacecraft, because they have to steer clear of the Sun in order to preclude overloading their sensors. The ATM had been designed to *stare* at the Sun and was therefore perfect for observing the comet during this most interesting period, but without the CMGs the ATM would not be able to aim its instruments with sufficient precision for comet observations. If the CMGs had to be shut down, therefore, only the Earth-resources work would be left, and this programme was suffering because winter weather in the northern hemisphere was impairing observations.

Systematic observations of the comet began on 23 November and a new camera had been flown up for the task – in fact, this was the back-up for the Ultraviolet Camera erected on the lunar surface by the Apollo 16 crew the previous year. On Christmas Day, Carr and Pogue made a spacewalk in order to service the ATM and photograph the comet. After pinning open another ATM cover, Carr inspected an X-ray telescope that had become inoperable when its filter wheel stuck between positions. He was to try to rotate it manually to its neutral (no filter) position. However, as the wheel was inside the instrument he would need to use a screwdriver to force it around. He shone his torch to inspect the mechanism, but 'slipped' as he inserted the screwdriver and the tool bent one of the blades of the shutter. He decided to wait until communications were restored so that the instrument's Principal Investigator could comment. The recommendation was to continue. After much fiddling in the confined space, he rotated the wheel out of the optical path and thus restored the telescope to partial operation.

On 28 December, the day that the comet made its perihelion passage, Lubos Kohoutek, its discoverer, was invited to Houston to talk to the Skylab astronauts. At only 21 million kilometres from the Sun the comet was well within the orbit of the planet Mercury. The ATM's White Light Coronagraph monitored it as it passed behind the Sun and the Skylab crew were the only human beings to witness it. There was no let up, however, because Carr was outside again the next day, this time with

Gibson, to photograph the comet as it re-emerged from behind the Sun and began to withdraw. The ATM was able to follow the comet until 4 January, when it moved too far from the Sun in the sky for the ATM to observe it. However, by mounting cameras in the scientific airlock the crew were able to track it into early February, at which time the programme was concluded.

BOILING POINT

Gibson vented his frustration on 20 December. In referring to the mission to date as "a 33 day fire drill", he warned that he was not going to allow himself to be rushed on one experiment simply because the schedule called for him to do something else. From now on, he would take as much time as was required to complete each job. It was the *quality* of the data that mattered, he insisted, not its volume. If the schedule had to be revised, then that was too bad.

Immediately after Bean's crew had returned, Neil Hutchinson, the lead flight director, had noted, "We sent up about six feet of [teleprinter] every day ... telling them where to point the solar telescope, which scientific instruments to use, and which corollaries to do. We laid out the whole day for them." It had been a steep learning curve for the planners, but "what we have done is we have learned how to maximise what you can get out of a man in one day".

As the new year approached, Carr expressed concern over how his crew's performance would be judged. At the heart of the issue, he felt that the planners had ignored him. "We had all kind of hoped – before the mission – that everybody had the message that we did not plan to operate at the SL-3's pace." Two days later, there was a very frank discussion over the open circuit. Dick Truly, the astronaut serving as CapCom, admitted that Carr's 'message' hadn't been communicated to the planners. The irony was, however, that except for the first few weeks Carr's crew had developed a pace which, if it were not for the exemplary performance of Bean's crew, would have exceeded the target set before Skylab had been launched. "We realise," Truly told Carr, "that these last couple of weeks, the work load that we've been putting on you is a level that you, very obviously, have handled with no problem." He then went on to say that if ever the crew found "a consistent gap that provides a little extra time" it would be "helpful" if they made this known so that more science could be achieved "per invested hour". This only served to confirm the crew's conclusion that the planners' metric was the sheer volume of data that could be secured and then represented as a statistic on a performance chart for the mission. From the crew's viewpoint, a three-month mission had to be treated like running a marathon rather than a sprint. It was crucial to settle into a *sustainable* pace. If Bean had been granted his extension his crew may well have 'burned out'. The final Skylab mission proved to be the most significant, not for the specific scientific data produced but as a case study for how to operate a space station as a long-term venture.

The new year brought thermal problems for Skylab. It was not that the airlock's coolant system had deteriorated, rather that the high inclination and high altitude of Skylab's orbit meant that for a few weeks in January it never penetrated the Earth's shadow, and the continuous sunlight severely taxed the thermal control system. The 8-month schedule starting in early May had been designed to preclude this. Although Carr's crew enjoyed a veritable heatwave for their final month, it was not as extreme as that endured by Conrad's crew in the early days. On the other hand, the solar scientists were delighted with the uninterrupted daylight. On 19 December, the ATM had documented the largest prominence for 20 years when an arch of plasma was blasted 500,000 kilometres into the corona. The Sun took about a month to turn on its axis, so in mid-January, as the previously active region reappeared, Gibson settled into a vigil in case the active site persisted. On 18 January he moved his sleeping bag into the airlock, where it was somewhat cooler, to be 'on call' for the ATM and Carr and Pogue picked up Gibson's chores to let him remain on the ATM. Two days later, Gibson was otherwise occupied when terrestrial observers spotted the early signs of renewed activity. He returned to the ATM and spent the rest of the day recording the ongoing low level of activity, but nothing developed. He also spent the next day doing the same for little reward, but that evening he announced, "I think we finally got one on the rise!" The solar flare lasted only half an hour, but for the first time the ATM was able to record the emergence of a flare from the continuous background activity; it was a fitting climax to the solar observing programme. As it happened, 23 January was an ominous day. Although the second CMG was being nursed by the controllers, it seemed to be on the verge of seizure. The engineers had noticed, however, that the unit suffered most distress when Skylab was turned to observe the Earth and performed best when Skylab was in solar-inertial attitude and the lubricant was heated by the Sun. However, as the mission was nearly over, it was not such a serious issue.

THE END

On 2 February, Carr and Gibson checked their spacesuits and the following day collected the ATM's film and several exterior experiments. After such a long mission, it took two days to pack everything for return to Earth and stow it in the Apollo. On 6 February, the Apollo's thrusters were used to raise Skylab's orbit in order to extend its life in case NASA managed to fund a follow-on mission.

"She's been a good bird," opined Bob Crippen, the CapCom on duty, as Carr undocked on 8 February.

"It's been a good home, Crip," Gibson agreed. "I hate to think we're the last guys to use it."

Carr made a fly-around to let his colleagues document the state of the station and then he withdrew.

The next day, the flight controllers reoriented Skylab with its axis aimed at the Earth, and its docking port furthest from the Earth, so that the gravity gradient would stabilise it; then, with its mission over, it was commanded to power down, having far exceeded its original goals. Carr, Gibson and Pogue were carefully monitored as they readapted to gravity. Their complete recovery showed that there was no cause to rule out a year-long mission. For that, however, NASA would need a new space station.

Although one of Skylab's solar wings was lost, the other was manually deployed and two layers of shielding were fitted to protect the Orbital Workshop's exposed skin from the Sun.

Table 3.1 Skylab flight log

	Launched		Returned		Pad	Days
Skylab	1330 EDT	14 May 1973	1237 EDT	11 Jul 1979	39A	–
SL-2	0900 EDT	25 May 1973	0949 EDT	22 Jun 1973	39B	28.03
SL-3	0711 EDT	28 Jul 1973	1819 EDT	25 Sep 1973	39B	59.46
SL-4	0901 EST	16 Nov 1973	1117 EDT	8 Feb 1974	39B	84.01

Table 3.2 Skylab EVAs

#	Date	Hours	Astronaut(s)	Objective
1	25 May 1973	0.6	Weitz	'Stand up' to try to free solar panel
2	7 Jun 1973	3.5	Conrad and Kerwin	Deploy parasol and ATM maintenance
3	19 Jun 1973	1.7	Conrad and Weitz	Retrieve ATM film
4	6 Aug 1973	6.5	Garriott and Lousma	Deploy A-frame shield
5	24 Aug 1973	4.5	Garriott and Lousma	Install CMGs
6	22 Sep 1973	2.7	Garriott and Bean	Retrieve ATM film
7	22 Nov 1973	6.6	Pogue and Gibson	Repair EREP
8	25 Dec 1973	7.0	Carr and Pogue	Photograph Comet Kohoutek
9	29 Dec 1973	3.5	Carr and Gibson	Photograph Comet Kohoutek
10	3 Feb 1974	5.3	Carr and Gibson	Retrieve ATM film

Table 3.3 Skylab primary time utilisation

Experiment series	Crew 1	Crew 2	Crew 3	Total
Solar astronomy	30	28	33	31
Biomedical	37	29	24	27
Earth resources	18	21	18	19
Astrophysics	10	10	9	9
Comet Kohoutek	–	–	10	5

Note that the observations of Kohoutek by the third crew represented a target of opportunity.

4

NASA's next logical step

SHUTTLE

In energy terms, the most expensive part of a space flight is attaining low orbit. Most of the mighty Saturn V's energy went into climbing up out of the Earth's deep 'gravity well'. Setting off for the Moon thereafter was relatively cheap. Even before Neil Armstrong took his 'small step' onto the lunar surface at Tranquillity Base, the Nixon administration had determined that cutting the cost of achieving low Earth orbit was, in the long term, the key to a sustainable space programme, and in the autumn of 1969 the *ad hoc* Space Task Group chaired by Vice President Spiro Agnew recommended that a reusable 'Shuttle' – officially, a Space Transportation System – be developed to provide routine access to low orbit at significantly lower cost than was achievable by using 'expendable' rockets. Looking a little further into the future, NASA urged that this Shuttle be used to assemble and service a 12-person space station in low orbit. Unfortunately, the cost of developing the new 'space plane' would be so great that by 1971 NASA was obliged to delete the station, or, at least, to push it well downstream. In 1972, once the Shuttle's configuration had been finalised, Nixon gave his consent to build it. Development was expected to take five years, and NASA hoped to have a fleet of vehicles in service by the end of the decade flying missions on a weekly basis.

WHITHER A SPACE STATION?

In developing its 'post-Apollo' strategy, NASA broadened its historically constrained international cooperation to include direct participation in the development and operation of human space systems. One of the objectives formally specified by the 1958 National Aeronautics and Space Act which commissioned the agency was to develop "cooperation ... with other nations", but so far this had yielded only joint work on satellites for space science research. In late 1969, therefore, NASA invited Canada, Japan and the European countries to consider whether they might wish to participate in the Shuttle's utilisation. Some in NASA were wary of cooperation to provide human space systems. With the obvious exception of the Soviet Union, no country possessed the engineering knowledge to support human space flight. It was inevitable, it was argued, that such cooperation would not only hold back the US development but also transfer technology to countries which, if they later decided, would be better placed to set up in competition by developing their own launchers and spacecraft. There was some merit in this argument, because some in Europe did indeed hope to become the third 'space power'. To preclude becoming dependent upon its international partners, NASA identified specific missions which required substantial systems to be developed. As long as these systems were isolated from the critical path and were defined in terms of 'clean interfaces', NASA would not risk compromising either its 'core knowledge' or its schedule. In keeping with the 'no exchange of funds' *modus operandi* which NASA had historically employed with international cooperation, the partners would donate their systems in exchange for flight opportunities on the Shuttle.

Seizing the opportunity, Canada offered to develop the Remote Manipulator System (RMS), with the result that Canada is now the world's leader in space tele-robotics. The situation in Europe was rather more complicated. In 1964, France, West Germany and Italy had signed up 10 countries to create the European Space Research Organisation (ESRO) to promote collaborative scientific research using satellites. The European Launcher Development Organisation (ELDO) was created to develop a family of rockets in order to free ESRO of dependence on American rockets. However, after five years it had become apparent that ELDO was unlikely to produce a viable launch vehicle in the near future, and it was at this point that NASA's invitation to participate in utilising the Shuttle was extended. Although France was sceptical of working so closely with NASA and preferred instead to continue the rocket development with the ultimate objective of establishing an independent human presence in space, Germany and Italy, frustrated by ELDO, were more amenable, and in 1972 they informally offered to develop an Orbital Manoeuvring Vehicle. This would operate as a 'tug', taking satellites released by a Shuttle in low orbit into a high orbit, retrieving others and returning them to a Shuttle to be serviced. However, the intention to use a liquid-propellant rocket stage was not well received by NASA, which was reluctant to fly such a volatile cargo. When the Department of Defense refused to rely on a foreign vehicle to ferry its payloads to their operating orbits, the US Air Force undertook to develop a solid-propellant transfer stage and the European offer was rebuffed.

In April 1975, ESRO and ELDO were merged to form the European Space Agency (ESA). ESA established two types of programme. The 'core' projects were to be mandatory and funded by all members according to a formula that would be reflected in assigning the development contracts. Other programmes would be funded only by those countries with a specific interest. This was sufficiently flexible for Germany and Italy to cooperate with NASA while the French developed the Ariane launch vehicle and a small reusable space plane which would establish an independent human space flight capability. ESA suggested to NASA that it develop a modularised facility incorporating pressurised and unpressurised modules to be integrated in 'pick and mix' fashion as required by a given mission and carried within the Shuttle's payload bay. NASA was receptive. Of course, for ESA this 'Research and Applications Module' (later renamed Spacelab) was a stepping stone, because pressurised laboratories would certainly form an essential part of any space station and so the investment in this technology would pay off when France's space plane (named Hermes) entered service.

When Spacelab was agreed upon as Europe's principal contribution to Shuttle operations, NASA was projecting weekly flights at an estimated cost of $10 million each. To accommodate this astonishing mission rate, NASA predicted the need for six Spacelab modules. In 1973, it was agreed that ESA would fund the development of Spacelab and donate the first module to NASA in return for an opportunity for its scientists to fly on that mission. The forecasted cost of $250 million would then be recovered by turning out the production copies which NASA would buy. In the event, the Shuttle failed to attain anything near this flight rate. NASA purchased the single Spacelab that it was contractually obliged to take and cancelled its options. It was a raw deal for ESA, as the cost of development was four times the projected budget and its only revenue was the $128 million for NASA's module. Even worse, the cost of flying proved to be so great that ESA eventually concluded that it could not afford to mount its own missions, so 'international' missions were mounted with US, Canadian, European and Japanese experiments and researchers. Nevertheless, as a result of its experience in building Spacelab, Italy is now the main supplier of pressurised space station modules.

Despite having been obliged to shelve its proposal for the Shuttle to assemble and service a modular space station, NASA conducted studies during the 1970s into possible missions and configurations for a station in order to be ready when an opportunity to seek funding presented itself. In late 1975 it invited industry to submit proposals for its 'Space Station System Analysis Study'. In 'Outlook For Space', published in March 1976, NASA's Administrator James Fletcher announced that proposals for large-scale projects would have to be assessed in terms of "Why should it be done, and what will it contribute?" A space station and a sample-return planetary mission were the only large projects that the report recommended for development in the near future. "A number of questions remain about the total role that humans can ultimately carry out in space," the report said. "A space station . . . forms the backbone of future human space activities and hence of many of the real opportunities to exploit space." A proposed modular design comprised a habitat for up to six people on 90-day tours of duty, two laboratories and a utility module supporting booms with solar arrays, at an estimated cost of $6 billion. In April, McDonnell and Grumman were each awarded $700,000 contracts for 18-month 'concept' studies and prompted to explore the scope for "evolutionary growth". One proposal for 1985 envisaged four modules docked end-to-end equipped to support a series of four-person crews on 90-day tours. Another proposal suggested that by 1990 a station for up to 12 persons could be assembled in polar orbit, where it would provide complete daily coverage of the Earth.

In May 1972, President Nixon and Soviet Chairman Kosygin had signed an agreement on cooperation in space. This Apollo–Soyuz Test Project resulted in the historic docking in 1975. Nixon's interest in this project had been as a symbol of the *detente* he was fostering with the Cold War adversary. In October 1976, the two space-faring nations began a series of meetings to consider further cooperation. The Soviets had established Intercosmos to look after its international space programme, and citizens from several fraternal Communist states were training to make brief visits to Salyut stations. On 6 May 1977 America and the Soviet Union signed an agreement to study the feasibility of docking the Shuttle with a Salyut in 1981. The Shuttle's development had fallen significantly behind schedule. Instead of being ready for orbital flight, the Approach and Landing Tests in which the test vehicle 'Enterprise' was carried aloft on top of the Boeing 747 'Shuttle Carrier Aircraft' were still several months away and the first launch was projected for 1979. In November, in Moscow, NASA considered experiments that could be undertaken during the joint Salyut mission. It offered to deliver a 10-tonne rack of experiments which would otherwise be awkward for the Soviets to launch and dock. Of course, this demonstration would strengthen NASA's case for using the Shuttle to assemble its own station. When Ronald Reagan moved into the White House in January 1980, however, the *detente* of the 1970s was abruptly terminated, the prospect of cooperation faded and the discussions were formally terminated in April 1985.

In 1977, Huntsville's engineers proposed modifying a Shuttle External Tank (ET) by installing a bulkhead that would isolate the upper half of the oxygen tank to serve as the core of a station that could support a crew of three astronauts for

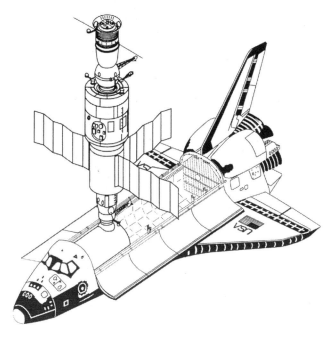

In 1977, NASA and the Soviet Union studied a joint mission in which a Shuttle would dock with a Salyut space station.

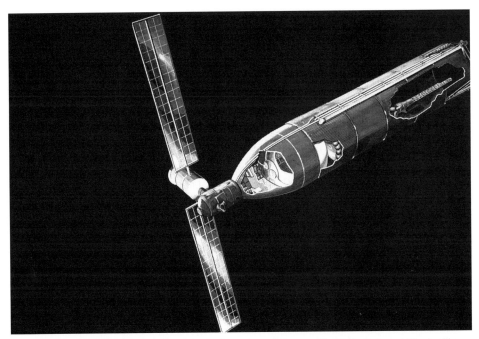

A 1977 Huntsville plan for a space platform based upon a Shuttle's External Tank. The pre-fitted tank would be launched with a partial load of propellant. A multiple docking adapter with an airlock has been mated on the nose, and a power module has been docked. (Courtesy of Marshall Space Flight Center.)

tours of duty of up to 90 days. In addition to proposing to use Skylab-proven systems, they advocated reclaiming Skylab 'B' from the National Air and Space Museum so that its systems could be refurbished. In this plan, the modified ET would be launched in a lightweight configuration with a partial load of propellant and the Shuttle would have an empty payload bay. Instead of being jettisoned in the immediate aftermath of main engine cutoff, the ET would be retained and delivered to a stable orbit. A later Shuttle would mount an androgynous docking module incorporating an airlock on the cap on top of the ET to provide access, and a solar panel would be deployed from another module to provide electrical power. In the fullness of time the 30-metre-long hydrogen tank could also be pressurised to expand the facility. This proposal's heritage in the Apollo Applications plan to convert a 'wet' stage was clear, and the rationale was the same: to put off-the-shelf technology to productive use.

SKYLAB REPRISE?

As for Skylab itself, its orbit was decaying much faster than predicted. With re-entry possible as early as 1982, NASA began to consider having a Shuttle deliver a propulsive unit to boost the station's orbit, both to gain time and to preserve it in case it could be put to use. In October 1977 NASA ordered Huntsville to develop a Teleoperator Retrieval System to be delivered in the Shuttle's payload bay and flown by remote-control to dock with Skylab. As the schedule stood at that time, it was intended that the fifth mission in early 1980 should undertake this task, but when it became evident that the Shuttle's development would slip even further it was advanced to the third mission, in 1979. Martin Marietta was awarded the $1.7 million contract, and in an effort to minimise costs the company proposed a design based on off-the-shelf hardware in the form of the thruster system used by the transtage of the Titan III launch vehicle, propellant tanks from the Viking Mars spacecraft and a left-over Apollo docking system to mate with Skylab. In May 1978 Huntsville awarded two contracts to study possible Skylab exploitation. The decision to refurbish it as an 'interim' station would not be taken until the reboost had been achieved. Although this robotic vehicle's development was driven by the effort to reboost Skylab, it was expected that such a device would have broader utility because it would be able to be flown by an astronaut on board a space station. However, by August 1978 NASA admitted that the continuing delays in Shuttle development offered only an even chance of making the first flight in September 1979. On 18 December 1978, with Skylab's orbit decaying even more rapidly in the run up to the solar maximum in 1980, the rescue plan was cancelled, and the station re-entered on 11 July 1979, scattering fragments across the Australian desert. In the event, the Shuttle did not make its first flight until April 1981.

BEGGS TAKES OVER

In contrast to Huntsville's strategy of starting with a modest tended platform that could be progressively expanded to form a continuously occupied station to serve as a laboratory in the Skylab tradition, Houston awarded Boeing a $400,000 contract in early 1981 to define a low-orbit Space Operations Centre that would:

- service satellites prior to dispatch to their operating orbits;
- refuel vehicles capable of ferrying satellites back and forth between low and geostationary orbit;
- repair satellites that had been retrieved;
- serve as a development centre for "telescience" and robotics;
- service free-flying experiment carriers; and
- prepare vehicles to venture into deep space.

In this plan, successive Shuttles would deliver 15-metre-long cylindrical pressurised modules and a truss with a pair of solar panels. With early approval, it was thought feasible to have a four-module 'interim' facility for a four-person crew by the end of the decade. A key objective was to evaluate different methods of assembling such a truss in order to determine the optimal degree of pre-integration, as a basis for future space construction. In early 1982, Boeing's contract was extended by five months to evaluate the possibilities of crews of eight persons serving tours of up to 90 days.

Given the increased crewing and the ongoing occupancy, the study proposed the addition of a laboratory module, and then considered the types of scientific research that might be undertaken. Boeing also looked at options for using an External Tank. One possibility was to store propellant in it; another option was to convert it into a 'hangar' for satellites so as to enable the crew to work on them in a shirt-sleeved environment. The case for in-space servicing of satellites derived from the short mean-time-between-failure of contemporary electronics technology. If a satellite was designed to be serviced, it might be cheaper to repair it than to replace it.

As the Shuttle moved from development to flight test, NASA needed another 'big ticket' programme to preserve its engineering base. In March 1981, newly elected President Ronald Reagan appointed James Beggs to head up the agency, and in his confirmation hearings Beggs insisted that "next logical step" was the construction of a permanent space station. Upon the Shuttle's first test flight in April, NASA was finally back in the spaceflight business. At the Paris Air Show in June, Beggs alerted possible international partners to the prospect of jointly developing a space station.

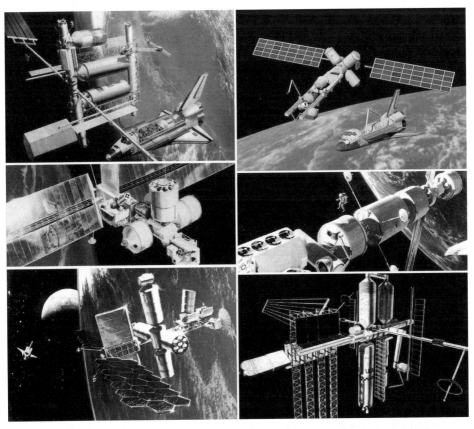

As the Shuttle began flying, a large number of 'modular' space station designs were investigated.

Beggs realised that if the space station was to be developed it would be necessary for the competing field centres to unite to support a single plan. All the interested parties inside the agency met in November 1981 for the 'Space Station Conference'. It was taken for granted that the space station would continue the international cooperation of the still-to-fly Spacelab, but the conference recommended that NASA start with the assumption that it would be a wholly-US programme. A multi-phased strategy was devised. Phase 'A' (the 'Concept Phase') would evaluate the possible missions for the station and recommend an appropriate architecture. Phase 'B' would assess hardware configurations to provide the capabilities needed to fulfil these missions. NASA invited its potential international partners to adopt a similar approach and to offer "space station requirements and concepts which could benefit NASA". International cooperation was a key part of Beggs's strategy, because not only would this translate into 'extra' funding in addition to whatever he could get from Congress once he had secured presidential backing, but the possibility of external support was likely to ease funding legislation through the Congressional committees that appropriated NASA's budget. The extent of the Department of Defense's involvement was, however, something of a complication. To secure the US Air Force's support for the Shuttle, Fletcher had to accommodate the military's requirements. It was far from clear whether the military would support NASA's development of a space station, but if so it might also impose such stringent security measures as to effectively prohibit foreigners from the station. In early 1982, having established guidelines for international cooperation, Beggs promptly sent Kenneth Pedersen, his Director of International Affairs, to (a) appraise Canadian, European and Japanese interested parties on the agency's deliberations, (b) invite them to conduct their own Phase 'A' studies and (c) propose missions that they might be able to contribute. The Space Station Task Force (SSTF) was created in Washington under the chairmanship of John Hodge, a former flight director for Gemini and Apollo missions, with its membership drawn from the Office of Manned Space Flight and the Office of Advanced Programmes. The SSTF was to ensure that control of planning for the station was exercised by Washington. In an effort to keep foreigners out of the crossfire between the warring field centres, all contacts with potential international partners were to be made through the SSTF. This arrangement provided the further advantage that it stood a better chance of keeping such discussions at the level of missions rather than hardware, and it would enable the technology transfer to be more strictly controlled. This was not simply a case of NASA being secretive, but certain American laws had to be upheld. When formalised by Beggs in May, the SSTF began to define realistic missions and set up studies of their system requirements and interfaces. Its guiding principle was that the missions should rely upon, or at least be significantly enhanced by, the presence of astronauts.

Japan created a task force in May 1982 to work with its National Space Development Agency (NASDA) and then recommend to the top-level Space Activities Commission how best to contribute to the space station. Japan had been too late for the development of Shuttle systems, but had contributed to a number of Spacelab missions and had strategic plans to establish an independent presence in

The core of a 'modular' space station proposed by Rockwell International in 1982.

space. Beggs met Erik Quistgaard, the ESA Director, in Paris in June, who agreed that ESA would study "utilisation aspects" of a space station, and several months later Canada's National Research Council agreed to do likewise. Beggs had his international support.

When Reagan went to Edwards Air Force Base on 4 July 1982 to welcome the Shuttle home from its final test and to declare the National Space Transportation System to be operational, Beggs had hoped that the president would issue a call for a space station, but Reagan simply urged the agency to "look aggressively to the future ... by establishing a more permanent presence in space." However, the United States was not alone in pursuing this objective ...

5

Success with Salyut

EARLY STRUGGLES

In July 1972, a few weeks after the new version of the Soyuz spacecraft was tested in its automated regime, the Soviet Union launched the second Salyut. Unfortunately, it was lost when the second stage of the Proton malfunctioned. Alexei Leonov and Valeri Kubasov, who had narrowly missed flying to Salyut 1, had been hoping to make the first 28-day tour, so they missed another opportunity.

Although the next space station to roll off the Khrunichev production line was configured for military reconnaissance, the large spacecraft designed by Chelomei had yet to undergo flight testing so the station would be serviced by a Soyuz. The switch in configuration was masked by slipping the Almaz vehicle into the Salyut series. However, as Salyut 2 refined its orbit on 14 April 1973 to accommodate the launch of its first crew, an electrical fault crippled it. Tass announced that its mission had been successfully completed. A 'scientific' station was launched on 11 May, but a fault immediately after separation from the booster sent it spinning out of control. Its nature was masked by labelling it Cosmos 557. This was yet another frustration for Leonov and Kubasov, who had hoped to be its first residents. A few days later, as the best-trained crew, they were reassigned to the recently announced Apollo–Soyuz mission scheduled for the summer of 1975.

Salyut 3, which successfully achieved orbit on 25 June 1974, was configured for military reconnaissance. Although its first crew spent a fruitful fortnight aboard it in July, the second crew's ferry was unable to dock a month later and was obliged to return to Earth in a hurry because the stripped-down 'ferry' version of the Soyuz had no solar panels and its batteries were good for only 48 hours of independent flight.

The vehicle that was lost in 1972 was a virtual copy of Salyut 1, with two pairs of small solar panels, but the scientific station that was written off as Cosmos 557 had a trio of rather larger panels on the narrower of the main cylindrical sections, and they were gimballed to face towards the Sun irrespective of the station's orientation. Launched on 26 December 1974, Salyut 4 was of this type. Its first crew served a very successful 28-day tour early in the new year. Unfortunately, the launch of the next

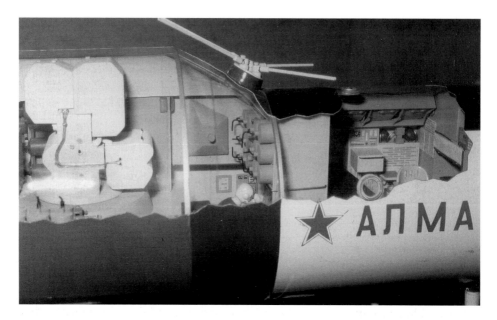

This cutaway model of the Almaz reconnaissance platform clearly shows that the main compartment (left) is dominated by the camera. A cosmonaut is at the control panel, which serves to isolate the crew quarters in the vehicle's narrower cylinder. (Courtesy of Dietrich Haeseler.)

crew on 5 April 1975 had to be aborted during 'staging' and the capsule made a forced landing. The backup crew was dispatched on 24 May and achieved the operationally significant milestone of reoccupying a station. On the original plan, Salyut 4's mission would have been completed prior to the Apollo–Soyuz flight in June but the Soviets now had a choice of either truncating the second period of occupancy and bringing the crew home before the international mission, or leaving them in space. In the event, they opted to fly the two missions in parallel and the station crew landed a week after the international mission, having spent 63 days in space. Owing to the delay in reoccupying the station, its environmental unit was exhausted and the internal wall panels were thick with a smelly green mold by the time it was finally vacated.

The run of bad luck returned to plague the reconnaissance platform launched on 22 June 1976 as Salyut 5. The first crew docked on 7 July, but they were forced out by an acrid odour on 24 August, 10 days short of the planned 60-day mission. After a method was devised to replace the station's air, another crew was dispatched on 14 October, but they were unable to complete the rendezvous. A third crew was launched on 7 February 1977. Although they docked successfully, they did not enter the station immediately but instead spent the night in their ferry while Salyut 5's air was first vented and then replaced. After a fortnight aboard the station, they returned to Earth. The following day, the station released a small descent capsule into which the cosmonauts had placed the film exposed by the first crew, whose

This view of Salyut 4 from an approaching ferry just prior to docking shows the tri-form arrangement of the new solar panels.

departure had been too rushed to save the results of their endeavours with the reconnaissance camera.[1]

By mid-1977, therefore, the Soviet space station programme had achieved mixed results. Although four of the 'scientific' type and three of the reconnaissance type had been dispatched, crews had managed to board only four stations and only two of these had been reoccupied. Despite having spent a total of almost 200 days in space, no crew had broken the 83-day Skylab record.

LIMITATIONS

Whereas Chelomei had envisaged supporting his Almaz station using the 20-tonne TKS logistics craft, the 7-tonne Soyuz crew ferry could not transport much cargo. Because the early Salyut stations had to be loaded with virtually everything required for their overall missions prior to launch, the operational life of a station was limited not only by the food and water available for the crew but also by propellant, which was consumed in overcoming orbital decay and in altering the orientation in order to perform experiments and optimise the output of the solar panels. The habitable life of a station was also limited by the ability of its environmental system to regulate the temperature and maintain a breathable atmosphere. If the crew exhausted the air, food or water, or the station consumed its propellant, or the thermal control system broke down, or toxins accumulated in the air or water supply, operations would have to be terminated.

Furthermore, the endurance of a mission was restricted by the in-orbit service life of the Soyuz spacecraft, which was mainly defined by the ability of the various gaskets in the propulsion system to withstand the extreme thermal stresses of the space environment.

What the programme required, therefore, was a second docking port and, in the continuing absence of the TKS ferry, a form of the proved Soyuz that could dock

---

[1] As an aside, this descent capsule was auctioned in New York in 1993.

This view of the rear of the Almaz reconnaissance platform (in model form) shows the solar panels in their stowed configuration and the docking system running through the twin-chambered engine. The film capsule that Salyut 5 ejected was built into this same assembly. (Courtesy of Dietrich Haeseler.)

automatically with an occupied station to deliver logistics, to pump fluids aboard through pipes and to allow dry cargo to be manually transferred by the residents. Automated rendezvous and docking was not new: the first Soviet docking, in 1967, had been performed by two Soyuz spacecraft flying in automated regime, and most of the approaches to the Salyuts had been automatic. Furthermore, a prototype of the cargo ferry had spent several months docked with Salyut 4 after it had been vacated. The primary objective of the Soviet medical space programme was to progressively extend the endurance record in order to establish that humans could survive in weightlessness for the duration of a flight to Mars. However, the Soyuz was limited to a two-month flight. If a crew was to serve on board a station for longer than this, their ferry would require to be periodically replaced. In addition to unloading cargo, therefore, long-duration crews would host a succession of visitors who would remain on board for a few days and leave in the expiring Soyuz, so that the residents always had a 'current' spacecraft available for their return.

THE AMAZING STORY OF SALYUT 6

Launched on 29 September 1977, Salyut 6 had a docking port at each end. Although this meant that it would be possible for two crews to undertake an in-space handover, this was not actually an objective at this stage in the programme. In fact, the plan was to occupy the station for about half of its 18- to 24-month service life, with 90-day, 120-day and 175-day missions.

Unfortunately, Soyuz 25 failed to dock at the front port in October and had to return to Earth. Yuri Romanenko and Georgi Grechko encountered no problems in docking at the rear port on 10 December, however, and a few days later they donned the EVA suits stored in the airlock compartment at the front and opened the port to inspect its docking mechanism, which they confirmed to be in good shape. This impromptu spacewalk was the first by a cosmonaut since January 1969, and *that* had been the first since Leonov's pioneering excursion in March 1965, so the Soviets were relatively inexperienced in the art of extravehicular activity at this point.

On 11 January 1978, Soyuz 27 slipped into the front port without incident, forming, for the first time, a structure in space composed of three spacecraft. Vladimir Dzhanibekov and Oleg Makarov thereby became the first space crew to visit an inhabited station. Although a ferry can be launched at any time that a station's orbital plane intersects the point at which a newly launched spacecraft attains orbit, ideal recovery windows for a Soyuz occur only at two-monthly intervals. This interval is due to the fact that the Earth is not a perfect sphere but is oblate, and the gravitational attraction of its equatorial bulge acts on a spacecraft's angular momentum in such a way as to make its orbital plane precess. The rate of this rotation is dependent on the altitude and inclination of the orbit. In the case of Salyut 6, it was 58 days. This arcane dynamic dictated the timing of the visiting flights. A visiting crew would be launched early in the 10-day window and be recovered towards its end. Visits exploiting ideal conditions could therefore be made at two-monthly intervals and last approximately a week. When Dzhanibekov and Makarov left on 16 January in Soyuz 26, they left the residents with a spacecraft that had sufficient service life to see them through the hoped-for 90-day mission.

Salyut 6 had expended much of its initial propellant load in climbing to its operating orbit. As the pipes for refuelling were incorporated into the rear port, it had not been possible to accommodate a tanker while this port had been in use. On the original plan, the crew would have docked at the front and a tanker would soon have drawn up at the rear, then departed to clear the way for the first visitors. Soyuz 25's failure to dock had upset the plan, but now that the rear port had been vacated it was possible to dispatch the tanker. Progress 1 docked on 22 January. It had an orbital compartment filled with 1,300 kilograms of dry cargo, including apparatus for experiments, but the descent module had been replaced by a cluster of tanks containing propellant.

The addition of the rear docking port had meant that it was no longer practicable simply to bolt a Soyuz propulsion module to the station. Chelomei's design had required the docking port to be in the rear because the return spacecraft was mounted on the front. His twin-chambered engine had been designed to fit around the rear transfer tunnel, and as this engine had proved itself on Salyuts 3 and 5, it was also utilised on Salyut 6. Rather than have the engine and its propellant tanks exposed as they were on the Almaz configuration, the entire assembly, together with other apparatus, had been encased in an unpressurised bay that formed an extension of the shell of the main compartment. On Korolev's 'scientific' stations the main engines had burned nitric acid and hydrazine fed by a hydrogen peroxide powered turbine and the thrusters had vented cold hydrogen peroxide. In Chelomei's configuration the tanks were pressurised by a bladder inflated by high-pressure nitrogen gas and the engines burned unsymmetrical dimethyl hydrazine (UDMH) in nitrogen tetroxide. Furthermore, the orbital manoeuvring engines and the attitude-control thrusters now used the same propellants and drew them from a common supply. The 'unified' propulsion system greatly simplified the task of replenishing the station's propellants. The first step – to verify the integrity of the pipes – involved pressurising them with nitrogen to check the hermetic seal of each manifold and valve within the flow path. This could be controlled either by Kaliningrad or by the crew and on this occasion the cosmonauts supervised the task in order to certify the system. The next step was to reduce the pressure in the station's propellant storage system. When the station's engines were primed, the bladders were pressurised at 20 atmospheres, and unless this pressure was relaxed the pump in the ferry would be unable to force fluid into the station's tanks. A 1-kW compressor had been incorporated into the station to drive the nitrogen back into its bottle. The fact that this nitrogen was stored at 220 atmospheres meant that this operation had to be undertaken in stages while the station was in sunlight over a period of several days. Although this was a slow process it was simple and reliable. Once the pressure on the bladder had been reduced to 3 atmospheres, the pump in the ferry, which was pressurised by a bladder at only 8 atmospheres, could feed propellant into the station. When the oxidiser had been transferred, the pipes had to be vented to vacuum to flush out any residue (if mixed, the hypergolic reactants would ignite), and the seals had to be reverified before the entire process could be repeated for the fuel. It took a week to pump aboard the tonne of propellant the ferry carried. Before departing on 7 February, the tanker acted as a 'tug' and boosted the station's orbit to overcome orbital decay, further improving the station's propellant margins. Since the earliest days of designing space missions it had been taken for granted that a spacecraft would be able to be refuelled in orbit. As fluids in weightlessness do not behave as they do on Earth, there had been a real possibility that the transfer would be problematic. Although this was the first time that such a transfer had been attempted, it was achieved without incident.

Another major milestone was achieved when Soyuz 28 docked at the vacant rear port on 3 March with Alexei Gubarev and Vladimir Remek, a Czech cosmonaut. The Soviet Union had invited fraternal socialist countries to nominate citizens to train as cosmonauts to fly under the auspices of the Intercosmos organisation. The visitors departed on 10 March and the resident crew followed six days later, having logged 96 days in space.

With this series of flights the space station programme had recovered from the early docking problem and had taken a tremendous leap forward. While claiming the 84-day endurance record set by the final Skylab crew, Romanenko and Grechko had played host to two visiting crews, one of which was an international mission, and, in receiving Progress 1, had installed new apparatus for the experimental programme, undertaken maintenance on the station's environmental system and refuelled its

A view of Salyut 6, the first of the two-ended stations, with a Soyuz docked at the rear, taken from a ferry heading for the front port. The dorsal solar panel is viewed almost edge on. Note the circular railing around the hatch on the airlock compartment at the front and the handrails on the narrower compartment.

propulsion system. In a very real sense, they left Salyut 6 in a better condition than they had found it.

After three months of dormancy, Salyut 6 was commanded back to life. On 17 June 1978 Soyuz 29 docked at the front port and Vladimir Kovalyonok and Alexander Ivanchenkov settled down to a 120-day mission. Soyuz 30 delivered Pyotr Klimuk and Miroslaw Hermaszewski, a Pole, on 28 June, and no sooner had they departed (in their own spacecraft) than Progress 2 arrived with various supplies and new experimental apparatus. In addition to mapping with a large-format camera and surveying natural resources with a multispectral camera, the main focus of Salyut 6's scientific research was materials processing, and a variety of furnaces were being shipped up by the Progress ferries for evaluation.

In marked contrast to Skylab, where the flight planners had attempted to schedule the astronauts' activities in fine detail, the cosmonauts worked a 5-day week of 9-to-5 shifts and were left to do whatever they pleased at other times. However, apart from watching the Earth roll by (which was actually fascinating) there was little to do except to volunteer for more work!

On 29 July Kovalyonok and Ivanchenkov ventured out to emplace cassettes of materials on the station's hull, this time by way of the side hatch in the airlock compartment at the front of the station. These cassettes were to be retrieved by a later crew and returned to Earth for analysis. Barely a week after Progress 2 left, Progress 3 took its place. After the docking failures that disrupted the programmes of earlier stations, ferries were drawing up at Salyut 6 as if following an efficiently operated railway schedule. After being loaded with accumulated rubbish, Progress 3 left on 21 August to clear the way for the arrival of Valeri Bykovsky and Sigmund Jahn of the German Democratic Republic a week later in Soyuz 31. On 3 September, the visitors departed in Soyuz 29. As the rear port had the only pipes for fluid transfer, the station would not be able to be replenished while the new spacecraft remained on the rear port. The solution was to have the residents undock and fly this vehicle around to the front. This end-swap constituted a significant operational overhead: it took several days to load the accumulated scientific results into the ferry and return the station to its automated regime as a precaution against being unable to redock, only to reactivate it after the manoeuvre, which actually involved just half an hour of independent flight. Although the swap could have been performed at any time, it was actually done on 7 September just before the landing window closed to simplify recovery in the event that the redocking failed. After unloading Progress 4 in October to restock the station for the next crew, the residents left on 2 November, having extended the record to 140 days.

The first two endurance missions had exceeded their targets. If a 6-month-long mission could be achieved before the station expired, its programme would be able to be successfully concluded. However, there was a problem. While Progress 4 had been in the process of pumping propellant into the station, the bladder in one of the tanks had developed a leak and there was a risk that corrosive fuel seeping back through the nitrogen pressurisation system might induce an explosion if the engine were to be used. Since Salyut 6 was otherwise in excellent condition, it was decided to try to drain the leaky tank.

After Vladimir Lyakhov and Valeri Ryumin docked at the front port in Soyuz 32 on 26 February 1979, Progress 5 slipped into the rear port on 14 March. Two days later, the complex was put into an end-over-end rotation; the centrifugal force of the rotation separated the nitrogen gas from the fluid in the fuel tank and the fuel was then drained into an empty tank in Progress 5. Finally, the faulty tank's valves were opened to vent the residue on the inside of the bladder and the tank was filled with nitrogen and sealed. While the engine had been out of action the orbit had decayed considerably, so it was raised by Progress 5 before it departed on 3 April. There were actually two fuel tanks in the system, and now that the risk of an explosion in the pressurisation system had been eliminated, the backup tank could be used. With Salyut 6 once more fully operational, Lyakhov and Ryumin settled down to their marathon mission.

An engine fault prompted Soyuz 33 to abandon its approach on 11 April and perform an emergency descent. If they had been able to dock, they would have taken away the older ferry. It was decided to send up another Soyuz without a crew to provide a replacement. In the meantime, Progress 6 arrived on 15 May and was retained until a few hours before the new Soyuz arrived on 8 June, at the start of the next recovery window. One week later, Soyuz 32 was loaded with 180 kilograms of accumulated scientific results and automatically returned to Earth, having extended the spacecraft's proven life to 108 days. Lyakhov and Ryumin swapped Soyuz 34 to the front of the station on 14 June, just before the window closed. Their tour was proving to be lonely. The first visting crew (which included a Bulgarian cosmonaut) had not made it to the station and the dispatch of the second visiting crew (with a

Hungarian cosmonaut) had had to be cancelled in order to send up the replacement spacecraft. Each visiting crew was to have relieved the residents of 50 kilograms of results (mostly bulky film packs from the Earth-resources camera), so if Soyuz 32 had not been used as a relief vehicle, the backlog would have posed a significant problem.

Progress 7's arrival on 30 June brought a major new piece of apparatus in the form of a 350-kilogram radiotelescope which the residents were to mount in the docking port projecting into the ferry's open orbital module, so that when the spacecraft drew away a 10-metre-diameter dish would unfold facing back along the station's axis. Once the observations were complete, the antenna was jettisoned in the expectation that it would drift away but the framework snagged on the rear of the station. Since Lyakhov and Ryumin were scheduled to retrieve the cassettes which their predecessors had put outside the airlock hatch, it was decided that they should inspect the antenna and try to cut it free to clear the rear port. On 15 August, while Lyakhov retrieved the cassettes, Ryumin took a long-handled metal cutter, made his way along the length of the station, peered over the edge and, upon judging it to be manageable, snipped the fouled struts and pushed the structure clear. The impromptu repair demonstrated how the programme was maturing. A few days later they placed the station in its automated regime and, having extended the record to 175 days, returned to Earth.

By this time, Salyut 6 was approaching the end of its nominal service life but its systems had been regularly maintained and it was in good shape. If it had proved impracticable to release the fouled antenna to clear the rear port the station's mission would surely have been declared to be complete (and by any measure it had been tremendously successful) but it was decided to send up one more long-duration mission. Progress 8 settled into the vacant station's rear port on 29 March 1980 and on 10 April Soyuz 35 brought Leonid Popov and – surprisingly – Valeri Ryumin, who volunteered for the back-to-back mission after Valentin Lebedev suffered an injury in March. Ryumin, therefore, had the odd pleasure of reading the 'Welcome' note which he had written before vacating the station the previous year. After a fortnight's maintenance, they started work using some new materials-processing furnaces. Progress 8 left on 20 May and a week later they received Valeri Kubasov and Bertalan Farkas on the postponed Hungarian mission. After the frustrations of the programme's early years, Kubasov had finally made it to a station; unfortunately Alexei Leonov, his former commander, had retired. The visitors left on 3 June in Soyuz 35 and the following day the residents transferred Soyuz 36 to the front port. Yuri Malyschev and Vladimir Aksyonov were then launched in an improved version of the spacecraft, designated the Soyuz-T, which had a simplified engine with a longer in-space service life, a better flight computer and a third seat (reinstated by mass savings in other systems). The Soyuz-T had been developed for Salyut 6's successor but the station's longevity had prompted this test flight. Three days later they returned to Earth; this was the first time that two visiting missions had been dispatched in the same window. Progress 10 arrived on 1 July and departed on 18 July to clear the port for Soyuz 37 with Viktor Gorbatko and Pham Tuan of Vietnam, who left at the end of the month in the old ferry. The residents then transferred Soyuz 37 to the front port (the end-swapping procedure was becoming routine). Soyuz 38 arrived on 19 September with Yuri Romanenko and Arnaldo Tamayo Mendez of Cuba. Progress 11 arrived as the station entered its fourth year in space and, to provide the station with a backup propulsion system, the tanker was left in place when Popov and Ryumin departed on 11 October, having set a new record of 185 days. In fact, Ryumin had spent 12 of the past 20 months on board Salyut 6. His second tour had been more sociable than his first; indeed, by hosting four visiting crews he had effectively been transformed into a hotel manager!

Could Salyut 6 sustain *another* crew? The power output of the solar transducers had been progressively diminishing, and some of the storage batteries had had to be replaced, but as long as power-hungry experiments were not operated in parallel the situation was viable. Although most of the elements of the environmental system had been designed to be serviced, and the longevity of the station derived from routine maintenance, the hydraulics of the degrading thermal regulation system had not been designed to be serviced. Could they be replaced? In November 1980 a repair crew was launched to find out. Konstantin Feoktistov, the veteran cosmonaut who had helped to design the station, had hoped to make this flight but the medics had grounded him. Leonid Kizim, Oleg Makarov and Gennadi Strekalov spent a fortnight refurbishing everything to which they could gain access and were even able to drain coolant from a part of the thermal regulation system in order to replace faulty valves and gaskets. Upon their return to Earth, they recommended that Salyut 6 would be able to support a 90-day mission hosting two Intercosmos visits. On 13 March 1981 Vladimir Kovalyonok and Viktor Savanykh docked at the front in Soyuz-T 4. They promptly unloaded Progress 12 (already docked) to clear the way for Vladimir Dzhanibekov and Judgerdemidiyin Gurragcha of Mongolia in Soyuz 39 on 23 March. On 26 May, a few days after Leonid Popov and Dumitru Prunariu of Romania returned to Earth in Soyuz 40, the residents returned home at the end of that same window, having made a relatively short 75-day tour.

A month earlier, on 12 April, NASA had finally managed to dispatch the Space Shuttle on its first test flight. During their many 'grounded' years, the agency's astronauts had looked on enviously as their cosmonaut counterparts had had success after success on board Salyut 6. In fact, apart from the docking failure of the very first mission, and the engine fault that had obliged Soyuz 33 to abandon its approach, Salyut 6's success had tremendously advanced the Soviet space station programme. Thanks to routine maintenance, it had lasted over twice as long as its design life and during this time had been inhabited for 684 days. Replenished by a dozen Progress ferries, it had sustained five long-duration crews, the first four of which had progressively extended the endurance record well beyond the half-year point. The resident crews had hosted eleven visiting crews. It had been expected that the Intercosmos visits would have to be split between Salyut 6 and its successor, but Salyut 6's longevity had enabled it to host virtually the entire series. Most impressively, Salyut 6 had given the Soviets valuable practice in ongoing flight operations and in overcoming disruptions to the schedule. Their real-time planning had put them in an excellent position to ramp up operations with its successor.

6

Reagan's go-ahead

TWIN-TRACK APPROACH

Having failed to secure presidential approval for the space station in 1982, James Beggs decided to refine the concept and resubmit it the following year. As part of its review of space policy, Reagan's administration had transferred policy-making from the Office of Science and Technology Policy to the National Security Council, which promptly set up the Senior Interagency Group (Space) under the chairmanship of Robert McFarlane, Assistant to the President for National Security Affairs. It was "to provide a forum to all Federal agencies for their policy views, to review and advise on proposed changes to national space policy, and to provide for orderly and rapid referral of space policy issues to the President for decisions as necessary." In October 1982, McFarlane set up a working group comprising senior members of the Department of State, the Department of Commerce, the Central Intelligence Agency, the Department of Defense and the Arms Control and Disarmament Agency to evaluate the case for the station. The Office of Science and Technology Policy had only 'observer' status. Unfortunately, it took until April 1983 for the working group to define its terms of reference, and even then it became impeded in the fine detail of NASA's plan. Beggs, however, had quietly launched a twin-track approach by seeking out individual members of the White House's staff who might be particularly supportive.

Meanwhile, in August 1982 the SSTF had hired eight aerospace companies to study the activities that the station ought to be capable of undertaking. This was another aspect of trying to define the station using a 'top down' rather than 'bottom up' approach. NASA was insistent that there be no direct contact between these parallel 'Space Station Needs, Attributes and Architectural Options' studies, in order to ensure that their analyses would be independent. In an 'International Orientation Briefing' on 13 September the SSTF formally invited its potential international partners to submit their Phase 'A' studies by March 1983 (the same time as its own study groups). In an effort to prevent US aerospace companies trying to force themselves on the agency by forming strategic alliances with foreign companies likely to be hired by eventual international partners, NASA prohibited them from discussing the station with foreigners. Once the proposals were in, the Concept Development Group set up by SSTF digested the options and formulated its plan. A significant theme to emerge from the US studies was the merit of a "distributed architecture" in which the station was assembled from modules, trusses and platforms, each of which was to serve a specific mission. Such an architecture would provide "cleaner interfaces" than would be possible with a single integrated structure. Realising that it was unlikely to secure a return to the Moon, NASA de-emphasised the station's utility as a jumping-off point for deep space missions. Nor did it endorse the concept of the orbital repair shop for satellites. Nor, indeed, did it accept the power-station-in-the-sky that some had advocated. Instead, the Concept Development Group favoured a microgravity laboratory to establish the potential of manufacturing-in-space applications; it would serve as a technology 'incubator' for future industries. With this decided, the SSTF hosted a workshop for potential users.

At the end of 1983, in a speech in Los Angeles, Beggs's deputy, Hans Mark, noted that NASA had "only two priorities". Firstly, "to make Columbia and her sister ships a truly effective system" and "to establish, in the next five years, the programme that will put us in space on a permanent basis". As a popular phrase of the time put it, the time had come to upgrade NASA's "visitor status" in space. The agency was developing an initiative, Mark added, to establish "a space station of some kind". Although he hoped work would begin in Fiscal Year 1984, he pointed out that "we have not yet persuaded the President that this is indeed the proper goal for this nation to pursue".

In fact, Reagan was perfectly willing to spend large sums of money as long as doing so strengthened America. In March 1983 he had launched the Strategic Defense Initiative (SDI), which the Press had promptly dubbed 'Star Wars' after George Lucas's film of that name. Its mission was to determine whether space-based interceptors and "beam weapons" could destroy strategic ballistic missiles long before they posed a direct threat to US territory. From NASA's viewpoint, this expensive programme was a major competitor for federal funding. Indeed, Caspar Weinberger, the Secretary of Defense, opposed NASA's space station because its assembly would use Shuttle launches that would probably be required for SDI development. Nevertheless, Beggs met Reagan on 7 April 1983 to discuss his prospects. Reagan's interest was in establishing America's leadership in space and it was evident that the Soviets had been far from idle over recent years with Salyut 6. Its successor was already in orbit and its first crew had pushed the endurance record to 211 days. Furthermore, the Soviets had demonstrated the viability of assembling a modular station. Of course, America's station would be bigger and better, and if it was opened up to friendly nations this would reaffirm America's role as the senior Free World partner in space activities. Given this background, the fact that Beggs had secured expressions of interest from Canada, Europe and Japan was a key consideration. On 11 April Reagan asked McFarlane to make a recommendation on whether he should approve a station proposal. By August, it was apparent to Gil Rye, the National Security Council member who was responsible for space policy, that SIG(Space) would not reach a consensus by its November deadline, largely because Weinberger had said that not only would he not back the proposal, but would actively oppose it. As a supporter of the station proposal, Rye had formed an independent, much smaller, group in May to make a recommendation as a fallback.

That summer, Beggs informed Congress that if a space station programme was approved for Fiscal Year 1985, then it ought to be practicable to award hardware development contracts within three years, and that the development would cost $6 to $8 billion. In July, the American Institute of Aeronautics and Astronautics hosted a Space Station Symposium. In giving the opening speech, Beggs emphasised to representatives of other national space programmes that international cooperation was "essential". Canada and Japan were robust in support of participation. West Germany and Italy were both enthusiastic but ESA was unable to commit. Japan was eager to undertake microgravity materials research in the expectation that the insight thereby gained would be transferred to benefit its industrial base. As the symposing progressed, the National Academy of Sciences said that microgravity science was unlikely to justify the cost of developing the station. Worse, it would divert funds from terrestrial science programmes. However, to be constructive, they listed the facilities that the station would have to include if it was to undertake "good science". A number of participants related how a permanently inhabited facility could perform research into biological sciences, pharmaceutical production and materials processing. After canvassing 30 potential user groups, the chairman of the Space Applications Board reported that, with the exception of materials sciences, there was little scientific requirement for a permanent human presence in space. Clearly, the Skylab lessons had been forgotten. Some of the potential industrial users said that the apparatus on the station should be operated remotely from Earth, using a "telescience" capability that had still to be developed, and that astronauts should board the station only to service the equipment. In addition to confirming the entrenched attitudes, the symposium highlighted the fact that, despite its best efforts to be inclusive, NASA had yet to identify (to use marketing slang) a 'killer application' that would – *in, and of itself* – justify the development of a space station.

THE LEADERSHIP IMPERATIVE

In October, the White House formally asked NASA to respond to Reagan's 1982 policy call for "leadership" in space. Beggs was "absolutely convinced" that the station should be "the next bold step in space". It was "an essential piece of our long range plan to reap the full commercial and scientific benefits of space". In a sense, the argument was self-evident. With launch costs so high, it clearly made little sense to repeatedly ferry heavy apparatus into orbit and back. The Continuous Flow Electrophoresis System which had flown on several Shuttle missions, for example, filled an entire cabinet on the orbiter's mid-deck and it was toting up to-orbit payload costs. It would make better sense to fly it up and install it on a station, use it intensively, and then return it once it was no longer required. A space station would completely change the economics of performing research in space, he said. By taking a sufficiently long view, the development and operating costs of the station could be seen to be a worthwhile investment. In addition to Rye, NASA had found an ally in

Craig Fuller, the Cabinet Secretary. On 1 December, Fuller presented the 'fallback report' of Rye's group to a meeting of the Cabinet Council on Commerce and Trade, representing the workers of America who would benefit from such a vast engineering programme. Reagan informed Beggs a few days later that he intended to endorse the station. Beggs had therefore made a successful 'end run' around Weinberger's opposition. As soon as the decision had been made, Reagan's political advisers recommended that he announce it in his State of the Union speech. On 18 January 1984, McFarlane convened a meeting of high-level executives (including Beggs) to discuss international cooperation, and not only endorsed Beggs's efforts but also recommended that the President issue the formal invitation in his speech. Reagan dispatched private letters to alert European, Japanese and Canadian leaders to his intention to invite their participation. On 25 January 1984, before the gathered Congress, the Great Communicator was in magnificent form. "We can follow our dreams to distant stars, living and working in space for peaceful economic and scientific gain." With the spirit roused, he made the commitment. "Tonight, I am directing NASA to develop a permanently manned space station." In an echo of Kennedy's historic call, he instructed that this be operational within 10 years. "We want our friends to help us meet these challenges and share in their benefits"; NASA was therefore to invite other countries to participate "so we can strengthen peace, build prosperity and expand freedom for all who share our goals". Some critics immediately argued that the State of the Union Address was an inappropriate time to announce such a venture, as this context would suggest to a foreigner that it was being done primarily for US domestic political reasons. Some US politicians, evidently having missed Beggs's statements in favour of partnership, decried such a blatant appeal for foreign help.

However, a Presidential commitment did not guarantee that funding would be forthcoming. On 1 February, when Beggs presented NASA's budget request for Fiscal Year 1985 to Congress and sought $150 million to launch the station's engineering design effort, the reaction was, as he put it, "favourable, but not unanimous". In fact, his schedule for $200–250 million in Fiscal Year 1986, $1.2 billion in 1987, and $2 billion per year thereafter had been met with undisguised scepticism. Beggs was able to sweeten the pill on 1 May, however, by pointing out that foreign hardware corresponded to commitments of $2 billion by ESA and $1 billion by Japan. Asked by the Press to justify a space station, Beggs unwisely suggested that its best use might reveal itself only once it was in service.

Neil Armstrong was critical of the proposal. He urged NASA to forgo a station in low orbit. Buzz Aldrin agreed with him: "The Solar System's most desirable 'space station' already has six American flags on it." The Shuttle was confined to low orbit, so a return to the Moon would require the development of a reusable propulsive stage to ferry large payloads to lunar orbit. George Keyworth, Reagan's Science Adviser, expressed a similar view and urged NASA to plan to use its station as a jumping-off point. Of course, the irony was that the agency had been obliged to delete the way-station mission from the station's rationale.

In early February, Beggs set off to assess the international response to Reagan's call to arms. "Based on our own scientific and technological capabilities," said

Having ordered NASA to build a space station, Ronald Reagan showed off an early conceptual model on a tour of Europe in June 1984. Prime Minister Margaret Thatcher was not impressed.

ESA's Erik Quistgaard, "we are considering ways to join forces with you for the advancement of this project." In fact, ESA had due cause to feel optimistic about the future. The first of the French-led Ariane rockets was now launching satellites on a commercial basis, a West German scientist had flown on the first Spacelab mission, and France had started work on a larger rocket to launch the Hermes space plane. Europe's strategic plan to be the third player in human spaceflight seemed to be on track. Participation in the station was regarded as an ideal means of addressing the learning curve. Italy's Aeritalia also had teamed up with Boeing to bid for the Orbital Manoeuvring Vehicle (OMV) – a 'tug' that was to be based at the station and used to deliver a succession of satellites to their operating orbits.

At the London Summit of the seven leading industrial nations in June 1984, Reagan displayed a large model depicting an early concept for the station. This had a cruciform of small modules around a central node at which the Shuttle would dock, a pair of solar panels projecting out to either side, and an open platform with berths where satellites would be checked out prior to being taken to their operating orbits by the OMV. It was well received by the Italian, West German and Japanese leaders, but Prime Minister Margaret Thatcher was unimpressed, believing that there was little scope for commercial manufacturing in space.

While it is the case that NASA's extensive series of studies had pared the station's mission down to a laboratory in space, in political terms it was far more than that. As an investment in the future, the rewards were expected to be both significant and diverse. The microgravity environment would facilitate unprecedented basic research into life and materials sciences. While this was undoubtedly true, pioneering research would be able to be undertaken only if state-of-the-art apparatus was made available, and it would require a great deal of equipment to address every discipline in a meaningful way, which in turn would require a *large* station. To NASA, the tremendous scope of potential activities was seen as a 'plus'. As a new "national laboratory", NASA argued that the station would be "one of the most powerful investigative tools" for scientific research to "open new windows on science". The logic was compelling. With appropriate funding, such a multidisciplinary orbital facility might indeed "become a spawning ground for innovation" and "a nurturing place for new technologies", and it just might even become "a centre for the inception and development of the advanced technologies upon which our nation's economic and social well-being depends in an increasingly competitive and sophisticated world". However, the enormous cost was the stumbling block. For NASA, the key was to push for the maximum return by expanding the range of (or at least not closing off) missions, but the trend was the other way. The danger was that by paring back the station's capabilities in order to minimise the cost of its development and assembly, that which remained would not be considered to be worth the start-up cost. As the NASA booklet 'Space Station – The Next Logical Step' put it in 1984, as the programme was getting started: "The space station is a vital part of our nation's investment in science and technology. The pace and quality of progress . . . will affect the competitive strength of our industries and have a strong impact on jobs and pay scales. These, in turn, will influence how well we live as individual citizens and as a nation." Also, as 'Space Station – Leadership for the Future' observed in 1987, by which time the basic design had been formulated, the station would:

- enhance capabilities for space science and applications;
- stimulate advanced technologies;
- promote international cooperation;
- develop the commercial potential of space;
- challenge the Soviet lead in space stations;
- contribute to American pride and prestige;
- stimulate interest in science and engineering education; and
- provide options for future endeavours in space.

When considering the merit of the station in retrospect, it is important to bear this wider context in mind. From the point of view of the Reagan administration, it was endeavouring to reinvigorate the American nation following what it thought of as the humiliation of the Carter years, and was willing to fund the military–industrial complex to buttress America from what it considered to be the increasing threat from the Soviet Union. A measly $8 billion spread over a 10-year period to affirm America's leadership in space was essentially a 'small change' item to Reagan. As a point of comparison, the reinvigorated Department of Defense was soon spending $250 billion *per year*. Reagan saw the space station as a symbol of the superiority of Western society in general and the US as its senior partner in particular, so the cost was not really an issue; but to Congress, which would have to find the cash, the cost was very important.

DOWN TO BUSINESS

Given the tremendous success that the Soviet Union was having with its most recent Salyut stations, it was apparent that NASA would have to move fast if it was going to catch up. On 15 February 1984, Beggs gave responsibility for the space station programme to the Johnson Space Center, which henceforth would manage:

- systems engineering and integration;
- business management;
- operations integration; and
- customer integration.

The SSTF's Concept Development Group sketched out a 'reference concept' capable of supporting the defined mission, which, once the political rational was stripped away, was "to enhance capabilities for space science and applications". As a result of consultations with potential users, it was evident that the station would need high-capacity communications, state-of-the-art data processing and tremendous power generation – 75 kW, in fact. The Gemini and Apollo spacecraft had used 'fuel cells' which create electricity by reacting hydrogen and oxygen to produce water as a by-product, but their utility was constrained by the supply of cryogenic reactants. In the long term, there was the prospect of developing a 'solar dynamics' power system in which solar energy would energise a pump to run a turbine, but as this technology had yet to be developed solar transducers, as on Skylab, were the only viable option. As transducers were inefficient, however, *large* arrays would be needed to provide sufficient power to run all the apparatus that SSTF hoped the station's users would supply. If such large panels were to be affixed directly to the pressurised modules, their presence would impede the Shuttle's access, so it was decided to put the panels at either end of a large truss. To overcome the power loss in such lengthy cables, a high-voltage low-current 'transmission line' technology would have to be used. With so much electrical apparatus on board, heat build-up would be significant, and a serious cooling system would be necessary to transfer the heat from the pressurised modules to large radiator panels mounted on the truss. Once the rest of the reference concept had been defined, it consisted of a pair of laboratory modules, two habitat modules, four linking 'nodes', and the truss for the solar panels and thermal radiators. The international partners could choose either to supply one of the baseline elements or to contribute additional facilities. The initial crew complement would be eight persons, and flight opportunities would reflect the contributions of the partners. Importantly, the modular design would enable the station to be expanded as and when appropriate, so this was seen only as the initial, not the final, configuration. The SSTF was careful not to close off future options. In order to provide "options for future endeavours in space" the nodes would have more berthing connectors than required by the baseline. For example, if the research led to commercial manufacturing, production modules would be able to be added, and, further downstream, once a case for 'free-flyers' could be argued, servicing platforms would also be able to added. The Space Station Programme Office was established in April 1984, with Phillip Culbertson as Associate Administrator for the Space Station.

The configuration that the Concept Development Group devised was for a 120-metre-long latticework 'keel' oriented with its axis pointing towards the centre of the Earth. The pressurised modules would be clustered near the lower end of the keel so that the 'gravity gradient' (which results from the fact that gravity is inversely proportional to altitude) would tug on the heavier end and therefore stabilise it, without it consuming propellant by firing thrusters. This arrangement meant that the keel's orientation would remain fixed with respect to the Earth as it pursued its orbit. It would be possible to install nadir-facing instruments at the base and astronomical instruments at the top, and thus facilitate astronomical and terrestrial studies in parallel with microgravity research. A perpendicular truss mounted half-way up the keel would carry the solar panels and thermal radiators, and so give the Shuttle unobstructed access to the module cluster. The design of the 'Power Tower' (as it was dubbed) was influenced by the National Space Transportation System's 'Shuttle-only' policy because the only way to assemble it would be to ferry its individual elements into space in the Shuttle's payload bay. Although the bay was 18 metres long and could accommodate a cylinder 4.5 metres in diameter, size was not the crucial constraint because the Shuttle was mass-limited rather than volume-limited. Furthermore, the vehicle's centre of mass was the dominating factor because it could not land if its centre of mass was too far forward. This was an issue even for payloads on one-way trips into space, because the payload could not be jettisoned for an emergency landing following an abort during the ascent. Although NASA hoped that the Shuttle fleet would eventually provide 'routine' and 'cheap' access to orbit, the operating overhead for the foreseeable future would be very high, so another aspect of the plan was the need to draw up the manifest to minimise the number of launches because the agency had not factored in the launch costs when calculating the cost of the station. Nevertheless, apart from the fact that it would not be cheap, using a fleet of Shuttles to assemble the station was precisely what NASA had proposed in 1969 when Nixon had ordered that the "expendable" Apollo-era technology be phased out. "The Shuttle is a central element of the space station endeavour," said Culbertson on 1 August. "With both systems operational, the US will have a capability to work in space unmatched by anyone." That would certainly be the case, because although the Soviets were clearly developing systems to assemble a modular station, they had shown no interest in a reusable space plane, so the Shuttle's cargo-carrying capacity would be a significant advantage in terms of *station operations*, which was why in 1972 NASA had accepted the need to delay a station in order to push ahead with the Shuttle. If it had pursued a station using Apollo technology then, as the Soviets were discovering, it would have found itself constrained by an inability to return bulk materials to Earth, which would have compromised a manufacturing-in-space facility. Despite the frustrations of putting the station off, developing the Shuttle had been the right decision.

On 14 September 1984, NASA issued a Request for Proposals to industry. The Power Tower was the Reference Configuration but the studies were free to suggest modifications. Beggs rejected advice to require including an 'automatic operation' option, it being an unnecessary complication, but sanctioned the incorporation of telerobotic systems and agreed that the station ought to be able to be utilised on an

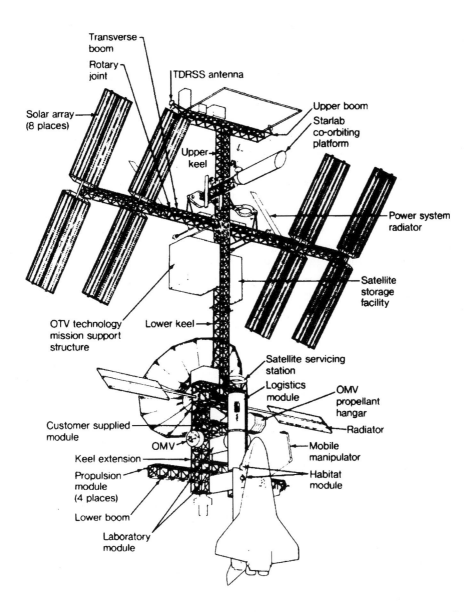

Transverse boom

Rotary joint

TDRSS antenna

Solar array (8 places)

Upper keel

Upper boom

Starlab co-orbiting platform

Power system radiator

Satellite storage facility

OTV technology mission support structure

Lower keel

Satellite servicing station

Logistics module

OMV propellant hangar

Radiator

Customer supplied module

OMV

Mobile manipulator

Keel extension

Propulsion module (4 places)

Habitat module

Lower boom

Laboratory module

After considering the range of possible missions, the Concept Development Group of the Space Station Task Force chose the 'Power Tower' as its Reference Concept for detailed planning. It was to be built around a 120-metre-long latticework 'keel' oriented with its axis pointing towards the centre of the Earth. The pressurised modules would be clustered at the lower end of the keel so that the 'gravity gradient' would tug on the heavier end and stabilise it. A perpendicular truss mounted half-way up the keel would carry the solar panels and thermal radiators, and so give the Shuttle unobstructed access to the module cluster.

intermittent basis as it was being assembled, rather than only when it was complete. Thirteen bids had been received by the November closing date and development contracts were issued on 1 April 1985. Instead of assigning the programme to a single prime contractor, NASA awarded four 'work packages' (WPs) with two contractors working in parallel under the supervision of one of the NASA field centres:

WP1   Marshall Space Flight Center
      Boeing and Martin Marietta
      $26 million
      - a common module to be delivered by the Shuttle that could be configured to serve as the station's laboratory, a habitat, and a logistics carrier
      - environmental control and life support systems
      - propulsive systems to hold the station's attitude and to counter orbital decay
      - the requirements for the satellite servicing facilities and the transfer vehicles ('tugs');
WP2   Johnson Space Center
      McDonnell Douglas and Rockwell International
      $30 million
      - the truss structure
      - the interface between the Shuttle and the station
      - guidance, navigation and control systems
      - crew habitability
      - thermal control system
      - communications and data management systems
      - the airlock
      - EVA systems
      - the remote manipulator;
WP3   Goddard Space Flight Center
      General Electric and the Radio Corporation of America (RCA)
      $10 million
      - attached and free-flying payloads
      - laboratory systems;
WP4   Lewis Research Center
      Rocketdyne and Thompson, Ramo and Wooldridge (TRW)
      $6 million
      - power generation, energy storage and distribution systems.

This three-year Phase 'B' study (officially for 'Definition and Preliminary Design') was to yield a specification that was sufficiently refined for rapid fabrication, launch and assembly. It was hoped to have the structure in orbit and operational in the early 1990s, in line with Reagan's 10-year stipulation.

MEANWHILE IN EUROPE

The Rome meeting of ESA's Council in January 1985 decided to develop a pressurised laboratory as its principal contribution to the space station. Although this module would rely upon the station for power and life support, ESA stipulated that it might later detach it, in order to integrate it into an independent European facility. This strategy would build upon Europe's expertise in developing Spacelab. ESA also decided to develop a 'tended' co-orbiting platform to be operated in association with the station.

NASA presented a workshop in Copenhagen, Denmark, in early 1985 to brief its international partners on its progress in defining the station. Canada, Japan and Europe each signed Intergovernmental Agreements to participate in the station, but it took a year to firm up the contributions and the contracts were not finalised until September 1988.

With both Europe and Japan promising to supply laboratories, NASA reduced its contribution from two such modules to one. The Shuttle's Remote Manipulator System had proved to be as crucial for productive operations in space as the Lunar Roving Vehicle had been to the Apollo astronauts on the Moon, and it was natural that Canada should offer to supply a larger and more dexterous manipulator for the station. As Beggs had hoped, international participation was spreading the cost of development.

In 1985, the French Space Agency (CNES) awarded the contract for the Hermes space plane to Aerospatiale and Dassault-Breguet. Because it was to ride the new Ariane V rocket into orbit, this 15.5-metre-long delta-winged vehicle's weight was limited to 17 tonnes, including 2.5 tonnes of propellant, but it would carry two pilots, up to four scientists and a 5-tonne payload in its 35-cubic-metre bay. Although Hermes would be less capable than NASA's Shuttle, it represented a considerable independent capability in terms of human space flight, which was ironic, because part of Beggs's rationale for inviting European cooperation had been to forestall direct competition with NASA's capability. Hermes was accepted as an ESA programme in October 1986. The 'Preparatory Programme' to define how it would fit into other European space programmes was to extend through to the end of 1987, after which the precise timescale for its development would be worked out.

SPACEHAB

Spacehab Incorporated had been created in 1983 to develop the Shuttle's potential for commercial microgravity research. There were 42 locker spaces on the mid-deck of the Shuttle's cabin but 80 per cent of this capacity was routinely assigned to equipment for the crew, and experiments that required crew oversight used most of the others. The list of educational and commercial users had grown faster than NASA could service it, so Spacehab set out to alleviate this backlog. With NASA's enthusiastic support, the company raised the venture capital required to place a contract with McDonnell Douglas for a pressurised 'mid-deck augmentation module'. In fact, the fabrication of the primary structure was subcontracted to Aeritalia, the acknowledged experts. Although the flat-topped module would occupy only the front quarter of the payload bay, it would *double* the habitable volume. With 60 lockers of experiments to tend, NASA decided to add a mission specialist specifically to oversee the module's operations. By October 1985 the company had received some 800 expressions of interest from potential users. This private investment matched the rapidly developing sense that microgravity research would soon be a lucrative business, and Spacehab sought to establish itself as the leading 'service provider' by packaging payloads and steering them through NASA's arduous certification process. In the fullness of time, of course, the company hoped to offer this service for payloads destined for the space station, too.

INDUSTRIAL SPACE FACILITY

As the prospects of microgravity research grew in the early 1980s, it began to look as if there would soon be a role for a large free-flying facility on which longer-term experiments could be performed or, better yet, applications run. The development in 1985 by McDonnell Douglas of a pallet to stand in the Shuttle's payload bay for 'Electrophoresis Operations in Space' (designed to refine biological products such as insulin) seemed to be a clear sign that sooner, rather than later, the company would seek to expand its orbital operations. Perhaps it would lease time on a semi-permanent multi-role platform? Perhaps by the early 1990s there would be an ever-increasing call for time on orbital platforms. In this lucrative service-provider market it would clearly pay to be the first in the field. It would also be of benefit to be seen to have close links with NASA.

Space Industries Incorporated (SII) had been set up in Houston in 1982 by Max Faget, the chief designer of the Mercury capsule. When Joe Allen resigned as an astronaut, he had joined the company. SII was therefore familiar with how NASA conducted business, but nevertheless set out to transform the way that microgravity payloads were developed. A decade before Dan Goldin made it the norm, SII was an advocate of a 'faster-cheaper-better' approach. In the mid-1980s it noted the future for a large orbital facility for long-running experiments, and proposed the Industrial Space Facility (ISF) as an automated materials processing factory to be deployed and tended by Shuttles. A docking system set in the front part of the bay would be linked to the mid-deck's airlock by a short tunnel. The ISF would have an end hatch and, once raised from the bay by the RMS and rotated, would be mounted on the docking system. After deploying a pair of solar panels, the commissioning crew would enter the module and activate its systems. When everything had been verified, the Shuttle would withdraw and leave the ISF to execute its predefined programme, perhaps with telerobotic assistance. When the programme was complete, the 'product' would be retrieved by another Shuttle, the applications serviced (or superseded upon the expiry of a specific lease) and a new batch of raw material loaded on board. On these servicing missions a logistics module would be carried in the payload bay. Unlike the space station that NASA was proposing, the ISF was to be self-contained. The design was sufficiently flexible to allow several modules to be joined together and, if necessary, a module could be returned to Earth for refurbishment. In the long term, ISF modules might be integrated into the space station as interim factories.

The ISF proposal was well received, Westinghouse, Boeing and Lockheed backed the engineering studies, and NASA announced an agreement in August 1985 that guaranteed two flight opportunities, so that SII could assure prospective clients that it had agency support. Furthermore, to obviate SII having to raise capital to cover launch costs, NASA introduced a 'fly-now-pay-later' deal whereby the company would reimburse the cost of flying from the revenues earned from renting time on its module. The company's close association with NASA had paid off. In reality, however, this revolutionary start-up deal had its origins in the Reagan adminis-tration's July 1984 call for commercialisation of space operations, and this was NASA's way of helping private ventures make commercial headway in their most crucial formative years.

In April 1989, however, NASA effectively killed off the ISF when a specially commissioned panel advised against leasing the commercial free-flying module for microgravity work. It added a caveat that if the development of the station was significantly delayed then it would be worth reconsidering the free-flyer as an interim vehicle.

REDESIGN

In March 1985, the prospect of users demanding accommodation for apparatus on the station prompted Culbertson to emphasise its need to host a large number of 'attached payloads'. As a result, a 'T'-piece was added to each end of the keel to

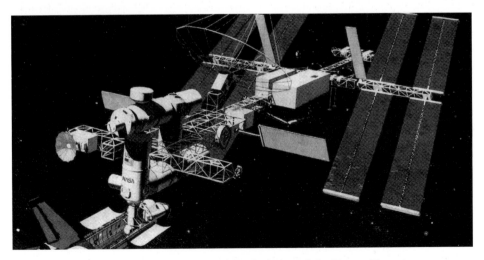

A Rockwell International artist's depiction of the 'Power Tower'.

Another artist's depiction of the 'Power Tower'.

Ron McNair plays his saxaphone aboard Challenger on STS-41B in 1984. He was killed when the Challenger was lost on mission 51L in January 1986.

carry the external packages. However, these trusses would have to be short to enable the robotic manipulator, which was to run along the length of the keel, to mount and service payloads at their extremities. In retrospect, this constraint was seen as a limiting factor in the Power Tower's design.

In April, it was realised that if the modules were clustered near the bottom of the keel, the ongoing accelerations as the station adjusted to irregularities in the Earth's gravitational field would 'pollute' the microgravity environment sufficiently to interfere with the research programme. The solution was to move the modules onto the station's centre-of-mass, which meant relocating them near the intersection of the vertical keel with the perpendicular solar panel truss. In the redesign, the single keel was superseded by a pair of keels joined top and bottom by long trusses that were long enough to host all of the attached payloads that the station was likely to attract. On 3 October, the Space Station Control Board formally recommended this 'Dual Keel' configuration for evaluation.

DISASTER

Having made nine Shuttle flights during 1985, NASA hoped to ramp up the pace in 1986, but the year started badly when its administrator had to take leave of absence to defend himself in an investigation regarding his previous position at General Dynamics.

On 28 January, 72 seconds into the mission of STS-51L, one of the Solid Rocket Boosters broke loose and ruptured the External Tank. As the stack disintegrated, Challenger was torn apart by the excessive aerodynamic loads. All seven astronauts were killed. Addressing the nation that evening, Reagan told the children who were later to have received a televised lesson from primary school teacher Christa McAuliffe aboard Challenger: "I know it is hard to understand, but sometimes painful things like this happen; it's all part of the process of exploration and discovery; it's all part of taking a chance and expanding Man's horizons. The future doesn't belong to the faint hearted, it belongs to the brave." At a memorial service a few days later, Reagan was at his best. "We remember Ronald McNair ... His dream was to live aboard the space station, performing experiments and playing his saxophone in the weightlessness of space. Well Ron, we will miss your saxophone, and we will build your space station."

7

Salyut 7's jinx

STRIVING FOR ONGOING OPERATIONS

Launched on 19 April 1982, Salyut 7 was a virtual copy of its predecessor. On the premise that what had worked so well once should work well again, the new station was expected to support four years of operations. The goal was to work towards in-orbit handovers so that the station could be occupied on a continuous basis, in the process eliminating the time that would otherwise be wasted in first deactivating and then reactivating its systems.

The extended life of the Soyuz-T reduced the need for crews to visit in order to exchange ferries, which would considerably relieve the hectic schedule that had been imposed by Salyut 6's long-duration missions. This was just as well, because if the crew complement was increased to three, Progress resupply ferries would have to be dispatched every two months. The reinstatement of the third seat opened the prospect of a biomedical researcher studying adaptation to microgravity by flying up with one crew and then serving with successive crews, eliminating the need for the engineering cosmonauts to endure such long flights.

Soyuz-T 5 delivered Anatoli Berezovoi and Valentin Lebedev on 13 May. The open-ended flight plan was dependent upon how well the station performed, but it was hoped to be able to exceed the 185-day record. On 25 May Progress 13 slipped into the rear port with propellant to replenish the fuel used as the station climbed to its operating orbit. The tanker departed on 4 June to clear the port for Soyuz-T 6, which arrived on 25 June with Vladimir Dzhanibekov, Alexander Ivanchenkov and Jean-Loup Chretien, a French cosmonaut whose flight marked the broadening of the Intercosmos programme. The visitors left on 2 July and Progress 14 arrived 10 days later. At the end of the month, the residents ventured outside to mount recently delivered cassettes on the hull. Progress 14 departed on 11 August. A week later Soyuz-T 7 docked with Leonid Popov, Alexander Serebrov and Svetlana Savitskaya, the first woman to fly in space since Valentina Tereshkova's pioneering mission in 1963. After a week, the visitors departed in the older vehicle and two days later, just before the recovery window closed, the residents flew Soyuz-T 6 around to the front. Progress 15 took the rear port on 20 September, departed on 14 October, and was replaced by Progress 16 on 2 November, so the transportation infrastructure was running smoothly.

Meanwhile, Salyut 7 was suffering problems. The early stations had relied upon ground radar tracking to supply data on their orbital path. Salyut 1's crew had spent up to 30 per cent of their time on tasks related to orienting the station for the job in hand. Salyut 4 had introduced 'Delta', a semi-automatic navigational system which monitored the rising and setting of the Sun on the local horizon to calculate the orbital period, and the 'Argon' computer combined the timings with radio-altimeter data to compute the orbital parameters, the station's progress around the globe and the times during which it would be within communication range of the ground sites, which greatly reduced the crew's workload. It had functioned perfectly throughout Salyut 6's marathon mission but, in August, Salyut 7's Delta failed. When it was realised that it had overwritten its memory and destroyed its programme, Lebedev keyed in the 325 six-digit sequence and managed to restore it. However, the system continued to misbehave. When it failed completely, it was decided to recall the crew prior to the ideal recovery window, and on 10 December they landed in darkness. Nevertheless, they had established a new endurance record of 211 days.

Another development was also afoot. After Salyut 6 had been finally vacated a 20-tonne spacecraft had docked at its front port. Its heritage was not revealed at that time (this had to await Mikhail Gorbachev's promotion of 'glasnost') but Cosmos 1267 was a test of Chelomei's heavy logistics ferry. It could deliver 3,600 kilograms of dry cargo and augment the station's power supply by 3 kW, and its descent capsule could return as much as 500 kilograms to Earth. On 10 March 1983 Cosmos 1443, which was similar, docked at Salyut 7's front port and assumed responsibility for manoeuvring the complex until the station's Delta could be replaced. On 20 April Soyuz-T 8 set off with Vladimir Titov, Gennadi Strekalov and Alexander Serebrov. The rendezvous antenna was ripped off during the separation of the aerodynamic shroud but Titov requested permission to continue with the rendezvous using data supplied by the radars of the ground tracking network. All went well until, at about a distance of 100 metres, they entered the Earth's shadow and he withdrew to avoid any possible collision with the 40-tonne complex. By sunrise the ferry had drifted too far away to set up another approach, so they had to return home.

When Soyuz-T 9 was launched in late June it carried a crew of only two, Vladimir Lyakhov and Alexander Alexandrov, and it was reported that they were *not* to try to break the endurance record. In fact, the plan was that they would hand the station over to the next crew in October, who would in turn hand it over to their successors early in the new year. Once the replacement Delta system was retrieved from the Cosmos 1443 "warehouse", it was installed to restore Salyut 7 to full operation. Almost all of August was spent unloading the rest of the cargo and setting up the new apparatus, then the accumulated results were loaded aboard Cosmos 1443's descent capsule (which, rather perversely, was equipped with couches) so that the spacecraft could depart on 14 August. Before the recovery window closed, the residents transferred their ferry to the front in order to clear the way for Progress 17's arrival on 19 August. Unfortunately, as oxidiser was being pumped aboard a pipe in the unpressurised engine bay at the rear of the station fractured, disabling the engine and one of the two sets of attitude control thrusters. It appeared that Salyut 7 was jinxed! After using its own engine to boost the station into an orbit somewhat higher than normal to stave off orbital decay, Progress 17 left on 17 September to clear the way for the first handover. However, the attempt to launch Titov and Strekalov on 26 September was aborted in the final phase of the countdown when a faulty valve started a fire that engulfed the rocket seconds after the escape tower lifted the capsule away.

Lyakhov and Alexandrov agreed to extend their tour, and also to undertake a spacewalk for which Titov and Strekalov had trained but they had not. The size of the station's solar panels was restricted by the stepped-cylinder shape within the aerodynamic shroud. A docked ferry augmented the supply but it was still limited. A pair of 'clip on' panels had been ferried up by Cosmos 1443 and a second pair had arrived in Progress 18 on 22 October. These were to counter the degradation that had been noted in the case of Salyut 6's panels towards the end of its life. Attachments had been bolted to the frames of the main panels to accommodate an extra strip down each edge. Lyakhov and Alexandrov ventured out on 1 November to mount the first pair. The 50 steps in the attachment sequence were daunting. Alexandrov took up position on a foot restraint at the base of the station's dorsal panel and Lyakhov handed him the folded panel. Once it was connected to the lower attachment point and the handle was inserted into its winch, Alexandrov, with some difficulty, managed to unfold it accordion-fashion alongside the larger panel. Upon reaching its maximum extension, the side strip automatically engaged the clamp at the far end of the larger panel. After turning the main panel through 180 degrees, they returned two days later and added the matching strip. As a pair, these new panels boosted the station's supply by 1.2 kW. Their addition meant that full advantage could now be taken of the materials-processing furnaces. Lyakhov and Alexandrov then settled down to await the recovery window later in the month. Progress 18 undocked on 13 November and the station was vacated on 23 November. Although they had not set a human endurance record, the frustrated handover meant that the orbital life of the Soyuz-T had been extended to 149 days. The task now was to decide what (if anything) could be done to overcome Salyut 7's propellant leak.

On 9 February 1984 Soyuz-T 10 docked at Salyut 7's front port with Leonid Kizim, Vladimir Solovyov and Oleg Atkov, a physician who was to perform ongoing biomedical tests using new apparatus to track changes in blood chemistry and make a recommendation as to when to terminate the open-ended mission. Progress 19 arrived on 23 February, boosted the station's decaying orbit and left on 21 March, and on 4 April Soyuz-T 11, with Yuri Malyschev, Gennadi Strekalov and Rakesh Sharma, an Indian cosmonaut participating in the Intercosmos programme, took its place. This marked the first time that six people had been on board simultaneously. The visitors left on 11 April in Soyuz-T 10. Two days later, the new spacecraft was transferred to the front to accommodate the arrival of Progress 20, which brought the tools for the repair task four days later. On 23 April, with Atkov sealed into Soyuz-T 11's descent capsule in case they were obliged to abandon the station, Kizim and Solovyov were to affix a frame curving in a 120-degree arc around the periphery of the engine compartment to provide a 'workstation' for one cosmonaut during the repair. A special platform on Progress 20's orbital module unfolded on command from Kaliningrad to accommodate the second cosmonaut. The tools were affixed to the workstation. With the preparations complete, the two men retreated. Three days later, Solovyov stood on the platform in the gap between the ferry's orbital module and the rear of the station, and Kizim strapped himself to the frame which wrapped around the engine bay. The first task was to replace a valve associated with the oxidiser leak. After opening the access panel Kizim utilised a pneumatic punch to open the thermal blanket that protected the bay. Nitrogen gas from a tank in Progress 20 was then fed into successive stages of the replenishment system to confirm that the location of the leak was, indeed, in the predicted length of pipe. The repair called for bypassing this pipe by replacing the nearest valves with units incorporating connectors for another pipe. The disconnection of the first valve was aggravated by the discovery that some of the bolts had been sealed with glue to prevent vibrations loosening them and one bolt, as a result, took an hour to release. When the station passed into the Earth's shadow they continued in the illumination of the lamps on their helmets because, with only 5 hours of life support in their backpacks and with almost an hour reserved for making their way back and forth along the station, there was little time to waste. Finally, they managed to install the first valve. It had been hoped to fit the new pipe too, but they were so far behind schedule that they were obliged to stuff the insulating blanket into the hole to protect the engine and hurry back to the airlock. Returning three days later, they quickly attached the bypass pipe to the valve. On 4 May they replaced the second valve and connected the bypass pipe. Having done all that they could, they restored the insulation and closed the access panel. Of course, the station's engine would remain unusable until the fractured pipe had been isolated. The tools for this task were still under development, so the final part of the repair operation would have to be done at a later date.

Its role as a foot restraint over, Progress 20 left on 6 May and Progress 21 arrived four days later. On 18 May, Kizim and Solovyov continued their record-breaking series of spacewalks by fitting the second pair of clip-on strips to each side of the lateral panel alongside the airlock hatch. This time, Atkov remained in Salyut 7 to rotate the panel while his colleagues fetched the second strip from the airlock to allow them to add both strips on a single excursion. The winch confirmed its poor reputation when its handle broke. Progress 21 left on 26 May and Progress 22 took its place on 30 May, delivering replacement batteries. After boosting the station's orbit Progress 22 left on 15 July, and three days later Soyuz-T 12 arrived with Vladimir Dzhanibekov, Svetlana Savitskaya and Igor Volk, who had piloted the Soviet equivalent of the DynaSoar spaceplane in atmospheric tests in the 1960s and was flying as part of his training for the Buran Shuttle. The big event of the visiting programme was a spacewalk on 25 July by Dzhanibekov and Savitskaya. Kizim, Atkov and Volk remained in Salyut 7, on one side of the airlock, and Solovyov sealed himself into Soyuz-T 11 on the other side. Savitskaya used the foot-restraint near the hatch and set up a workstation to test an electron-beam welder. Before retreating, Dzhanibekov retrieved some of the exposure cassettes and deployed others. The visitors departed in Soyuz-T 12 on 29 July, their mission having lasted longer than the standard week-long Intercosmos visit.

On a spacewalk outside Salyut 7, Svetlana Savitskaya stands on a foot restraint near the airlock to test a welding tool on a platform attached to the hull.

On 8 August Kizim and Solovyov went to complete the repair to the engine using a pneumatic clamp delivered by Soyuz-T 12. Dzhanibekov and Savitskaya had been assigned to do this during their extended visit but their hosts had insisted that it was their responsibility. A video tape had been shown on the uplink to illustrate how to swage the fractured pipe in order to seal it. Progress 20 had left long ago, so they had to work without its anchor. Driven by air at 250 atmospheres pressure, the clamp completely crushed the broken pipe. Once the new seals had been verified using nitrogen they reset the thermal insulation. The curved anchor was left in place but the tools were returned to the airlock just in case they were needed again.

What Lyakhov had initially described as "a slight leak" had taken fully 24 hours of EVA to repair. In contrast to the spacewalks by their predecessors to affix the first pair of the clip-on solar panels, however, this impressive achievement was hardly mentioned in the media (Tass said only that the engine had been "serviced"). It was subsequently admitted that when designing the station to be serviceable, the prospect of replacing valves in the engine bay had been thought to be the "absolute worst case" scenario, and a probable 'show stopper'. Progress 23 arrived on 16 August to refill the station's tanks and departed on 26 August. Having set a 237-day record, Kizim, Solovyov and Atkov returned to Earth on 2 October.

Repeatedly, cosmonauts were demonstrating their ability to overcome serious engineering problems. Although some observers criticised the amount of time that crews spent maintaining the systems (arguing that it was time lost to science) this narrow viewpoint did not recognise the value of the knowledge derived from learning to sustain operations. Compared to the remarkable success of Salyut 6 the problems were very frustrating, but in effecting repairs the programme was making significant progress towards the ultimate goal of establishing a permanent presence in orbit.

Salyut 7's jinx struck again on 11 February 1985 when radio contact was lost. With the station in free-drift and its radar out of action, it would be impossible for an automated ferry to dock with it to provide an independent manoeuvring capability. On 1 March, Tass reflected that Salyut 7 had been in orbit for 34 months and said that it had "completely fulfilled" its "planned programme", a statement that had an ominous sense of finality. Nevertheless, Vladimir Dzhanibekov and Viktor Savinykh set off on 6 June in Soyuz-T 13 to see what they could do. Dzhanibekov was making a record fifth flight. On Soyuz-T 6 he had made a manual approach from a distance of 1,000 metres, so he was the obvious choice to try to dock with an uncooperative station. Savinykh had trained with Vladimir Vasyutin and Alexander Volkov to attempt a record-breaking 10-month mission in 1985, and had been selected to accompany Dzhanibekov because he had trained (along with Vasyutin) to affix the last pair of clip-on solar panels and if they managed to board the station and return it to full operation, that was his task. Soyuz-T 13 flew a propellant-efficient 2-day rendezvous. The transition from the transfer orbit to the point at which a straight-in approach was made had to be performed without the benefit of the radar transponder. Vladimir Titov's success in flying Soyuz-T 8 in close to the station without a radar was about to yield an unexpected dividend. Although the ground had been able to manoeuvre Titov to within a kilometre of the station, he had found estimating range and closure rate difficult as he had moved in to dock. At a range of 3 kilometres, Dzhanibekov activated an optical sight incorporating an image intensifier that would show the station's position and orientation while in the Earth's shadow, and a range-finder that would measure the separation and compute the closure rate. He paused 200 metres away and studied his target: luckily, although it was in a very slow precessional roll, it was not tumbling. A fly-around inspection showed no sign of catastrophic damage. It was evident that since the three solar panels (which were offset from one another) had ceased to track the Sun, the station had suffered some sort of a power failure. Dzhanibekov positioned Soyuz-T 13 in front of the station and then closed in, using the optical sight to maintain the alignment while Savinykh called out the range and rate, and docked at the first attempt a few moments before entering the Earth's shadow.

The standard procedure was to establish the electrical and hydraulic links, test the hermetic seal of the tunnel, equalise pressure with the station, and then open the hatches. However, because there was no power they could not tell if the station was pressurised so opening the valve was the moment of truth: there *was* air in the station. They sniffed it for toxic fumes. Satisfied, they swung back the probe and drogue assemblies and peered into the void (with the porthole covers closed, it was dark). Using torches, they ventured in. Dzhanibekov estimated that it was −10 °C.

The water pipes had frozen up, condensation had frosted the wall panels and peculiar weightless icicles had formed. Even wearing arctic clothing, fur coats and fur boots the cosmonauts had to return to their ferry every hour or so to warm up. They had to retreat there to eat and sleep. As long as they could not power down their ferry, time acted against them, because unless they could restore Salyut 7 to life within 10 days (the limit of their powered-up ferry) they would have no option but to abandon it and return to Earth. As the power supply had failed and the batteries had run flat, it did not take long to locate the fault: the switching system that controlled the feed from the solar panels to the batteries had failed. Ironically, the batteries had drained their remaining charge in running the motors which had kept the panels facing the Sun. The panels were still delivering power, so if the batteries could be recharged it should be feasible to restore the station to full operation. Two of the batteries were ruined, but the other six ought to be serviceable once recharged. The culprit switch was isolated and jump leads run from the panel lines to the recharger. Soyuz-T 13 was then used to orient the station so that its irregularly positioned solar panels collected as much energy as possible. A few hours later, with the first battery recharged, the telemetry link was reactivated to enable Kaliningrad to examine the state of the other systems. The main fault had been overcome, but there was a long way to go before Salyut 7 would be usable as a platform for scientific research. At the end of the week, the temperature finally climbed above 0 °C and the ice began to thaw, but it was several more days before the humidity fell sufficiently for the heaters to be safely switched on. This was just in time because Soyuz-T 13's resources were close to their limit. Once again, human intervention had saved a crippled station. The irony of the crisis was that if cosmonauts had been on board when the recharging switch had failed they could have pre-empted the draining of the batteries. The lesson was clear: although an operational station was a self-evident prerequisite for human inhabitation, maintaining a crew on board was likely to be crucial to keeping a station operational.

Progress 24 arrived on 23 June with propellant, water, air, clothes, three replacement batteries and a new water heater (to replace the original, which had been split by the ice) and departed on 15 July, taking away the damaged items. Although the next resupply ferry was successfully placed in orbit on 19 July, a fault soon afterwards led to its being assigned the anonymous designation of Cosmos 1669. Its cargo was too valuable just to be written off, however, and the flight controllers eventually managed to overcome the problem. It then followed the standard 2-day rendezvous and docked without incident. Among its cargo were two spacesuits to replace the originals which had been deemed unsafe, due to having been frozen. These new semi-rigid suits offered improved peripheral vision and increased mobility. The third pair of clip-on solar panels (which had been delivered by Progress 21) were fitted on 2 August. Cosmos 1669 left at the end of the month.

Vasyutin and Volkov's arrival in Soyuz-T 14 on 18 September reformed the crew for the long-duration mission. Georgi Grechko accompanied them on the way up and returned to Earth with Dzhanibekov in Soyuz-T 13. This handover was notable not only because it was the first time that one crew had relieved another but also because it was the first time that a cosmonaut launched as part of one crew had

This magnificent view of Salyut 7 with Soyuz-T 14 at its rear was taken as Soyuz-T 13 departed. Note the 'clip on' solar panels that spacewalkers had installed alongside the main panels.

returned as part of another. Vasyutin, Volkov and Savinykh were to return in March 1986, by which time (as planned) Savinykh would have been on board for 10 months. Given a previous statement that a physician would accompany crews on endurance missions, it would turn out to be unfortunate that a physician was not flown in this case.

Cosmos 1686's arrival on 2 October marked the first time that a TKS docked with an inhabited station. Tass noted that it was "similar in design" to Cosmos 1443 but did not say precisely how it differed. In fact, its descent capsule had been heavily modified to incorporate a bank of instruments, and so it was no longer recoverable. Vasyutin's crew had trained specifically to use the instruments, and the cargo included a deployable framework truss (a package too bulky for a Progress ferry) which they were to erect outside. In its 25 October report, Tass said that the cosmonauts were "in good health and feeling well", but over the next few days Vasyutin progressively succumbed to illness. This had begun with "a slight uneasiness" that had been aggravated first by an inability to sleep and then by a loss of appetite. Initially the crew had kept this news to themselves, but when it became evident that this was a serious illness they reported it and Vasyutin described his symptoms. The medics at Kaliningrad told him to relay the biomedical test data via the telemetry stream so that they could assess his condition. By 17 November Vasyutin's condition had deteriorated to the point that he was suffering periods of acute pain and the crew was told to return home, which they did on 21 November. Vasyutin was found to have developed a prostate infection which had manifested itself as an inflammation and a fever. This was the first time that a flight had been terminated due to illness. In retrospect, considering the number of cosmonauts who had spent long periods in space, it was remarkable that this had not occurred earlier.

As 1986 dawned, the Salyut 7, Cosmos 1686 complex continued in orbit in its automated regime. It was clearly capable of supporting another crew, and some experiments were still to be completed, but in the longer term it was decided to move on.

Table 7.1 Salyut occupancy

Station	Main expeditions (days)	Total
Salyut 1	23	23
Salyut 3	16	16
Salyut 4	30, 63	93
Salyut 5	49, 18	67
Salyut 6	96, 140, 175, 185, 13, 75	684
Salyut 7	211, 149, 237, 168, 50	815
		Total = 1698

8

'Space Station Freedom'

RECOVERY

As it absorbed the shock of losing the Challenger crew, NASA worked to ensure that the next mission would be ready, in case it proved feasible to go-ahead with its launch on schedule in March. This decision was met with incredulity by outsiders, but it was in keeping with the standard practice of not closing off options. It was soon evident, however, that the fleet would be grounded until the formal investigation submitted its report. A commission was formed on 3 February under the chairmanship of William Rogers, Secretary of State in the Nixon administration. After interviewing 160 people, examining 6,000 documents and setting up 35 *ad hoc* panels to investigate specific issues in depth, the commission published its 'Report of the Presidential Commission on the Space Shuttle Challenger Accident' on 6 June. In response, Reagan reversed the 'Shuttle-only' policy of the National Space Transportation System. As NASA came to terms with the prospect of the fleet being grounded for several years while the many problems identified by the commission were solved, it became apparent that it would not be feasible to continue to uprate the Shuttle's main engines, and this further constrained payload masses. At the end of 1986, a study group considered easing the pressure on the Shuttle by using expendable launchers (particularly the Air Force's new Titan IV heavyweight) to place some elements of the space station in orbit. However, the delicate Shuttle stack imposed a comparatively mild ride, and the hardware would have to be strengthened to withstand the increased dynamic loads of an expendable launch vehicle. Furthermore, retrieving hardware in low orbit and delivering it to the station would significantly increase the amount of spacewalking activity (which the agency was attempting to reduce) so the study came down *against* using expendables to ease the schedule. The modularity of the design and the manifest would simply have to reflect the reduction in the Shuttle's capacity.

PARTNERS

Canada's Space Station Committee decided that participation in the station would "demonstrate and enlarge the capabilities of Canada's space industry, assisting it to

An artist's depiction of the Mobile Servicing Centre proposed for the space station by the Canadian Space Agency.

expand employment and markets". Canada was one of the first countries to contract to have communications satellites launched by the Shuttle, and Spar Aerospace was manufacturing satellites under licence from the Hughes Corporation to build up its expertise. The Institute for Advanced Research concluded that taking the lead in remote-manipulator technology would pay a rich dividend and it would enable Canada "to take responsibility for an integral part of the Station", so in 1986 Canadian Astronautics initiated Phase 'B' studies for a 'Mobile Servicing Centre'.

Meanwhile, ESA's Council had decided in January 1985 that its primary contribution would be a multifaceted research facility, collectively known as the Columbus programme. In June, MBB-ERNO was appointed as prime contractor to conduct Phase 'B' studies for:

- a pressurised microgravity research laboratory to be permanently attached to the station;
- a small pressurised research module (with integral power, life support, thermal control and attitude control) to serve as a semi-automated free-flyer that would periodically dock with the station to be serviced;
- an automated platform in a polar (Sun-synchronous) orbit that would be periodically serviced by visiting Shuttles; and
- a space 'tug'

with the attached module study subcontracted to Aeritalia, the free-flying module to Dornier, the polar-orbiting platform to British Aerospace and the space tug to Aerospatiale in France.

Although the Thatcher government was reluctant to participate in developing human spaceflight facilities, the British National Space Centre was formed on 26 January 1985 and the Department of Trade and Industry set up a study in 1985 to assess whether the UK ought to participate in the Columbus programme. When the 11-volume report was published in 1986 it recommended that Britain take the lead in the development of the polar platform because it would enable British industry to expand its expertise in the construction of automated spacecraft. The ability to visit this platform to replace or repair apparatus would mark a significant operational advance. As a logical extension of what was already underway, it was expected that the polar platform would be the first element of the Columbus programme to become operational. In addressing a symposium by the British Interplanetary Society in mid-1986, Roy Gibson, the director of the BNSC, said that in addition to seeking prime contractorship for the polar platform Britain should play a significant role in the attached module. Gibson resigned in July 1987 when the government rejected this. Despite Thatcher's close ideological relationship with Reagan, Britain's involvement in the station was minimal.

Japan's National Space Development Agency was eager to mark its entry to the realm of human spaceflight. The 'core' of its sophisticated laboratory was to be a large pressurised module with an exposed platform attached to its far end for unpressurised payloads that would be serviced by manipulators controlled from within the laboratory. A scientific airlock in the main module would enable small items to be transferred in and out, thereby obviating the requirment for spacewalks to service the external experiments. The laboratory would also have a berthing mechanism to accept a small logistics module. Japan's long-term strategy, as formulated by the Space Activities Committee, was to develop a powerful launcher and a small space plane with which to ferry its own astronauts and logistics to the station, and to have an independent orbital facility in service by 2010.

In June 1988, two years of negotiations with Canada, Europe and Japan resulted in an Intergovernmental Agreement specifying the governmental commitments involved in participating in the development of the station. A trio of bilateral

Memorandums of Understanding were also signed between NASA and the individual space agencies specifying the technical issues. As a result of these agreements, the $16 billion which the station's overall development was expected to cost NASA would be supplemented by $4 billion from ESA, $2 billion from Japan and $1 billion from Canada. Furthermore, the international partners agreed collectively to cover 25 per cent of the station's operating costs during 25 years of service.

DUAL KEEL

Although the Shuttle fleet had been indefinitely grounded by the loss of the Challenger, planning for the station continued and the Dual Keel was finally accepted as the revised Reference Configuration by the Systems Requirements Review in March 1986. The Baseline Configuration was released in May, specifying the station's major characteristics. With some revisions to both the configuration and the assembly sequence, the Critical Evaluation conducted during August and September confirmed the design. It had assessed the Baseline Configuration both from the viewpoint of the Shuttle's capacity and the station's assembly sequence, operations and safety, and had concluded that the small nodes and tunnels that linked the modules had to be replaced by larger "resource" nodes that could accommodate the command and control equipment that was originally to have been located on the truss, as this would significantly reduce the spacewalking overhead of maintaining the station. Using nodes rather than by direct interconnects would also improve safety in an emergency. Furthermore, nodes would simplify reconfiguring the space station in the event that this ever proved necessary. The Critical Evaluation also suggested revisions to the assembly sequence to reduce the spacewalking tasks during the early stage of assembly, to require fewer launches, and to facilitate scientific utilisation during assembly.

The principal features of the revised Baseline Configuration were therefore:

- a pair of 110-metre-long keels joined at each end by 44-metre-long cross-pieces;
- two 13.5-metre-long pressurised modules to serve as the habitat and NASA's laboratory;
- two 7.25-metre-long logistic modules;
- an Earth-normal atmosphere and a closed-loop environmental control and life support system that processed hygiene waste and urine to yield potable water;
- solar panels delivering 75 kW power output, buffered by nickel–hydrogen batteries;
- five locations for attached payloads, a telerobotic servicer and a facility for servicing free-flyers;
- a gyroscopic attitude control system;
- four propulsive modules for orbital reboost and backup attitude control;

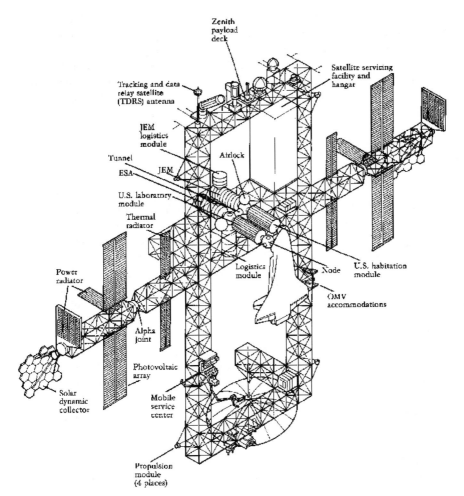

Zenith
payload
deck

Satellite servicing
facility and
hangar

Tracking and data
relay satellite
(TDRS) antenna

JEM
logistics
module

Tunnel   Airlock

ESA   JEM

U.S. laboratory
module

Thermal
radiator

Power   Logistics   Node   U.S. habitation
radiator   module   module

                                                      OMV
                                                      accommodations

Alpha
joint

Photovoltaic
array

Solar
dynamic   Mobile
collector   service
                         center

Propulsion
module
(4 places)

After rejecting the 'Power Tower', NASA opted for a more elaborate 'Dual Keel' concept.

- international modules provided by Europe and Japan;
- the remote manipulator system; and
- a 150-metre-long truss to support the cluster of pressurised modules, the robotic servicer, the propulsive modules, solar panels and thermal radiators.

The new configuration had five rather than seven pressurised modules, with NASA providing a microgravity research laboratory, the habitat and a logistics module, ESA providing a laboratory, Japan providing a laboratory with an integrated external pallet and logistics module, and Canada providing the remote manipulator. In addition, NASA and ESA would each provide free-flying platforms. As part of the downstream enhancement programme, the solar panels were to be supplemented by a more efficient 'solar dynamics' system.

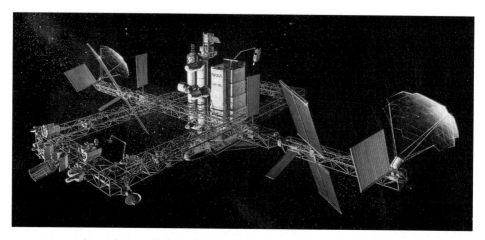

An artist's depiction of the completed 'Dual Keel' space station.

On 8 March 1986, having headed NASA from 1971 to 1977, during which he started the development of the Shuttle, James Fletcher was reappointed as the agency's administrator – as a "safe pair of hands" to oversee the recovery from the loss of the Challenger. In June, he appointed associate administrator Andrew Stofan to lead the newly created Office of Space Station in Washington. Stofan would have two deputies, a chief scientist, a senior engineer and six divisional directors, and would be responsible for turning the station from a dream into hardware. In December, Fletcher accepted the revisions to the work package that Stofan had recommended in light of the Critical Evaluation. After the completion of the definition and preliminary design in January 1987, the station was subjected to a detailed cost analysis. To limit costs during the assembly period, it was decided to phase in the 'habitability' status, with the initial 'tended' status being upgraded to 'permanent habitability' once the Baseline Configuration was complete. The schedule called for the hardware contracts to be issued in May 1987, and for the launch of the first element in 1993, so even though it would not be finished "within a decade" as Reagan had stipulated, at least it would have been started.

In early 1987, NASA realised that the Shuttle was unlikely to achieve a flight rate sufficient to deliver the truss elements for the dual keel, so this massive structure was deleted from the Baseline Configuration and redesignated as an enhancement that would be added only once the capacity of the solar panel truss to support attached payloads was exceeded. The National Research Council of the National Academy of Sciences undertook an independent review in September. It endorsed the Baseline Configuration, but emphasised that the nation's long-term goals in space had to be clarified before committing to the addition of the keel structure. In fact, capacity for mounting attached payloads was *reduced* by a budgetary review that shortened the solar power truss in order to reduce the number of launches. In addition to tailoring the station as a cluster of laboratories for microgravity research, this meant that future expansion of the station's mission with astronomical, terrestrial and solar studies would be more difficult.

'LIFEBOAT'

Before the loss of the Challenger and the grounding of the fleet, NASA had intended the station to include a "safe haven" to which astronauts could retreat in the event of an emergency such as a power failure or a partial depressurisation. If this was insufficient, a Shuttle would be launched to retrieve the crew. On 24 June 1987, however, Fletcher told Congress that it would be unwise to rely upon the Shuttle to return the station's crew in an emergency. He requested $3 million for a study of a 'Crew Emergency Return Vehicle' to be delivered by Shuttle and used as a 'lifeboat'. The obvious solution was a large Apollo-style capsule, but there was a case for developing a completely *new* vehicle based on 'lifting-body' technology. In August, as NASA was about to issue a request for proposals, Congress ordered a review because the projected development costs had increased dramatically. It was a sad reflection upon the agency that it could not reintroduce Apollo-era technology in a cost-effective manner. However, the *form* of the lifeboat was a side-issue that did not really influence the station's design.

In October 1989, Houston issued a request for proposals for this vehicle, now called the 'Assured Crew Return Vehicle'. It awarded a pair of $1.5 million, 6-month contracts to evaluate design and mission profiles. One candidate was the the Langley Research Center's HL-20, a lifting body shape with more than a passing resemblance to the Soviet BOR-4 'mini-Shuttle'. After reviewing the proposals, contracts for $1.5 million were issued in 1990 to Lockheed and Rockwell International to refine their concepts, development schedules, costs and risks with a view to beginning hardware development in 1992. In the fullness of time, the chosen vehicle configuration was redesignated the X-38.

PHASE 'C'

The Phase 'C' Development Contracts were issued in December 1987. The goal was to flesh out the Baseline Configuration by 1989 and conduct the final Critical Design Review by 1992 in order to 'freeze' the design. As earlier, there were four work packages for the design, development and testing of specific elements of the station; element integration; engineering support; associated software systems; and user operations. The 10-year-long contracts totalled $6.7 billion, this being $5 billion for the Baseline Configuration and $1.5 billion to provide options for enhancing the capability of these elements.

WP1   Marshall Space Flight Center
      Boeing ($1.6 billion)
      ● the habitation module
      ● NASA's laboratory
      ● the logistics elements
      ● structures for the nodes to link the other modules

- those parts of the environmental and thermal regulation systems built into the pressurised modules.

In addition, Huntsville would have overall responsibility for management, systems engineering, integration and operations as well as logistics support for these elements.

WP2   Johnson Space Center
      McDonnell Douglas ($2.6 billion)
- integrated truss structure
- Mobile Servicing System transporter
- attitude control and reboost propulsion systems
- thermal control systems
- Shuttle docking systems
- airlocks and EVA aids
- guidance, navigation and control systems
- data management systems
- communications and tracking systems.

Houston's responsibility also included supplying the internal systems for all NASA's pressurised modules, and overseeing the development and integration of the laboratory's experiments.

WP3   Goddard Space Flight Center
      GE Astro-Space ($895 million)
- two mounts for attached payloads, and an instrument pointing system
- integration of the flight telerobotic servicer
- planning NASA's role in satellite servicing facilities and defining the interfaces for this facility
- the station's information system software
- a 'tended' automated free-flying polar platform to carry experiments in a Sun-synchronous orbit.

However, these facilities were subsequently either cancelled or reassigned to other centres.

WP4   Lewis Research Center
      Rockwell International's Rocketdyne Division ($1.6 billion)
- the 75 kW electrical power system, including power generation, storage, distribution and associated software, for both the station and the polar platform
- proof-of-concept testing for a 'solar dynamics' power system.

After four years of having its requirements defined, the station was finally moving.
On 12 April 1988, Fletcher announced that the launch of the first element would have to be slipped by one year due to a budgetary shortfall. The first three launches would start the assembly of the truss and add the docking, power, propulsion, guidance and control systems. Nevertheless, by advancing the laboratory to the

fourth assembly launch, it would be feasible to achieve the 'tended' milestone in 1995 as planned. The laboratory would be launched "with useful capability" and its remaining racks would be delivered on a later flight. The addition of the habitat in 1996 would achieve the 'permanent habitability' milestone. In a process that was euphemistically labelled "reprogramming", contractors were told to schedule their subcontracting in accordance with the reduced annual budget.

THE PRICE OF FREEDOM

With the station seemingly progressing well, Reagan decided to assign it an identity and on 18 July 1988 he named it 'Space Station Freedom', a name that matched his antipathy towards the Soviet Union. Thus, the station became a symbol of the Cold War, just as had Apollo two decades earlier.

Nevertheless, the Senate was sceptical and it rebuffed the request for $967 million to initiate station construction and assigned just $200 million. "If this is the ultimate outcome," Fletcher warned indignantly, "it is a setback for NASA, and a blow for this country's future in space. The space station is the key to our major goals of the future. It has been studied, analysed and weighted against alternatives. . . . To suggest a further delay will cost us technologically and economically." Worse, it would cause international partners to question NASA's credibility. After a vigorous debate, the House awarded $902 million. Although this incident would turn out to be just the first of a series of confrontations through which Congress endeavoured progressively to scale down the station, arguably with a view to improving the programme, NASA sought to ensure that the station would be capable of growth, both in size and capability, so that it would be able to operate well into the twenty-first century.

In announcing NASA's $13.3 billion budget request for Fiscal Year 1990, which included $2 billion for the station, Fletcher warned that the agency was "as taut as possible". Furthermore, the station had been redesigned to minimise costs, so "there is simply no room for further trimming, or shaping, or cutting; we are either going to build it, and build it right, or not build it at all". Congress was not impressed, however, and Fletcher resigned on 8 April 1989. Dale Myers, his deputy, saw the agency through to the appointment of astronaut Richard Truly in July. Truly then appointed Bill Lenoir, another astronaut, as associate administrator with instruction to develop a plan for the consolidation of the Offices of Space Flight and Space Station. He also reassigned Richard Kohrs from deputy programme director of the National Space Transportation System to utilise his experience to oversee the design, development and operation of the station, which Truly described as "the linchpin of this nation's future in space".

Congressional antipathy towards 'big ticket' engineering for human spaceflight was demonstrated in July 1989 when Reagan's successor in the White House, former Vice President George Bush, celebrated the twentieth anniversary of Apollo 11's lunar landing with a clarion call of his own. "In 1961, it took a crisis, the 'space race', to speed things up," he reflected. "Today we don't have a crisis, we have an opportunity. I'm not proposing a 10-year plan like Apollo; I'm proposing a long-range continuing commitment." Looking beyond the station, which was "for the 1990s", he said that America's return to the Moon would be "for the end of the century". This would be a journey "back to the future, and this time back to stay". In the longer term there should be "a manned mission to Mars" as "a journey into tomorrow". The station was the "cornerstone" of this vision. NASA responded by appointing Aaron Cohen, director of the Johnson Space Center, to formulate options to pursue this 'Space Exploration Initiative'. It soon became apparent, however, that Congress was in no mood even to fund studies and the initiative stalled.

CRISIS

In July 1989, NASA formed a Configuration Budget Review Team, which in turn established three Control Boards to study options and make recommendations for "rephasing" the development of the station to live within a severely constrained budget. As certain aspects of the Baseline Configuration would have to be deleted to cut the short-term costs of the station, the team's guiding principle was not to do anything that would prevent such features from being reintroduced once the station was operational.

One trade-off was the propulsion system. Martin Marietta had been awarded a $139,000 contract in 1985 to study a "propellant scavenger system" that would ride the Shuttle's External Tank, drain the residual propellant as soon as the ET was jettisoned and then separate before the tank was destroyed upon re-entering the atmosphere. A tug would be waiting to retrieve the scavenger and take it to the station. It was estimated that (depending on the Shuttle's payload) such a system might be able to top up the station's tanks with 11 tonnes of propellant per flight for attitude control and orbital reboost. Such a scheme would be a step towards improving the overall efficiency of the station's infrastructure. Later, Houston studied a system to electrolyse water offloaded from the fuel cells of a visiting Shuttle to generate hydrogen and oxygen to further top up the station's propellant. Like the scavenger, this electrolyser would make efficient use of in-orbit resources. As the propulsion system was modular, with elements mounted on the truss, the reviewers decided to fit hydrazine thrusters instead of a hydrogen–oxygen system. Hydrazine thrusters would be cheaper to develop but they would be more costly to operate in the sense that they would not offer the efficiencies of the scavenger or the electrolyser. The cryogenic system would be cheaper in the long term but the budget could not carry the up-front cost, so reclamation systems were cancelled. The operational costs would be significant. It was estimated that ferrying up the hydrazine would correspond to a half of a Shuttle's payload capacity per year (that is, with five flights per year, an overhead of some 10 per cent) but this operating cost would be accounted for in another budget. A similar trade-off was the decision to make the environmental control and life support 'open loop' rather than 'closed loop' systems. Without an on-board water-recycling system, the Shuttle would have to ferry up water. Another 'savings' was eliminating the second airlock that was to have provided redundancy and also served as a storage bay for the spacesuits and other EVA apparatus. One of the nodes would now be modified as a contingency airlock despite the fact that it was ill-suited to the purpose. It was also decided to postpone the development of a new spacesuit. The Shuttle's spacesuits were flown repeatedly, but they were serviced between missions. Because this work would have to be done on the station, it had been hoped to redesign the suit to make it simpler to service. Furthermore, it was recommended that spacewalks be limited to times when a Shuttle was in residence.

When all of these measures were totted up there was still a budgetary imbalance, so the review reluctantly decided that the crew complement would have to be reduced from eight to four astronauts and the power supply reduced from 75 to 35 kW, both of which would seriously reduce the station's utility as a "national laboratory" for microgravity research. The mood had certainly changed since Reagan's call to demonstrate America's leadership in space.

The simplest way to reduce NASA's annual spending was to cut the Shuttle's flight rate, so it was recommended that only five assembly flights be launched per year. However, most of NASA's budget was consumed by 'fixed costs'. Reducing the launch rate would not only slip the station's completion date, but in the long run it would *increase* the cost because the salaries of the flight operations staff had to be paid whether the Shuttle flew or not. As always, the Congressional focus was on *annual cash flow* rather than overall programme cost. In October, NASA told Congress that it proposed to retain the March 1995 target date for the launch of the first element and pace the assembly to live within its projected year-on-year budget. As a result, the completion date would probably slip by 18 months to August 1999, which was five years beyond Reagan's target.

In early October, Kohrs appointed Robert Moorehead as deputy director of Programme and Operations for the Space Station. On 8 November, Truly finally consolidated the Offices of Space Flight and Space Station. "The division of responsibilities is clear", assured Lenoir, who was in charge, "and the mechanisms to assure coordination are in place."

Despite having relocated many of the station's utilities into the nodes to reduce the need for spacewalks, Congress was still sceptical, so in January 1990 an External Maintenance Task Team was established to review the station's maintenance and repair. The plan called for five spacewalks per week for maintenance of the station. In its report in July, the team suggested that the use of telerobotics might enable the rate to be reduced to one outing per week, especially if the work was streamlined, which would cut the annual spacewalking time from 3,200 to 500 hours. However, given that no Shuttle astronaut had ventured out since 1985 it was recommended that a series of trials be conducted to assess maintenance tasks. After a meeting of programme managers in March, Truly made his report to Congress.

In refusing the Fiscal Year 1991 request, Congress demanded a "restructuring" to slash $6 billion from the cost of the station in the 1991–1996 timeframe. NASA observed tongue-in-cheek that it would be able to save $221 million by retaining the imperial measurement system (like other federal agencies, it was obliged to make the conversion to the metric system). On 5 September 1990 the White House formed the Advisory Committee on the Future of the US Space Programme under the

A Rockwell International artist's depiction of a Shuttle docked with a space station 'node'. Note the observation cupola on the node. The Shuttle has a logistics module to be transferred to the station. Note the wafer of cryogenic stores at the rear of the bay to enable the Shuttle to remain in space for an extended period.

chairmanship of Norm Augustine, chairman and chief executive of Martin Marietta. He was given four months to make "a serious no-holds-barred" review of the space programme, then make recommendations for how it might be improved.

"We have a lot of challenges," Truly admitted in an employee 'pep talk' on 21 September, but he was confident. "I think Space Station Freedom is strong. I believe when this budget process is complete, you will find we have gotten good support."

When Congress announced the 1991 budget it was bad news. Not only was the station $579 million short, the Assured Crew Return Vehicle 'lifeboat' and the Orbital Manoeuvring Vehicle 'tug' had been "zeroed out". The message was blunt: "the budget crisis is only beginning". NASA was given 90 days "to implement a revised space station design and assembly sequence [with] an incremental approach". Specifically, it was to develop a plan assuming "an out-year growth in the station funding profile of approximately ten percent per year with a peak year funding level of $2.5 to $2.6 billion".

"Everything is fair game," warned Kohrs as the investigation got underway on 6 November. Suggestions included integrating the laboratory and habitation modules; delivering the truss in pre-integrated segments rather than installing the subassemblies by hand; increasing the power in 18.75-kW increments rather than in 35-kW increments; and deleting various items such as the mobile transporter, the astronaut working platform and the aft docking facilities. The driving force was to cut the number of launches, and hence the cost, *without pushing the completion date beyond the end of the century*. Another possibility was to shorten the main pressurised modules so that the Shuttle could lift them in a fully-outfitted form instead of launching them only partially outfitted and completing the process of installing their racks in orbit. The truss could also be shortened, pre-integrated and launched with its subsystems in place. This would reduce assembly EVA by more than 50 per cent. Moreover, the habitation module could be pushed down the manifest. Instead of achieving 'permanent habitation' a third of the way through the assembly, this could be delayed until the station was complete, but this meant that it would be *five years* before a crew would be able to move in. Although this would reduce the early cost, it would make the station less productive because the laboratory (which would be launched at an early date) would be usable during the assembly process only while a Shuttle was in place. NASA had originally wanted a crew of 12, but Reagan had allowed eight. In its 1989 review the agency had reduced the crew to four until the station was finished. Shortening the habitation module meant that the capacity would not be able to be restored until the station was enhanced by the addition of a *second* habitat, which was not a foreseeable prospect.

The 'permanent habitability capability' configuration as depicted in September 1990 by a Marshall Space Flight Center artist.

The international partners were frustrated by NASA's lack of consultation in this review. They argued that drastic changes to the infrastructure broke the terms of the 1988 agreements which defined the utilities that would be supplied to their modules. The reductions in crew and power would undermine their plans to operate their laboratories on an ongoing basis. In fact, it seemed likely that the international partners would have to take turns flying their astronauts. The result was criticism that the agency was treating its international partners as if they were subcontractors, and scepticism of Congress's commitment to the programme. In November ESA announced that it would withdraw from the programme if the redesign meant the launch of its Columbus laboratory would be postponed beyond the millennium.

A faction in ESA was actually arguing for the attached module to be cancelled in order to focus on the free-flyer that would ultimately be serviced by the Hermes space plane. With two Hermes spacecraft, it was initially estimated that the programme would be capable of four 12-day missions per year, but in mid-1991 the crew was reduced to three people, the payload capacity was reduced to 1 tonne, the unpressurised payload bay was deleted and the flight rate was reduced to a single 10-day mission per year. Despite this diminishing capability, the projected cost had escalated. Even if development proceeded according to plan, it would not enter service until 2015. If, as seemed ever more likely, the vehicle was downgraded to 'technology demonstrator' status, this would bode ill for ESA's dream of achieving independence in human spaceflight.

When Augustine published his report in December 1990 he said that the 90-day limit on the "restructuring" effort was unwise. "Redesign is simply too important to take less than whatever time may be needed for a thorough reassessment and the establishment of a configuration that can earn stable long-term funding and support."

The "restructuring" report was delivered to Congress in early March 1991. Its revised design would be cheaper, smaller, easier to assemble in orbit and need fewer launches. "This new design for Space Station Freedom accomplishes every major goal we set for ourselves when we kicked off this effort last November," assured Lenoir. "We took the directions from Congress and the Augustine Commission recommendations to heart. ... The programme we are announcing today addresses each and every one of their requirements."

The key changes were:

- shortening the truss by about 30 per cent and switching to pre-integrated segments (to reduce the *in situ* element-to-element connections by 80 per cent, remote manipulator operations by 50 per cent and spacewalks by 50 per cent);
- shortening the modules by 40 per cent and outfitting them prior to launch (thereby reducing the number of experiment racks in NASA's laboratory from 28 to 12 and deleting the individual crew cabins from the habitation module;
- reducing the crew complement to four people;
- slipping the 'permanent habitability' milestone right to the completion of the assembly process;
- cutting the power supply from 75 kW to 56 kW;
- cancelling the Flight Telerobotic Servicer; and
- cutting the data transmission rate to one-sixth of its initial specification.

The restructured programme placed the first element's launch in early 1996, and after six launches the 'tended' milestone in mid-1997 would enable docked Shuttles to use the laboratory for periods of up to a fortnight. At that stage, one solar power module would supply 18.75 kW, half of which would be available for science. The 'permanent habitation' milestone would be achieved by the 17th launch, but a crew would not be able to be left on board until an Assured Crew Return Vehicle (funding for which had yet to be approved) was in place.

The shortening of the truss by about 30 per cent prompted the cancellation of the facilities for large attached payloads, but hardpoints on the truss would still be used for small payloads. Martin Marietta had been assigned a $297 million contract in July 1989 to develop the Flight Telerobotic Servicer whose multiple dexterous robotic arms were to have assisted in the station's construction and maintenance without requiring astronauts to venture outside. Cancelling these two projects effectively deleted WP 3 from the programme.

Huntsville was now responsible for:

- the design and construction of the pressurised shells for NASA's laboratory and habitation modules;
- the logistics modules to resupply the station and provide temporary *in situ* storage;
- the nodes to link the laboratory and habitation modules; and
- subsystems internal to the pressurised modules, including environmental control and life support, thermal control, electrical power distribution and communications.

Houston would provide technical direction for the design and development of crew-related subsystems in Huntsville's package and take responsibility for the design, development, verification, assembly and delivery of:

An artist's depiction of Martin Marietta's Flight Telerobotic Servicer which was to augment the station's main robotic manipulator. Its multiple dexterous robotic arms were to assist in the station's construction and maintenance without requiring astronauts to venture outside.

- pre-integrated truss assemblies;
- propulsion units;
- Mobile Servicing System transporter;
- node design and outfitting;
- external thermal control subsystems;
- data management systems;
- guidance, navigation and control systems;
- communication and tracking systems;
- airlock and EVA systems; and
- Shuttle docking systems.

Most of the command and control systems would be located in the resource nodes. Huntsville would provide their shells and Houston would outfit them to perform their specific functions. The pre-integrated truss assemblies would serve as the station's backbone, providing structural and positional integrity for the solar power units, thermal regulation units, propulsion units in relation to the pressurised module cluster. The Mobile Servicing System that was to assist with the assembly and maintenance of the station would carry Canada's remote manipulator. The EVA systems included:

- the spacesuit extravehicular mobility units (EMU);
- provisions for communication;
- physiological monitoring and data transmission;
- crew rescue and equipment retrieval provision; and
- EVA procedures.

The guidance, navigation and control systems would have to:

- provide attitude and orbital maintenance;
- support the pointing of the solar panels and thermal radiators;
- provide attitude information to other systems and users; and
- accomplish periodic orbital reboost manoeuvres.

The traffic management system would control traffic in the station's vicinity, including Shuttle docking operations and trajectory determination of objects with orbits intersecting the station's orbit. The communications and tracking system had to provide:

- communication with astronauts during spacewalks;
- communication with Shuttles and any other associated vehicles, such as tended free-flyers;
- space-to-ground communication through NASA's Tracking and Data Relay Satellite System to ground data networks;
- internal and external voice communication through the audio subsystem;
- internal and external video requirements through the video subsystem;
- management of communication and tracking resources and data distribution through the control and monitor subsystem; and
- navigation data through the tracking subsystem.

The data management system had to connect on-board systems, payloads and operations for data acquisition and distribution, data storage and data processing, with user interfaces to permit control and monitoring of systems and experiments. The external thermal regulation system would provide primary cooling and heat rejection in order to control the temperature of the electronics and other hardware located throughout the facility.

The Lewis Research Center was responsible for the station's end-to-end electric power system. This involved specifying the system architecture and providing the solar power generators, batteries and power management and distribution facilities. In the 'tended' configuration there would be a single power module providing 18.75 kW, but the final configuration would have a trio of modules.

Vice President Dan Quayle's National Space Council endorsed NASA's "restructuring" report on 21 March. The following day, Lenoir announced, "We have a phased approach and we're within budget guidelines." In fact, almost all of the $8.3 billion total saved was achieved by deferring operational expenses, the *development costs* would be trimmed by only $2 billion. "Let's give it to the engineers to build," Truly pleaded as he presented the plan to the House. Democratic Representative Barbara Boxer was dismissive, saying that the pared back station had become just "a garage in space with nothing in it, and nothing going on around it". On 15 May the House appropriations subcommitee voted down the funding request for Fiscal Year 1992. Suddenly, even the station's supporters began to wonder whether it would ever become a reality. The subcommittee's Democratic chairman, Bob Traxler, argued for the $2 billion requested for the station to be given to social and environmental programmes, and not just left within NASA's bailiwick. "I was very disappointed in ... the Traxler subcommitee," admitted NASA's deputy administrator, J.R. Thompson, in a masterpiece of understatement. Remarkably, the full committee ratified this decision on 3 June. "Cancellation of Space Station Freedom would cause a total disruption to the planning of all of America's manned space endeavours," Truly told the Committee on Science, Space and Technology the next day. Part of his concern was for his agency's reputation with its international partners. "If we renege on these international obligations, America's word will be in question and our ability to negotiate international science and technology agreements will be seriously eroded." Republican Dick Zimmer of New Jersey was unrepentent: "I have reluctantly concluded that instead of furthering our goals in space exploration, the space station will actually impede America's progress" because it would "absorb all the available dollars" and so starve small but "highly promising projects". He was scathing concerning the station: "Its original mission has been so drastically curtailed, its cost has run up so far, and its schedule has slipped so much that its objectives can be better achieved with less money on smaller, more mission-orientated spacecraft." However, it was *Congress* that was throwing obstacles in the way of the programme and imposing the expense of a succession of redesigns, notionally with a view to saving money. On 6 June, the House overturned its appropriations committee and voted the station $1.9 billion in Fiscal Year 1992, and on 10 July the Senate restored the $100 million that the House had trimmed, thereby giving NASA its full request. In September, the Senate acknowledged that

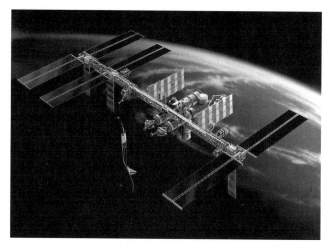

The 'permanent habitability capability' configuration as depicted in September 1991 by a Marshall Space Flight Center artist. Note that the nodes have been replaced by lengthy docking systems.

"the space station's redesigned configuration meets virtually all the conditions set forth in last year's conference report".

In November 1991, the new 'tended' configuration passed its integrated Preliminary Design Review. The next milestone was a Critical Design Review of the station's components and systems in 1992, followed by the integrated Critical Design Review in early 1993 which would mark the formal transition from design to construction, as Phase 'D'.

In 1991 the White House's Office of Space and Technology Policy paired up with the Space Studies Board of the National Academy of Sciences to conduct a joint review of the prospects for science on the station. They concluded that life sciences research would be severely undermined by the absence of a large centrifuge, and that while the merit of the materials-processing was not at all straightforward to assess objectively, the results may well not repay the tremendous investment. On 6 December 1991 NASA signed a Memorandum of Understanding with the Italian Space Agency, Agenzia Spaziale Italiana (ASI), and placed a contract with Alenia Spazio to build a trio of 4-metre-long pressurised modules to enable the Shuttle to ferry logistics to the station. Alenia had built the pressurised shell for Spacelab and it was to build the Columbus laboratory. Once the Shuttle had docked, its manipulator would swing this Multi-Purpose Logistics Module onto one of the nodes, where it would remain until recovered by a later Shuttle. This eased the pressure on NASA by enabling it to delete the large logistics modules that would otherwise have eroded its rapidly diminishing budget. In return, NASA was to deliver an Italian laboratory to expand the station and Italian astronauts would thus be able to take their turn on the station. For simplicity, the Italian laboratory would utilise the same shell, and its life sciences facilities would include a 2.5-metre-diameter centrifuge to overcome the recent criticism of the Baseline Configuration.

As Richard Kohrs explained in his introduction to a 'Media Handbook' that NASA produced in 1992: the station was "essential for advancing the human exploration of space". He then listed the contributions it was to make: "Continued progress in the human exploration of space requires ... multi-year studies of human adaptation, testing of life support systems, and experience in building, maintaining and operating a large manned space system." Seemingly as an afterthought, he added, "Freedom will also serve as a permanent Earth-orbiting laboratory". So, whatever benefit the station might offer to America and to the Free World, it was crucial for NASA's vision of its own future. As Kohrs explained it, the station's primary mission was to "provide a permanent outpost where we will learn to live and work productively in space". While it was to be "an orbiting research base with essential resources of volume, power, data handling and communications", its main theme was to be "long-duration studies of human physiology and well-being in space". Such research was a prerequisite to "long-range human exploration". In other words, in the exploitation of opportunities offered by a permanent facility in orbit, NASA intended to be its own best customer. The agency's position at least had the virtue of consistency – since its inception NASA had been lobbying to use a station in low Earth orbit as a starting point for voyages into deep space. As NASA officials had often noted, in terms of energy expenditure, low orbit is over halfway to anywhere in the Solar System.

FREEDOM'S FINALE

In response to a request by the White House, Truly announced his resignation on 12 February 1992. During his tenure as NASA's administrator, he had often clashed with Vice President Quayle's National Space Council. The agency was shocked a few days later when Congress ruled against the Advanced Solid Rocket Motor that had been ordered in the aftermath of the loss of the Challenger. Without this upgrade the Shuttle would not be able to lift the shortened modules unless they were sent up partially outfitted, and laying on additional flights to deliver the remaining racks would cost more in the long term.[1] Even so, in announcing Daniel Goldin's appointment as the new administrator on 11 March, Bush was optimistic: "Dan's a leader in America's aerospace industry and a man of extraordinary energy and vitality." Goldin promptly set about relieving the field centres of their historic powers in order to concentrate it in Washington. When he announced cutbacks that would result in thousands of layoffs, agency morale plummeted.

The European Space Agency was doing no better. By the spring of 1992, Hermes was late, over budget and 2 tonnes overweight. The radical solution was to strip it of its payload capability, thereby reducing its role to a crew transfer vehicle. One option was to use it as the station's lifeboat. In June NASA engineers flew to Paris for the

---

[1] The goal of the $3.8 billion Advanced Solid Rocket Motor was to increase the Shuttle's payload into a 28-degree orbit from 25 tonnes to 29.5 tonnes. It had been a troubled programme, however, and the initially planned 1996 test flight had long-since been pushed back to 2000.

In March 1992 Daniel Goldin was recruited from a senior management position at TRW to lead NASA.

feasibility study. In October, contracts were awarded to Spazio SpA in Italy and Aerospatiale in France to conduct further studies. Meanwhile, the Columbus free-flyer was put "on hold", which was bad news for British Aerospace. In November, however, ESA all but officially redefined Hermes as a technology demonstrator, thereby setting the stage for its eventual cancellation.

On 30 July 1992, Congress voted an annual station budget of $2.25 billion for Fiscal Years 1993 and 1994, with an increase to $2.27 billion in 1995. Representative Tim Roemer, an Indiana Democrat and long-term station critic, insightfully pointed out that the station had been cut so drastically from its original concept that it was now "in search of a mission". It survived a Senate motion on 9 September advocating its cancellation, when the Arkansas Democratic Senator Dale Bumpers dismissed it as "a bottomless money pit". The strategy was clear: by progressively eroding the station's range of possible missions its opponents were (to introduce a military term) defeating it 'in detail' so that as soon as it could be demonstrated that whatever remained could not possibly justify the development cost, the final blow would be dealt.

The Critical Design Reviews of the station's components and systems (that is, the individual work packages) in 1992 cleared the way for the integrated review of the 'tended' configuration in early 1993, and once this had confirmed that the design was essentially complete the contractors would be released to start manufacturing flight-rated systems. It is worth examining this configuration (which Kohrs wryly noted was "the product of years of planning") in some detail.[2]

Phase 1
The objective of the first phase of the assembly process was to provide sufficient elements of the station to achieve the 'tended' milestone.

Flight 1
The first Shuttle would carry a set of integrated components to provide the 'cornerstone', at the starboard end of the truss. The packed payload bay would contain:

- two pre-integrated truss segments;
- the starboard integrated equipment assembly;
- an 'alpha' rotary joint assembly;
- the starboard photovoltaic solar power module;
- the mobile transporter; and
- an unpressurised berthing mechanism.

How the truss should be assembled had been a major issue in designing the station. While some argued for a self-deploying structure, others had maintained that it would be better to use proved technology; that is, to assemble it by hand. In order to assess an astronaut's abilities, rod-and-pin construction kits had been tested on spacewalks. Jerry Ross and Sherwood Spring had repeatedly assembled and disassembled the ACCESS and EASE structures in November 1985, and had been filmed so that their work could be subjected to time-and-motion analyses. They had proved that spacewalkers could indeed perform construction in weightlessness, so proposals to develop self-deploying frames were shelved. In May 1992, Tom Akers and Kathryn Sullivan had built another frame to evaluate the fidelity of the Weightless Environment Training Facility. They reported that manipulating the structural elements was rather easier in space than it had been in the tank, which was encouraging. However, considering the unprecedented size of the station's truss (it was to be 66 metres long with a 4.9×3.7-metre-wide hexagonal cross-section) the designers compromised on seven prefabricated segments that spacewalking astronauts would manually connect together. In addition to providing unpressurised storage for the 'orbital replacement units', the interior of the truss would ultimately serve as a 'corridor' for spacewalkers. Instead of having the astronauts install the various systems that were to reside in the truss it was decided to 'pre-integrate' and verify them prior to launch, so that the spacewalkers would need only to hook up batches of plug connectors.

Each power module had a pair of 'wings' connected by 'beta' gimbal assemblies

---

[2] It is worth examining in detail because the assembly strategy is very different to that pursued on the International Space Station.

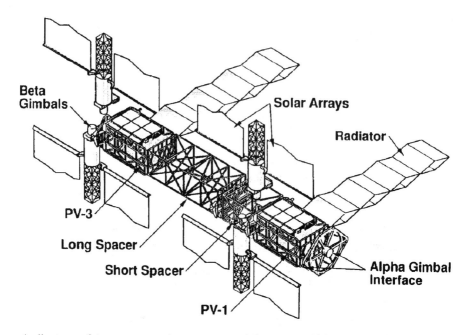

A diagram of two space station power modules separated by a truss 'spacer'. Each module has storage batteries to store the electricity, a radiator to shed excess heat, and a pair of 'beta' gimbals to rotate the booms which support the solar transducer arrays. The 'alpha' gimbal rotates the entire assembly around the main truss's axis to track the Sun.

which were free to rotate about the wing's primary axis. Barely six months after Reagan had announced his decision to build a station, STS-12 had tested the deployment mechanism for a large solar panel that Lockheed had developed for this eventuality. In its pre-deployed form the concertina array was just 75 millimetres thick, but it could be drawn out by a telescoping mast which then provided structural support. Under ideal illumination, each module would generate 18.75 kW. In addition, the 'alpha' rotary joint would enable the power modules to rotate around the truss's primary axis. With two degrees of freedom, the panels would be able to be maintained facing the Sun for maximum power output. Nickel–hydrogen batteries were mounted in the frame of the solar power module to sustain the station while in the Earth's shadow.

Once the cornerstone structure had been assembled, it was to be left in a 'horizontal' orientation with its primary axis aligned perpendicular to the velocity vector.[3]

---

[3] Because the space station will use nadir-inertial attitude in order to maintain its orientation with respect to the Earth, the direction of the velocity vector is defined as 'forward'.

Flight 2

The second Shuttle would dock at the unpressurised berthing mechanism, to unload its payload of:

- the third pre-integrated truss segment; and
- two propulsion modules.

The self-contained propulsion modules would be mounted on the truss. The hydrazine thrusters would serve in a backup capacity to the gyroscopes of the primary attitude-control system, and be responsible for maintaining the station's operating altitude. This mission would activate the power generation, propulsion and S-Band communications and tracking systems.

Flight 3

The third Shuttle's payload would consist of:

- the fourth pre-integrated truss segment;
- the starboard thermal regulation system radiator unit;
- the UHF and Ku-Band communications and tracking system; and
- the remote manipulator.

The thermal regulation system radiator unit would be mounted on the truss in such a way that it could shed excess heat, which was carried from the pressurised modules by a network of pipes. In fact, the system had two independent 'loops', one internal and the other external. Each pressurised module had an air-to-water heat exchanger with low- and moderate-temperature water loops, and would be responsible for transferring its unwanted heat to one moderate- and two low-temperature external ammonia loops. Each external unit had a trio of radiators, and was be free to rotate on the 'aft' edge of the truss mounted halfway between the module cluster and the solar power modules. Each radiator comprised a large number of strip-like sections which were joined edgewise. The ongoing operations of the Shuttle had enabled systems intended for the station to be tested. In fact, just 40 per cent of the systems *designed* for microgravity had worked first time, the remainder had needed attention to rectify unforeseen defects. When a strip-like section of a radiator was tested as the Station Heatpipe Advanced Radiator Experiment in 1989, it was foiled by the formation of bubbles. When one end was heated, it was meant to vaporise liquid ammonia which was then expected to flow along the 15-metre-long pipe by thermally-induced convection, whereupon a fin at the far end was to radiate the heat to space and in so doing recondense the ammonia, enabling it to flow back to the evaporator. It took several iterations of the design before the radiator element worked satisfactorily. However, the fact that it contained no mechanical apparatus meant that it would not draw power from the station. Furthermore, because the radiator elements were independent, this technology should provide a highly redundant thermal control system that ought to operate for years without maintenance, thereby reducing the station's operating costs.

Flight 4

The fourth Shuttle's payload would supply:

- the fifth pre-integrated truss segment;
- various EVA aids; and
- miscellaneous apparatus.

The Crew and Equipment Translation Aids (CETA) were to be fitted inside the truss to enable spacewalkers to move bulky apparatus rapidly along its length in order to facilitate repair and maintenance tasks. Jerry Ross and Jap Apt had assessed a number of potential 'transfer aids' in 1991 on STS-37. In fact, that test marked the first spacewalk since the loss of the Challenger, an interval of five years.

Flight 5

The fifth Shuttle's payload would supply:

- the port node and equipment racks;
- a pressurised docking adapter; and
- a cupola.

With more than half the length of the truss in place, the focus was now on making the station habitable. The node was to be fastened to the truss in a position just 'below' and just 'aft' of where the centre would be once it was complete, with its axis alongside the truss. The 'floor' of the square-cross-sectioned internal compartment was designed to match the Earth's surface. For consistency, the 'local vertical' was to be the same for all the pressurised modules. The evidence from Skylab was mixed, and it was far from certain that weightless astronauts really required this frame of reference. "A pleasing environment", the agency assured, "enhances crew productivity, and bestows a feeling of well-being."

In addition to a berthing mechanism on each end, the 5.2-metre-long node had four other berths around its periphery so that it could form the station's 'hub'. Once its connectivity was exhausted, another node would facilitate further expansion. In addition to the station's controls, this first node was to provide the life support system, fluid management and the core of the thermal regulation system. In an effort to minimise the station's logistics overhead, regenerative life support systems were to condense water vapour from the atmosphere and process urine and water from the hygiene systems to make it potable. A regenerative molecular filter was to cleanse the air of carbon dioxide and other toxins. With the 'loops' of water and oxygen utilisation 'closed', only top-up nitrogen would require to be ferried up by the Shuttle in high-pressure canisters. Of course, as part of their regular service cycle the Shuttles would have to provide food and clothing and remove trash and solid human waste. When mounted on the outboard end of the node, the cupola would provide wrap-round visibility at the station's port side (other cupolas would be installed later to cover other aspects). This first cupola would have a console for operating the remote manipulator. To this point, Shuttles had been docking at the unpressurised berthing mechanism on the 'cornerstone' of the truss, but once this Shuttle had added the node, it was to dock at its pressurised adapter, permitting the astronauts to board it and activate the internal systems.

Flight 6

The laboratory was to be mounted on the node's radial berth so that it ran 'below' the truss and faced 'ahead'. As a consequence of being scaled down so that it could be launched in a fully fitted form, the 8.4-metre-long module would contain 12 racks of apparatus for microgravity experiments, but a multidisciplinary programme was to address life and materials sciences. Its functionality was segmented into adjacent banks of racks so that suites of apparatus could be superseded as the research progressed. The launch configuration would represent only the initial interest of the scientific community. The scientific payload was housed in the racks that formed the laboratory's 'walls', and the equivalent of 6 racks of utilities were built into the 'floor' and 'ceiling'. The racks could be angled out for servicing, or removed to be replaced.

In addition to the crew's physiological adaptation to space, the life sciences research was to include gravitational biology with animals. A suite of generic furnaces and separators would facilitate the development of semiconductors, eutectic alloys and proteins as well as metal sintering and electrophoresis of active biological agents. It was far from clear, however, that the results would overcome the National Science Foundation's scepticism. In August 1992, NASA invited potential users to a workshop held in Huntsville to explain in detail the station's capabilities and services. Giving the keynote address, Goldin referred to the station as "NASA's tenth research facility", by which he meant that it would be part of the agency's organisation, not just a spacecraft.

The addition of NASA's laboratory module would have achieved the 'tended' milestone. In addition to continuing the assembly, later crews would be able to 'linger' for up to a fortnight to perform experiments.

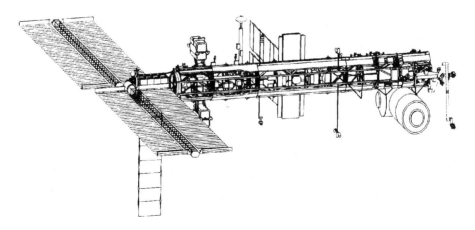

A diagram showing the 'tended capability' configuration of the space station.

An artist's depiction of the multiple-armed Special Purpose Dexterous Manipulator proposed by Canada to augment its Space Station Remote Manipulator System.

Phase 2

The second phase of the assembly would start by adding:

- the airlock;
- the special-purpose dexterous manipulator; and
- the mobile servicing system.

Mounted on top of the node, the airlock would provide spacewalkers with direct access to the truss, within which were the EVA aids. The remote manipulator's special-purpose dexterous 'hand' was to undertake delicate operations on attached payloads, in so doing reducing the spacewalking overhead of station operations.

The next four flights would add (in order):

- the sixth pre-integrated truss segment;
- UHF antenna;
- the port thermal control system;
- the seventh pre-integrated truss segment;
- three 'dry cargo' berthing mechanisms;
- two propulsion modules;
- an 'alpha' joint assembly;
- the port photovoltaic solar power module;
- the starboard node; and
- the outboard (starboard) truss 'spacer' element.

The second node would be connected in-line with the original, alongside the truss. Japan's Experiment Module with 10 racks of science apparatus would then be mounted on the first node, aft, opposite NASA's laboratory, and ESA's Columbus would be mounted on the second node alongside its Japanese counterpart.

The next two flights would add:

• the outboard (starboard) photovoltaic solar power module; and
• the Japanese exposed facility and the pressurised and exposed sections of the experiment logistics module.

The habitation module and racks containing further life-support system would then be added, with the module being mounted on the second node projecting forward alongside NASA's laboratory. It would provide facilities for a crew of four. The wardroom, toilet and the shower were to be inset into the walls of the square-cross-section interior compartment and the floor and ceiling would house utilities and storage lockers. The galley, located near one end of the module, adjacent to the wardroom, would provide an oven, a refrigerator, a freezer and a trash compactor. As with Skylab, there would be a large porthole in the wardroom so that the crew could savour the view as they congregated for meals, meetings or uplinked TV entertainment. Unlike Skylab, however, in this "descoped" form of the habitat the crew would not have individual cabins; they would have to attach their sleeping bags to the wall. The crew would work in pairs in order to sustain 24-hour operations. It would, however, be difficult for one shift to sleep soundly while the other made use of the facilities, and although they would meet for meals, one shift would be breakfasting while the other had its supper. As the station was expanded further, crew members would probably seek out their own sleeping niches in the logistics modules.

Adding the 'lifeboat' would mark the 'permanent habitability' milestone and complete the Baseline Configuration.

Table 8.1 Space Station Freedom milestones

	Tended capability by 6th flight	Permanent habitability capability by 17th flight
Power	1 unit, 18.75 kW	3 units, 56.25 kW
User power	11 kW	30 kW
Command uplink	70 kbps	70 kbps
Data downlink	43 Mbps	43 Mbps
Attached payloads	2 sites	4 sites
Resident crew	0	4 (expandable to 8)
Dedicated research crew	4 (living on Shuttle)	2 (on 24-hour basis)
Life support	Via Shuttle	Regenerative water loop

An artist's depiction of the completed module cluster forming the core of Freedom. NASA's laboratory and habitation modules are fitted with docking systems. They are mated to a pair of linking nodes, which also support the Japanese (shown with its external pallet) and ESA laboratories.

Table 8.2 Space Station Freedom element dimensions

Space Station Freedom element	Shape	Length (m)	Width/Diameter	Mass (tonnes)
Truss and equipment	Hexagon	66	3.7 × 4.9	146.0
US laboratory	Cylinder	8.4	4.4	15.5
Habitat	Cylinder	8.4	4.4	16.2
Columbus	Cylinder	11.8	4.5	17.0
Japanese module*	Cylinder	17.0	4.2	32.8
Connecting nodes, etc.†	Cylinder	5.2	4.4	46.0
Solar panels	Rectangle	34.0	11.9	7.9
Overall	–	108	–	281.4

* The mass of the Japanese module includes the exposed facility and the experiment logistics module.

† The dimensions of the connecting node pertain to the node itself, but the mass is for two nodes and the associated cupola, airlock and pressurised docking adapters.

Phase 3

By mid-1992, the launch of the first element was scheduled for early 1996 (having slipped 12 months in the years since the 1989 budgetary crisis) and 'permanent habitability' (now at the end of the assembly process) had slipped into 2000.

An artist's depiction of the truss structure with mounts for attached payloads.

Thereafter, as Shuttles sustained the station's operations they would add the third node, the fourth solar power module and Italy's life-sciences laboratory. Another pair of nodes would be added later to link the ends of the habitat and NASA's laboratory, thereby exploiting the station's modularity to offer scope for further expansion – certainly the agency could not be accused of coming up with an inherently limited design.

Addressing the National Space Club on 24 June 1992, Goldin was combative. "We can light up the sky with the inspirational work of Space Station Freedom, or we can stand by and watch the greatest technological 'bonfire' ... if it is cancelled. We have waited long enough. To keep the next generation of benefits from space flowing back to Earth [we] must have a permanent presence in space. We need Space Station Freedom and we need it now." He reflected that despite 30 years of space flight America knew very little about how the body adapts to weightlessness because, apart from Skylab, no NASA mission had exceeded a fortnight. The Russians had spent longer in space, but their data was woefully inadequate because the on-board research facilities were far too basic. "The only place to learn about operating for long periods in space, is in space." There is a progressive loss of bone mass in weightlessness. "The rate of bone loss in space is ten times as great," he noted. "On Earth, we call this disease osteoporosis. Twenty million American women suffer with it, so finding how to counteract it could bring relief to those women." In a country that focuses all too often on the short term, he noted that his was one of the few federal agencies that was dedicated to the future. About $2 billion of its budget was devoted to the station. "It sounds like a lot," he observed, "until compared with the $6.3 billion that we spend on pet food each year, or the $4.3 billion we spend on potato chips, or the $1.4 billion for popcorn." He reminded his audience that every time the US had pushed its frontier, it had discovered unimagined riches. Why should space be different?

In 1984 Reagan had been told that the station would be able to be built for $8 billion "within a decade". Unfortunately, by 1993 some $11.4 billion had been spent. The station had set the unenviable precedent of having spent its entire assigned budget without delivering a single piece of flight-worthy hardware. In effect, the budget had funded the salaries of the managers and engineers who had first designed and then repeatedly redesigned the entire programme. The agency's most recent estimate was that station's development and assembly would cost some $30 billion.

9

Mir

Although work remained to be done on board the Salyut 7, Cosmos 1686 complex to finish the programme abandoned by the Soyuz-T 14 crew, the Soviets decided to move on and placed a new vehicle into orbit on 20 February 1986. It was to serve as the 'base block' for a large orbital complex which would be assembled from modular components. Rather than simply designating it 'Salyut 8', it was decided to name the new vehicle 'Mir' (which can be translated, depending on context, as 'new world', 'peace' or 'community'). Its basic characteristics were defined by the Proton that would carry it, but it incorporated several advanced features. In addition to having an axial docking port, the new forward transfer compartment had a ring of four radial ports. Only the axial ports were equipped with the antennae required for automatic dockings. The modules destined for the radial ports would first dock on the axis and then use a small robotic arm to grasp a fixture on the multiple docking adapter to swing themselves around to engage the side port. Mir had been heavily computerised to relieve the crew of routine tasks. Instead of a trio of solar panels it had two larger ones, each of which was capable of producing 4.5 kW. Because the laboratory modules were to be added later, the base block served the role of a habitat, with improved facilities for the crew, including a pair of sleep compartments, each with its own porthole, and a separate toilet cubicle with a wash basin.

The reason for Mir having been inserted into the same orbital plane as its predecessor became clear on 13 March when Leonid Kizim and Vladimir Solovyov launched aboard Soyuz-T 15 and it was promptly announced that after they had commissioned Mir they would transfer to Salyut 7. As the chief cosmonaut Vladimir Shatalov put it, this was to be "the most complex flight".

A new version of the Soyuz had been developed for use with Mir, but the requirement to transfer to Salyut 7 meant that in this case a Soyuz-T had to be used. The new ferry would use a new type of rendezvous radar transponder called 'Kurs'. The older 'Igla' required the station to turn to face its docking port towards the approaching vehicle. This had been reasonable for a Salyut, but it would be impracticable once Mir had a cluster of heavy modules on its nose. Using Kurs, a

The Mir 'base block' in its pristine configuration.

spacecraft would be able to make its initial approach with the station in any orientation, then fly around to line up to the appropriate docking port to make a straight-in approach. In addition to the Kurs, Mir had an Igla transponder on its rear so as to guide in Progress cargo ships, so the Igla-equipped Soyuz-T 15 made its initial approach to the rear and Kizim then flew it around to the front, using a laser rangefinder to make his final approach. Progress 25 arrived on 21 March with a full load of propellent to replenish Mir's tanks – which had been virtually drained by the climb to its operating orbit – and departed on 20 April to clear the way for Progress 26, which arrived a week later with science apparatus.

When Kizim and Solovyov undocked on 5 May, Mir was trailing behind the Salyut 7, Cosmos 1686 complex, so they lowered their orbit slightly to catch up with their second target with which they docked the next day, thereby achieving the first transfer from one space station to another. Having accumulated almost 24 hours working outside Salyut 7 on their previous tour of duty (during which they had fitted a bypass in its engine) they were well qualified to tackle the external activities left by their predecessors. On 28 May, they retrieved cassettes from the hull beside the airlock and set up a workstation on which they erected a 15-metre-long lattice and pin framework truss ferried up by Cosmos 1686. A similar truss was to be erected on Mir soon, so verifying the deployment procedure and determining the dynamic characteristics of this prototype had been the primary reason for returning to Salyut 7. They finished the other experiments over the following weeks, then loaded some 400 kilograms of scientific apparatus into their ferry. Having returned Salyut 7 to its automated regime, they left on 25 June. By this time Mir had drawn ahead, so they repeated the procedure of lowering their orbit in order to catch up, and then followed the Igla approach with a fly-around to dock manually on Mir's front port just as if such flights were routine.

It had been hoped to launch the first science module during Kizim and Solovyov's tour but its fabrication was running late, and once the cosmonauts had exhausted the scope of the limited amount of apparatus that had been ferried up or salvaged from

A cutaway of the Mir 'base block' showing the spherical docking compartment, control panel, main compartment with work table and personal cabin, and rear access tunnel. The attachment at the rear is the steerable antenna for geostationary relay satellites.

Salyut 7, they got bored and were obliged to return to Earth on 16 July, abandoning the plan to hand over the station to their successors for continuous occupation.

While Soyuz-T 15 had been away, Soyuz-TM 1, the first of the Kurs-equipped spacecraft, had been launched without cosmonauts and had paid Mir a brief visit to verify the systems. This new configuration could carry an extra 250 kilograms into space and could return with 120 kilograms of compact cargo. Relaxing the dynamical constraints meant that it could return at any time of the day – a development that would greatly simplify station operations.

As Yuri Romanenko and Alexander Laveikin docked Soyuz-TM 2 on Mir's front port on 8 February 1987, Tass reported that they hoped to set a new endurance record. They started by unloading Progress 27, which was already in place, then Progress 28, which took its place on 5 March. On 5 April, the first module, referred to as 'Kvant' (Quantum), began its final approach to Mir's recently vacated rear port, but as it closed within 200 metres the spacecraft's Igla radar lost lock and it started to diverge from the straight-in approach. Romanenko and Laveikin hastily raced from one porthole to another in search of it and reported that it almost struck one of Mir's solar panels. After rectifying the fault, Kaliningrad set up a second rendezvous on 9 April and this time the approach was perfect. However, when the command was sent to retract the docking probe to achieve a hard docking the latches in their collars failed to engage. In effect, the 22-tonne vehicle was dangling off the rear of the station by its capture latches, an arrangement that was tenable only as long as neither vehicle attempted to manoeuvre. On 11 April the cosmonauts exited one of the vacant radial ports and made their way along the 13-metre-long base block's hull. Upon peering over the rim, Laveikin reported in amazement that a cloth sack was jammed between the two collars. Once Kvant had extended its probe he retrieved the obstruction, then watched Kvant retract its probe to complete the docking. The newcomer was actually a composite, with the engine block of a TKS ferrying a 5.8-metre-long pressurised module. The next day, the TKS withdrew, exposing a docking port and antennae for the Igla and Kurs systems.

It was not revealed at the time, but Kvant had been designed for use with Salyut 7. It is noteworthy that whereas vehicles like Cosmos 1443 were just temporary adjuncts because they had to be jettisoned to release the front port for the next ferry, Kvant's double docking ports would have enabled it to become a permanent part of that station. The evolutionary nature of the development of the elements for the construction of a modular orbital complex is thereby evident. Kvant was to have been sent up in the second half of the Soyuz-T 14 crew's mission, after Cosmos 1686 had left. With the decision to scrap Salyut 7, the module had been reconfigured for use with Mir. As the Progress tankers would have to dock with Kvant, pipes had been fitted on its hull to enable fluids to be pumped into the base block. Mir would no longer be able to fire its main engines, and Kvant had no engines (which was why it had had to be delivered by a propulsive unit), so the station would now be reliant upon Progress ferries for major orbital manoeuvres.

In addition to a batch of X-ray and ultraviolet telescopes and an electrophoresis system for purifying biological agents, Kvant had environmental systems, including an electrolyser to reclaim water and oxygen from urine, and a filter to remove carbon dioxide from the air supply (a regenerative unit which would be more efficient than the use-once-and-discard lithium hydroxide canisters of the base block). It also contained six electrically driven 'gyrodynes', two on each of the Cartesian degrees of freedom, with which to orient the station without consuming propellant. Once the computer had been reprogrammed to take account of the module on the rear, the cosmonauts began to refine its ability to make attitude changes with the gyrodynes. When Progress 29 slipped into Kvant's port on 23 April to form the first orbital complex of four vehicles, the computer's 'mass model' was revised again. It would have to be updated for each configuration as the complex was expanded, so that the computer would know how to calculate manoeuvres.

As part of its cargo, Kvant had delivered a third solar panel for the base block. Two spacewalks would be required to install it, as it was in two segments. It had been hoped to do this immediately, but when they had gone out to see why Kvant could not dock the medics monitoring Laveikin's biomedical telemetry had observed what appeared to be a heart irregularity, so they had asked for the installation of the new panel to be delayed, firstly to early May, then to late in May and then again to 12 June. The docking adapter was too cramped to accommodate both men and the bulky package, but Soyuz-TM 2 was at the front port to provide an escape route in case they were unable to re-enter the base block, so the panel was loaded into the orbital module and Mir's adapter module was depressurised with the docking tunnel open. With the upper docking port open, Romanenko secured himself on a workstation and Laveikin passed out the stubby cylindrical package. They mated it with a motor mount already present on the base block. On 16 June, with the second package mounted on top of the first, they deployed the 10-metre-long truss (similar to that tested by Kizim and Solovyov outside Salyut 7) to extend the arrays. Finally they connected the colour-coded cables to sockets on the base block to link the new panel into the electrical system. The new panel boosted the power to 11.5 kW, which was sufficient to operate a furnace, the X-ray telescopes and the gyrodynes simultaneously. Unfortunately for Laveikin, the biomedical telemetry confirmed that his heart developed 'extra systolic activity' when he exerted himself. On 24 July, after Progress 30 had been and gone, Soyuz-TM 3 brought Alexander Viktorenko, Alexander Alexandrov and Mohammed Faris of Syria as Mir's first international visitor. They had originally been scheduled to visit the Soyuz-T 14 crew on board Salyut 7 once Cosmos 1686 had departed. At this point, Vladimir Shatalov made the public announcement that Laveikin was being recalled and that Alexandrov would take his place and serve with Romanenko for the remainder of his record-breaking mission. Laveikin therefore bade his comrade farewell on 29 July. Ironically, although the improved biomedical monitoring system had been sufficient to reveal the irregularity in his heart, it had been insufficient to produce a firm diagnosis and when specialists examined him upon his return to Moscow they pronounced him fit to fly!

On 31 July, Romanenko and Alexandrov flew Soyuz-TM 3 around to the front port and settled into what became a 6-month slog, visited only by a succession of cargo ships until Soyuz-TM 4 arrived on 23 December with their relief in the form of Vladimir Titov and Musa Manarov. Valeri Poliakov, a physician and the deputy head of Moscow's Institute of Medical Biology, was to have flown up to assess Romanenko's state of health prior to his return but the military had insisted that Buran pilot Anatoli Levchenko fly for experience. Because the two groups of cosmonauts were separate, when Levchenko appeared in the hatch Romanenko had to ask him who he was. When Soyuz-TM 3 departed on 29 December it carried three cosmonauts who had launched at different times: Romanenko, who, by having spent 326 days in space, had extended the record by fully three months; Alexandrov, who had spent a mere 160 days in space; and Levchenko, who had barely adapted to weightlessness. Titov and Manarov, who had been assigned a 12-month mission, had trained to commission the first of the modules intended for a front radial port. On 12 April 1988, 'Cosmonaut Day', deputy flight director Viktor Blagov reassured them that this would be dispatched on time. Meanwhile, they had a complex spacewalking assignment. Because one of Kvant's X-ray telescopes was suffering intermittent faults it had been decided that the cosmonauts should attempt a repair. Replacement parts and special tools were delivered by Progress 36 in May together with a video showing what had to be done. On 9 June, Soyuz-TM 5 delivered Anatoli Solovyov (not to be confused with Vladimir Solovyov), Viktor Savinykh and Alexander Alexandrov (a Bulgarian researcher, not to be confused with the Soviet cosmonaut of that name who had recently visited Mir). In fact, this was the second Bulgarian mission. Alexandrov had backed up Georgi Ivanov, who had been launched in Soyuz 33 in 1979 and been prevented from reaching Salyut 6 when the ferry's engine misfired. Since that was the only Intercosmos mission to be aborted, this visit to Mir had been laid on as a consolation. The primary purpose of this mission was to replace the station's ferry, so once this had been transferred to the front port, the residents set off on 30 June to inspect Kvant's balky telescope. They manoeuvred the 40-centimetre-wide cylindrical detector and their kit of tools out through one of the vacant ports, along the length of base block, across a handrail to Kvant, and along another rail to the unpressurised compartment at its rear, whereupon their task was made difficult by the absence of restraints. They paused to rest when they became tired, so the schedule slipped, but they managed to cut through the 20-layer thermal blanket protecting the unpressurised compartment to gain access to the telescope. Unfortunately, the 2.5-metre-long instrument had not been built to be serviced by engineers wearing space gloves and the small bolts proved difficult to release. The procedure recommended after the hydrotank simulations had assigned 20 minutes to this task, but it actually took an hour and a half. Hoping to catch up, the cosmonauts were astonished to discover that the apparatus incorporated some embellishments that had not been present on the development unit used for training. Pushing on regardless, they sawed through several bolts, wiped away an unexpected resin deposit, removed a number of screws and cut stainless steel clips to expose the detector, but when a special tool was inserted to release the clamp that held the detector in place its tip sheared off. After a vain effort to prise off the clamp, they deployed a new thermal insulation blanket over the hole to protect the apparatus and then returned with the replacement detector. The remainder of the operation would have to wait until a better tool could be sent up.

Soyuz-TM 6's arrival on 31 August brought a hastily formed crew. As a veteran of Salyut 6 and Salyut 7, Vladimir Lyakhov, the commander, required no specific training. Earlier that year, Glavcosmos chairman Alexander Dunayev had announced that an Afghan was to visit before Soviet military forces finished their withdrawal from Afghanistan. As a result, Abdul Ahad Mohmand received only six months' training. It was usual for a foreigner to receive two years of training, but the fact that he was fluent in Russian assisted the assimilation process. In fact, Mohmand had taken over from Mohammad Dauran when he had fallen behind in training by having his appendix removed. Titov and Manarov were already on Mir when Mohmand entered training, so they had no idea who he was when he floated through the hatch. The third seat was taken by Valeri Poliakov, who was to assess the ability of the residents to exceed Romanenko's endurance record. Rather than make a brief 'house call', Poliakov was to remain on board when Lyakhov and Mohmand left, and the fact that a Frenchman was to accompany the handover at the end of the year meant that Poliakov would have to serve with the next crew, so this hastily arranged mission offered him a once-in-a-lifetime opportunity to try to set a record himself.

On 6 September Lyakhov and Mohmand departed in Soyuz-TM 5. The orbital module was jettisoned according to plan, then the solar glare as the spacecraft passed over the terminator into sunlight prompted the primary and the backup infrared horizon sensors to provide conflicting signals to the computer, which inhibited the imminent retrofire manoeuvre. For years, recovery window constraints had required the spacecraft to have crossed into sunlight at least ten minutes before retrofire to provide the sensors with sufficient time to verify its orientation and stability. However, despite the relaxation of this requirement for the Soyuz-TM variant, passing through orbital dawn at the critical moment had upset the system. Seven minutes later, as Lyakhov was interrogating the computer to discover what was wrong, the sensors cleared up and finally presented the signal to confirm that the spacecraft was correctly oriented, so the still-waiting computer lit the engine. The spacecraft was now far beyond the planned de-orbit point, however. If permitted to continue, the burn would result in a landing several thousands of kilometres downrange. Deciding to descend nearer the normal recovery area on the next orbit, Lyakhov immediately shut off the engine. In fact, the de-orbit burn was rescheduled for two orbits later. The engine lit on time, but shut off six seconds into the 230-second burn. Lyakhov restarted it, but 50 seconds later the computer decided that the craft was out of alignment and shut it off again. The situation had suddenly deteriorated significantly. Recognising that the track had drifted too far west to try again on the next orbit, Valeri Ryumin, now flight director, announced that they would have to stay in space for another day. Overnight analysis determined that on the second de-orbit attempt the computer had used the incorrect programme; for some reason it had selected part of the rendezvous sequence that it had flown on its way to Mir. Having established that the engine had not really malfunctioned, the tension in Kaliningrad evaporated. In space, however, the situation was dire. Without its orbital module the spacecraft was rather cramped for two spacesuited figures. It was of little consolation to know that it would have been much worse if the third seat had been occupied. There was no point in trying to return to Mir, because their Kurs antenna and the docking unit had been discarded with the orbital module. The descent module had not been designed for prolonged independent operation. Lyakhov lamented that the toilet was in the lost module and he suggested that, in future, the module should not be jettisoned until after the de-orbit burn had been completed, even though doing so would require extra propellant. With communications restored, a new computer programme was read up and Lyakhov keyed this manually into the computer. Although the Western media reported that they were "stranded in orbit", the situation was actually manageable. A television crew was on hand to record the pre-dawn twilight landing and the two cosmonauts immediately scrambled out to greet their recovery team. Despite the improving technological maturity, spaceflight was evidently still a risky business.

Titov and Manarov returned to Kvant on 20 October, reopened the thermal blanket and used the new tool to release the clamp without incident. Working rapidly, they extracted the faulty detector, inserted the new one and affixed a simple clamp to lock it into position. In contrast to their previous spacewalk, by the time they replaced the thermal insulation they were an hour ahead of schedule. Clearly, working effectively in space was primarily a matter of having the right tools.

Unfortunately, the large module had not materialised by the time of the handover in December, so the task of commissioning it passed to Alexander Volkov and Sergei Krikalev. In fact, if the schedule picked up and the next two modules were launched six months apart, they would commission them both. The handover lasted three weeks, and Frenchman Jean-Loup Chretien, who had visited Salyut 7, not only became the first foreigner to fly twice but also the first to make a spacewalk. In fact, it had been decided that there would be no more 'free rides'; from now on international researchers would be charged a fee to defray the costs of participation, including Progress shipments. By the time that Titov and Manarov finally returned to Earth they had extended the endurance record to 366 days. By February 1989 it became clear that Volkov and Krikalev would be unlikely to see the second module. The real-time nature of the planning process was frustrating for all concerned, but the primary goal of the programme was to maintain a permanent presence in orbit and the crew simply had to undertake such work as was practicable.

The process of expanding the complex would be complicated by the need to control the attitude of an asymmetric configuration. In a linear train the centre of mass was at least on the major axis, but with a large module projecting out to the side in an 'L' shape it would be outwith the structure. When such a module was expected, the Soyuz would need to be at the rear, but Mir would not be able to be resupplied while both axial parts were occupied. These operational limitations meant that the first new module would (sooner rather than later) have to be moved off the axis onto a radial port. With the 20-tonne module projecting out to one side the complex would be difficult to manoeuvre, so experiments which required Mir to assume a specific orientation would be impracticable. Furthermore, because an engine designed to deliver impulse axially would not be able to push through the offset centre of mass, it would rotate the complex, so Mir would not be able to adjust its orbit until the opposite module was in place. Two factors were therefore crucial to the next phase of expansion. As soon as possible after the first module was swung out to the side, the attitude control system would have to be fine-tuned to manage the asymmetric configuration (it would be impracticable to receive incoming spacecraft until Mir could be stabilised). As soon as possible thereafter the second module would have to be received and swung around to create a 'T' shape. Ideally, this phase of the expansion would be accomplished over a period of a few months. It was therefore vital that the first module should not be launched until the second was also nearing completion. With the prospects in decline it was decided at the end of March not to dispatch the next crew until both modules were ready to go, so Volkov, Krikalev and Poliakov vacated Mir on 27 April. This was particularly disappointing for Poliakov, because he had hoped to stay in space and serve with his third pair of cosmonauts, and thereby extend the record to 18 months. It was decided that when Mir was reoccupied, crews would focus on expanding the complex and would fly fairly brief tours of duty; the recently set 366-day record would probably stand for some time.

When Progress-M 1, the first of the Kurs-equipped tankers, slipped into the front port on 25 August it was revealed that Mir had the plumbing to enable it to be replenished from either end. This eliminated the requirement for crew ferries to swap ends in order to keep the rear port free. Mir was reactivated when Soyuz-TM 8 delivered Alexander Viktorenko and Alexander Serebrov on 8 September. The plan was for the first module to dock on 23 October and swing onto the dorsal radial port two days later, with the second module following in early February 1990. In fact, the first launch was delayed to 26 November in order to replace some chips in its Kurs system. The spacecraft, unimaginatively designated Kvant 2, was a TKS with Chelomei's descent capsule replaced by a large airlock. Unfortunately, it soon ran into difficulty. The outermost three segments of one of its two solar panels unfolded properly, but the innermost did not. In fact, because the innermost panel had not locked, the outer part of the 10-metre-long panel was free to swing, which made manoeuvring awkward. After analysing the telemetry the flight controllers put the vehicle into a slow roll, then commanded the motor that rotated the solar panel to cycle back and forth to cause the centrifugal force to straighten it. On 2 December, the day after Progress-M 1 left, Kvant 2's Kurs was overloaded when it was still 20 kilometres out, and thinking that it was closing too rapidly it aborted its approach. Actually, even if it had been able to close in, a docking would have been risky because Mir's gyrodynes dropped off-line. After a revision of software, Kvant 2 made another approach four days later. This time Viktorenko locked Mir stable using Soyuz-TM 8's thrusters and the docking was achieved without incident. Two days later, Kvant 2 deployed its robotic arm, grasped a fixture on the hull of Mir's multiple docking adapter and swung itself up onto the dorsal port. It consisted of two primary compartments and the airlock. In addition to internal cargo it incorporated a second toilet, a shower, a second electrolysis unit and another set of six gyrodynes. The airlock was stocked with two of the latest Orlan-DM spacesuits and a backpack-style manoeuvring unit and its outer hatch was extra-wide so that a cosmonaut wearing the backpack could readily pass through. The 7 kW provided by Kvant 2's solar panels gave Mir a much needed boost. Progress-M 2's arrival on 22 December produced the first five-vehicle orbital complex and brought its mass up to about 72 tonnes.

With Kvant 1 in place on Mir's rear port, Kvant 2 first docked at the front axial port and then swung itself around to the dorsal port, leaving the orbital complex in a difficult-to-control asymmetric configuration. Note the loose thermal blankets on the Soyuz-TM 9 spacecraft.

On 8 January 1990, with their ferry now on the front port in case of trouble, Viktorenko and Serebrov exited a vacant port in the multiple docking adapter and made their way to Kvant 1 in order to install a bulky star tracker on each side of its unpressurised instrument compartment to assist in orienting the expanded complex. Three days later, they moved the radial drogue from the upper to the lower port, ready to receive the next module. On 26 January they used Kvant 2's airlock for the first time, and put a framework receptacle just outside the hatch to serve as a restraint when they tested the manoeuvring backpack. This was done by Serebrov, on 1 February. He remained tethered as a safety measure, and Viktorenko stationed himself by a winch to reel him in. Although the test went well, the backpack was not intended for use on Mir, it was being tested for use on Buran. The arrival of Soyuz-TM 9 with Anatoli Solovyov and Alexander Balandin on 13 February provided the residents with a welcome change of pace. When the visitors left a few days later, they did so in the old ferry. It was a poor deal for the residents, however, because some of the thermal blankets on Soyuz-TM 9's descent module had become detached soon after achieving orbit and were now protruding from the collar between the descent and orbital modules like petals on a flower. An inspection by the departing visitors indicated that unless something was done to remove or fix these blankets they would block the field of view of a horizon sensor that would be needed when setting up the retrofire orientation. It was decided to send up some new tools and have Viktorenko and Serebrov attempt to reattach the blankets on an impromptu spacewalk. There was no hurry, however, as long as something was done before the ferry had to be used. Much of March and April was devoted to refining the computer's ability to control the asymmetric configuration using a dozen gyrodynes. As these trials consumed much more propellant than expected, an old-style tanker was sent up to replenish the base block's tanks.

As the new module, named 'Kristall', started its approach on 6 June, one of the attitude control thrusters misfired and the automated system aborted the approach. The second attempt four days later was perfect. Once Kristall had swung itself down to the lower port, the second phase of Mir's expansion was finally complete. Significantly, a small compartment at the far end of the module had a pair of androgynous docking ports, so Buran would be able to dock. In addition to a high-resolution mapping camera and a package of multispectral sensors, Kristall's main payload was a bank of materials-processing furnaces. Not long afterwards, it was announced that the funding to complete the final modules would be withheld until it could be shown that the apparatus in Kristall could produce an economically viable result. This effectively meant that the final phase of the expansion (which at that time envisaged the modules being launched in late 1991 and early 1992) would be postponed indefinitely.

With Soyuz-TM 9 now on the front port, Viktorenko and Serebrov ventured out on 17 July to inspect its loose thermal blankets. Exiting by a vacant port on the multiple docking adapter would have provided a more direct route, but this would have meant making their way past the antennae on the orbital module *en route* to the descent module. Since there were no handholds, they might easily have damaged apparatus mounted on the ferry. It had therefore been decided to exit through Kvant

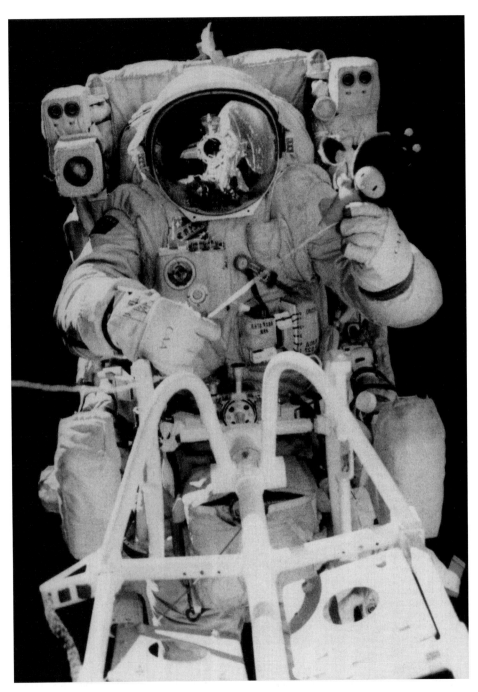

Alexander Serebrov, wearing an Orlan-DM suit, tests the Astronaut Manoeuvring Unit (Russian acronym YMK) backpack.

The addition of the Kristall module restored a sense of symmetry to the Mir complex.

2's airlock. A 7-metre-long folded metal ladder had been flown up and they were to angle the ladder across the orbital module to provide access to the descent module. The ladder was also to serve as a work platform.

Unfortunately, they were in such a hurry to start that they released the hatch before the airlock had fully depressurised. The other hatches on the complex opened inwards but the airlock opened outwards in order to allow spacewalkers to load the compartment with equipment without having to leave clearance for the hatch to swing in. The hatch was to be opened in two stages. At first it should be opened a few millimetres to crack the hermetic seal, and it would then be retained by a catch. Only after the airlock had fully depressurised was the catch to be released. In this case the catch released immediately and the residual pressure made the hatch flip open far beyond its designed limit. The cosmonauts expressed surprise, but set about their spacewalk oblivious to the fact that the hinge had been damaged. Each cosmonaut had a pair of short safety tethers with mountaineering clamps that he alternated between a series of handholds. Their progress was slowed by the need to hold in place in darkness while dragging the bulky equipment. It took almost two hours to reach the multiple docking adapter, and the ladder then proved awkward to deploy.

Upon finally reaching the descent module they inspected the exposed heat shield for any sign of damage. The preferred option was to use spare pins to clip the loose blankets into position, but efforts to reattach them failed because the blankets had shrunk and would no longer reach the locking ring at the base of the descent module. The next option was to roll up each segment and clamp it to the orbital module where it would not block the field of view of the sensors mounted further aft. They managed to secure two of the blankets in this way but the third was so badly torn that they were unable to make as tidy a job as they hoped; this blanket was not so important, however, because it was clear of the sensors. Having been out for six hours, they did not have time to dismantle the ladder as they returned. Unhindered by tools, they made better progress. Having been on their life-support systems for so long they could not afford to stop, so in orbital darkness they double-checked each transfer of their safety lines in the illumination of their helmet lamps. Back in the airlock they found that no matter how hard they tugged the hatch it would not close the last few millimetres, so they left it ajar and passed into Kvant 2's middle compartment as an emergency airlock. By the time this compartment had been pressurised they had been in their suits for over seven hours, which was uncomfortably close to the absolute limit. They set out on another impromptu spacewalk on 26 July to recover the ladder, which could not be left in place as it might swing across the front port after Soyuz-TM 9 departed, but first they used a television camera to show the engineers that a bolt had been pulled partly out of one of the hinges and an aluminium cover plate had been twisted up. This appeared to be what was preventing the hatch from closing. Upon returning, a thorough inspection of the hatch revealed that part of the bent plate had snapped and become stuck in the hinge. Once this was removed they were finally able to close the hatch, but doing so was a struggle so it was decided to have the hinge plate replaced by the next crew.

Soyuz-TM 10 delivered Gennadi Manakov and Gennadi Strekalov on 3 August, and a week later Viktorenko and Serebrov left for home in their patched-up ferry. The new crew's main assignment was to extensively rewire the complex using power cables that were delivered by Progress-M 4. When they tried to replace the airlock hatch's hinge plate on 30 October they discovered that the hinge pin was badly bent, so they were told to leave it as it was; the entire hinge would have to be replaced. When Viktor Afanasayev and Musa Manarov relieved them in December they were accompanied by the first fee-paying visitor, Toehiro Akiyama, who was the Tokyo Broadcasting System's chief foreign news editor. After celebrating the new year, Afanasayev and Manarov replaced the hinge assembly on 7 January 1991 without difficulty and then set off to affix a small bracket to the base block. On 23 January, after Progress-M 6 had delivered it, they installed a 12-metre-long telescoping crane on this bracket. Operated by a pair of hand-cranks, the crane could transfer a load of 750 kilograms between any two points on the left side of the complex but care had to be taken to steer clear of all the solar panels. It was 'parked' against Kvant 2 so that the cosmonauts could ascend it on their way back to the airlock, and then slip down it at the start of the next spacewalk. On 26 January they used the crane to swing a pair of mounts for solar panels from the airlock to Kvant 1, where they were affixed one either side of that module.

One of the frames mounted on Kvant 1 by spacewalking cosmonauts in preparation for relocating Kristall's solar panels is visible in this view of the Mir complex.

When Progress-M 7 appeared on 21 March, its Kurs reached the conclusion that it had deviated from the straight-in approach to the rear port and it aborted. This was the first time that any of the Progress ferries had suffered a problem. Upon its second attempt two days later, the controllers watching the downlink from the spacecraft's docking camera were surprised to observe that even though the Kurs thought that it was properly aligned it was actually misaligned, and once again it passed perilously close to the station. Afanasayev and Manarov undocked their own spacecraft on 26 March, manually flew it to the rear, and let its Kurs system attempt an automatic approach; it drifted off course, implying that the fault lie in the Kurs antenna on Kvant 1. Having localised the fault, they then docked manually. Progress-M 7's successful docking at the front two days later confirmed that there was nothing wrong with its Kurs. The most likely cause of the problem was that a cosmonaut had disturbed Kvant 1's antenna during the recent spacewalk. The possibility of replacing the damaged antenna with that at the rear of the base block (which had been redundant since Kvant 1 docked) was considered, but it was deemed simpler to send up a replacement. In the meantime, the rear port would be unusable by automated vehicles. Upon inspecting the Kurs antenna on 26 April Manarov reported that the parabolic dish was missing!

Mir's first female visitor was Britain's Helen Sharman, who flew up with Anatoli Artsebarski and Sergei Krikalev on 20 May and left a week later with Afanasayev and Manarov. Artsebarski and Krikalev had drawn a heavily construction-oriented flight plan calling for as many as eight spacewalks. First, however, they had to transfer Soyuz-TM 12 to the rear port to free the front for Progress-M 8, which arrived on 1 June with the tools to repair the damaged antenna, which they duly did on 25 June; it was a delicate task involving the use of a dentist's mirror to inspect part of the mechanism, all while wearing a spacesuit. On a second excursion three days later, they deployed several experiments including one from the University of California. On 15 July, they made a start on their primary task by placing a platform on Kvant 1's 'roof'. They returned four days later with a large box containing tubular rods and sleeve joints that would form a 14-metre-long truss. They had assembled the first segment in the airlock and once this had been mounted on the platform a further two segments were added. They noted that it was difficult to perform such work in the ever-changing illumination of the daylight pass, but was straightforward in the illumination of their helmets during orbital darkness. More truss segments were added on 23 July but the final (twentieth) was not attached until 27 July. To mark their achievement they placed the Soviet Union's 'Hammer and Sickle' flag on its end. During 24 hours of external activity they had amply shown that manual construction work was feasible.

The flag-raising ceremony proved poignant, because in early September, in the aftermath of the failed coup to oust Mikhail Gorbachev, there was speculation that Mir might be sold to NASA. Upon hearing this, Artsebarski and Krikalev asked whether *they* were to be included in the deal as sitting tenants. This amusing anecdote exposed a real debate about the future of the orbital complex. In the harsh economic realism that accompanied the political chaos it appeared likely that Buran and its Energiya rocket would be cancelled, and with little prospect of generating a real financial return, not only would Mir not be completed, but it might even be *abandoned* in orbit. Apart from jesting that they seek to establish communications with NASA, Artsebarski and Krikalev had little option but to continue their programme.

When Soyuz-TM 13 slipped into the front port on 4 October it delivered Alexander Volkov and, for the first time, *two* guest cosmonauts: Franz Viehboeck of Austria was making a commercial flight, and Takhtar Aubakirov's flight marked the expansion of the Intercosmos programme to the individual Soviet republics. As a Kazakh, Aubakirov was representing the republic from whose territory every Soviet mission originated. In fact, he had been assigned to the visiting mission scheduled for November but in July the crisis in funding had prompted the decision to combine the two visits and thereby save a rocket for later use. One consequence of this was that because Alexander Kaleri could not fly as originally planned, and also because Aubakirov could not serve as Mir's flight engineer, when Afanasayev left with the visitors on 10 October Krikalev was obliged to remain on board with Volkov and so serve two back-to-back tours. As Soyuz-TM 13 swapped ends on 15 October, Krikalev had the satisfaction of observing that his repairs to Kvant 1's Kurs antenna had been successful and Mir was back in full service. On 25 December, Gorbachev resigned. A few days later the 'Hammer and Sickle' on the Kremlin was taken down and the Soviet Union was superseded by the Commonwealth of Independent States. The flag on Kvant 1 was now providing 'top cover' for a State that no longer existed.

As 1992 dawned, Boris Yeltsin, the President of independent Russia, created the Russian Space Agency and appointed Yuri Koptev to manage all civilian space operations and take control of existing launchers, spacecraft and Mir. In space, the routine continued almost as if nothing had happened and on 19 March Soyuz-TM 14 arrived with Alexander Viktorenko, Alexander Kaleri and Klaus-Dietrich Flade, who was making a commercial flight for unified Germany. When Krikalev (whom the Western media had dubbed "the last Soviet citizen") finally returned to Earth on 25 March, he found that things had changed more than he could have thought possible. Mir had already been in space longer than any of its predecessors, and longer than its own nominal life expectancy, so Viktorenko and Kaleri devoted the first part of their tour to overcoming its deterioration. Four of Kvant 2's six gyrodynes had failed, so on 8 July they ventured out with a pair of heavy-duty cutters to slice through the thermal insulation to affix two new ones so that the station would be able to continue to control its orientation effectively. On 29 July Soyuz-TM 15 arrived with Anatoli Solovyov, Sergei Avdeyev and Frenchman Michel Tognini. While this was France's third flight, it was the first arranged on a commercial basis. In fact, the earlier visits had been so productive that France marked this mission by announcing that it wished to revisit Mir at roughly two-year intervals. In the post-Soviet financial crisis such fee-paying missions were seen as international acknowledgement of the orbital complex's value, and thus played a significant part in countering those who argued that Mir had become a costly irrelevance that should simply be abandoned. However, the crucial issue for those in favour of continuing operations, and indeed for resuming the expansion, was whether it would be feasible to sustain the seven-year-old base block long enough to fulfil such 'advance bookings'.

Solovyov and Avdeyev's primary assignment was to finish the construction work begun by Afanasayev and Krikalev. A thruster unit was to be mounted on top of the truss. This 700-kilogram payload was so bulky that it could be delivered only by a Progress whose central module had been adapted to carry it. Because the craft had to dock at the rear port, it had not been possible to send it up until Kvant 1's Kurs had been confirmed as fixed. There was relief all round when Progress-M 18 docked without incident on 18 August. On 3 September they installed the thruster unit. A pivot had been incorporated into a segment of the truss one-third of the way along its length. They tilted down the upper part so that its tip came to rest over the hatch in the ferry from which the unit would emerge, locked the truss in position, and then retrieved the 'Hammer and Sickle' as a souvenir of a bygone age. Next they employed a special ratchet in the ferry to slide the payload out of its compartment. Four days later, they ran an umbilical along the truss's length, mated one end to the

At a time that NASA was being criticised for relying upon astronauts to assemble its space station, the construction of the Sofora girder on Kvant 1 and the installation of the thruster unit on its end showed that spacewalkers could serve as 'orbital hard hats'.

thruster unit and the other to a plug on Kvant 1, then attached several metal braces to the unit so that it could be connected to the platform at the end of the truss. After another four-day rest, the thruster unit was connected and the truss was rotated to a position 11 degrees beyond 'vertical', to place the side-mounted thrusters in the same plane as the base block's roll-control thrusters. It was expected that an 85 per cent saving in propellant usage would result from controlling the roll of the complex with this unit because its thrusters were farther from the complex's axis. This would reduce the frequency with which the base block required topping up. It represented a trade-off, however, because the self-contained unit would have to be replaced once it had exhausted its cold-gas supply.

The European Space Agency signed up in November to make two visits in 1994 and 1995 totalling 6 months, which would enable it to make a start on the research that it intended to undertake in its Columbus laboratory when the much-delayed Space Station Freedom became operational. Later that month, Germany signed an agreement to use Mir as a base from which to test technologies in which it hoped to establish a lead, one of which was a manoeuvrable free-flying robot to rendezvous with, examine and repair satellites.

An Energiya artist's depiction of the Mir complex with the Kristall module swung back onto the axis so that Buran could dock at the androgynous port at its end.

When Soyuz-TM 17 arrived, Mir was 'full up', with Soyuz-TM 16 on the Kristall module, Progress-M 17 on the rear and Progress-M 18 on the front, so it stood off and waited while Progress-M 18 departed.

Upon drawing alongside Mir on 26 January, Gennadi Manakov and Alexander Poleshchuk saw the unprecedented sight of the station with a ferry in place at either end. However, their Soyuz-TM 16 was fitted with an androgynous docking system and it docked at the end of the Kristall module, increasing to six the number of vehicles in the complex. During their tour their primary task was to install motors on the solar panel mounts which had been affixed to Kvant 1 two years previously. On 19 April the first motor was swung from Kvant 2's airlock to Kvant 1 and, after a struggle, successfully installed. As the cosmonauts made their way home they noted that one of the crane's cranking handles had come loose and drifted away, so the installation of the second motor was postponed until Progress-M 18 delivered a new handle. They mounted the second motor on 18 June. When Soyuz-TM 17 arrived on 3 July *all* the docking ports were occupied, so the new arrivals filmed Progress-M 18's departure from the front port then took its place. Frenchman Jean-Pierre Haignere accompanied the handover.

Vasili Tsibiliev and Alexander Serebrov had barely settled down to work than they were asked to extend their tour beyond the end of the year because the factory that made the rocket engines would not deliver until it received payment, and cash was not available. This kind of production bottleneck threatened to become a regular feature in the new corporate-oriented rather than state-oriented economy, and it would tax the patience of programme managers and cosmonauts alike. Nevertheless, by this point Mir's future as a microgravity laboratory for fee-paying clients seemed to be secure.

10

An International Space Station

SITUATION TRANSFORMED

As the Cold War 'thaw' continued, Presidents George Bush and Mikhail Gorbachev signed an agreement on 31 July 1991 to symbolise the rapprochment between the two superpowers with a cosmonaut flying aboard the Shuttle and an astronaut visiting the Mir space station.

In August, Gorbachev was ousted by a coup aiming to reverse his reforms and when Boris Yeltsin climbed onto a Red Army tank in front of the besieged parliament, the Russian people had found their new popular leader. The failure of the coup initiated a process which, by the end of the year, had transformed the Soviet Union into a Commonwealth of Independent States. The Cold War had become obsolete almost overnight. Yeltsin, the President of Russia, established the Russian Space Agency on 25 February 1992, headed by Yuri Koptev.

At their first summit, in Washington on 17 June 1992, Bush and Yeltsin agreed "to give consideration to" a joint mission. In effect, this was Bush's way of welcoming democratic Russia into what he had dubbed the 'New World Order', and it built upon the agreement he had signed with Gorbachev the previous year. The next day, Dan Goldin and Koptev ratified a one-year $1-million contract by which NASA and NPO-Energiya – a quasi-independent industrial conglomerate (headed by Yuri Semenov) which ran the Mir space station – would consider how Russian space technology might be used in conjunction with the space station programme. One option was to adapt the Soyuz as an interim lifeboat. NASA was impressed by the vehicle's reliability, its ability to dock automatically, and its ability to descend either on land and at sea, and associate administrator Arnold Aldrich had visited Russia in March to be briefed on its specifications preparatory to an astronaut riding the Soyuz to serve a tour of duty aboard Mir. When Goldin met Koptev on 7 July he signed a contract for a 3-year study of the feasibility of the lifeboat option. The primary concern was the spacecraft's orbital life. It was powered down while docked with Mir and cosmonaut tours were timed to ensure that no ferry spent more than 180 days in space. If it was to serve as a lifeboat, the Russians would have to prove that it could survive several years in space and still be powered up in an emergency.

A second option was to install a Russian docking system in a Shuttle which would not only enable a succession of astronauts to serve aboard Mir but would also allow NASA to make a start on some of the science projects planned for Freedom. A further possibility, once the station was operational, was to utilise Progress spacecraft for logistical support, so there was evidently considerable scope for collaboration. On 20 July Goldin and Koptev agreed to study the feasibility of the Shuttle visiting Mir. After the US Senate's 63 to 34 vote in support of the station on 9 September, Goldin was exuberant. "This vote reflects the commitment by members in the Senate to invest in America's future." On 5 October, Bush and Yeltsin signed the 'Human Spaceflight Cooperation' protocol to formalise a series of joint 'Shuttle–Mir' missions.

At the end of the year, NASA announced plans to consolidate the management of its programme and create a contractor-led integration team to assure the station's successful manufacture and orbital assembly. The Joint Vehicle Integration Team was to be based in Houston and draw its members from Boeing, McDonnell Douglas, Rocketdyne (the three prime contractors) and Grumman (the engineering and integration contractor). "These changes will reduce overhead costs," Aldrich assured.

NEW BROOM

When Bill Clinton moved into the White House in January 1993, his priority was to eliminate the budget deficit that had built up in the Reagan and Bush years. Despite the fact that the final Critical Design Review was about to clear the way for hardware fabrication, on 18 February Clinton ordered Goldin to develop a rather more cost-effective station plan. With the station facing development cost overruns, and with the likelihood of inflated operating costs, he requested a range of options, each of which was tailored to a specific budget ranging from $5 to $9 billion spread over five years. Cost estimates depend upon who makes them, when they are made, what they regard as associated expenditures, and the period of time covered. NASA had argued that the station's development cost would be amortised over a 30-year operating life, but with the Shuttle's low flight rate and high operating overhead the station's operating costs began to dominate the equation so Clinton ordered that costing include only 10 years of operations, and gave NASA three months to respond. On 9 March, as the agency launched its fourth station review in as many years, Goldin formed the Redesign Team under the chairmanship of Bryan O'Connor, a former astronaut serving as deputy associate administrator. The team had specific objectives: (1) to develop a cost-effective capability for significant long-duration research in materials and life sciences; (2) to bring near-term and long-term annual funding requirements within the constraints of the budget; (3) to continue to accommodate and encourage international participation; (4) to minimise technical and programmatic risk; and (5) to recommend how the Russian hardware might be encompassed.

O'Connor promptly set out to define these objectives and reconfigure the hardware to support them. In addition to the station's hardware configuration and assembly, O'Connor's remit included the management of the programme, so everything was open to reconsideration. Because the review would assess the possibility of other spacecraft operating in conjunction with the station, Goldin invited the Russians to attend. On 4 June the Redesign Team submitted its report. "The Team has done a tremendous job," said Goldin. "They have developed three technically viable space stations. Each provides for international cooperation, establishes a fully-capable space research centre in orbit that will enable high priority science, technology and engineering research, and in every case will do so for significantly less money than the current Space Station Freedom Baseline Configuration." Clearly, cost was the driving factor, but this was no time for unrealistic promises. Even so, none of the proposals came close to the budgets that Clinton had specified; in fact, they all exceeded the highest stated figure. Goldin was nevertheless unrepentent: the cost estimates were accurate and provided a solid basis for reaching a decision because they compared "all options on an apples-to-apples basis". NASA had not been asked to recommend one option over the others but rather to characterise the strengths and weaknesses of each in an unbiased manner. In the interests of impartiality, NASA had neutrally labelled the options 'A', 'B' and 'C'. On 7 June, the report was forwarded to an Advisory Committee chaired by Charles Vest of the Massachusetts Institute of Technology, who was to consider the options and make a recommendation to the White House. Significantly, Vest endorsed the fact that the cost estimates were all higher than Clinton had specified.

THE OPTIONS

NASA regarded Option 'A' as a case of "modular build-up" as it reduced the cost of assembling Space Station Freedom while preserving the international agreements and the requirements of users. Although it would be built on the basis of already designed technologies, these would be simplified or repackaged if doing so would be more cost-effective, and particularly if it would reduce the number or complexity of spacewalks. However, the most significant savings would be made by reducing the number of launches needed to assemble the station. There was, in fact, a sub-option using a propulsive stage designed by Lockheed to deliver military satellites (in the context of this submission, this Titan Launch Dispenser was referred to simply as Bus-1). In a modified form Bus-1 could provide the station with ongoing guidance, navigation, control and periodic orbital reboost. It also had the advantage that it was cleared for flight on the Shuttle. Adapting Bus-1, it was argued, would eliminate the development cost of the propulsive unit designed specifically for the station. In both cases, assembly would involve 16 Shuttle flights (one less than the nominal plan) at a rate of five launches per year with four milestones. The differences between this and the Baseline Configuration of Space Station Freedom were that the truss had been shortened (by at least three segments); the nodes had been eliminated (despite earlier safety concerns which had prompted their inclusion) and their systems placed in NASA's laboratory; the thermal radiators and power modules had been simplified and the 'alpha' joints had been deleted (i.e. power generation would be less than optimal); the mobile transporter had been deleted (i.e. the remote manipulator would be static); the regenerative environmental systems would be elided (although the operational overheads would be increased); the airlock would be simpler; and certain modifications would reduce the station's maintenance overheads significantly (particularly those involving spacewalks).

Beginning in October 1997, three Shuttle missions (one per month) would install a photovoltaic solar power module (whose transducers had recently been upgraded to deliver 23 kW) and three sites for attached payloads in order to create a 'power facility' to which a Shuttle with a Spacelab could dock for 'utilisation' flights lasting up to 20 days. The next assembly mission, in April 1998, would attach a core-cum-laboratory with support systems and five racks of scientific apparatus. It would incorporate berthing mechanisms for the logistics carrier and the international laboratories and a mount on which to install the remote manipulator. This would be the 'tended' milestone. Utilisation would continue at the rate of four missions per year. By December 1999, after eight flights, the international laboratories, the cupola and the second solar power module would be in place. The addition of the airlock, the habitation module, the third (and final) solar power module and a pair of Soyuz-based lifeboats would complete the station in September 2000.

Option 'B' was similar to the planned Baseline Configuration, and it preserved the incremental strategy for evolutionary growth. Also, because the design was mature, it would preclude the redesign costs that would result from making substantial changes. The savings would derive from a lower flight rate. Although assembly would start in October 1997 (as for 'A') the process would not be complete until December 2001. However, a dozen utilisation and logistics flights would be interspersed with the 20 assembly flights. The 'power facility' milestone of a photovoltaic solar power module, the starboard truss, one site for an attached payload, all the major subsystems – including the control moment gyroscopes, the orbital reboost propulsion modules, S-Band communications, and thermal regulation – would be achieved by the second mission. The next six flights would add a node, the airlock, the second solar power module, another two sites for attached payloads and NASA's laboratory with 16 racks, thereby achieving the 'tended' milestone by December 1998. It would take a further nine flights to complete the port truss with two more solar power modules and add the cupola, the remote manipulator and the Japanese and European laboratories to achieve the 'international' milestone by March 2001. The final three flights would add the second node, another three sites for attached payloads, the habitation module and the two Soyuz-based lifeboats to finish the station. Although the truss would be one segment shorter than the Baseline Configuration, the primary hardware changes were simplified data management, communications and tracking, environmental, thermal regulation and life support systems. It was clear that this was the option that NASA would prefer.

Option 'C' offered a 'radical' solution in which most of the station's systems would be integrated into a single large module that would be sent aloft as a one-off Shuttle configuration by decommissioning Columbia (the oldest vehicle in the fleet) and cannibalising its aft fuselage. Although NASA referred to this as a "single launch" option, a series of Shuttle flights would be required to add the international modules. Nevertheless, it was estimated that the station would be able to be assembled in only ten launches – which was how most of the saving would be achieved. The redesign would impose a significant delay (which would be costly) and assembly would not be able to start until September 1999. On the other hand, because the pace would be rapid, the station could be completed by early 2001.

The "core module" of this configuration was to be 6.6 metres in diameter (the same as Skylab) and 20 metres long (much longer than Skylab's main pressurised compartment). It would be able to house unpressurised apparatus in 3-metre-long bays at each end. The pressurised compartment, which would form seven "decks" and have 40 racks for scientific equipment built into it, would have a volume of 725 cubic metres. For launch, this module would be protected by a 28-metre-long aerodynamic shroud mounted on the aft fuselage of the decommissioned Shuttle orbiter. This combination would be mated with the External Tank in the usual manner (that is, by way of the aft fuselage, with a small strut near the top end). The ascent to orbit would be controlled by a specifically reprogrammed Shuttle general-purpose computer. The module would deploy solar panels. However, as these were to be on fixed mounts, solar-inertial attitude would be required for maximum power generation (as had been so on Skylab). Thermal regulation would be by both conformal and deployed radiators. In addition to the docking system for the Shuttle, there would be seven berthing mechanisms (two of which were for lifeboats). As there would be no airlock, spacewalks would be feasible only when a Shuttle was in attendance, via the airlock integrated into the docking assembly in the payload bay – a constraint that might prove serious in an emergency. Because it was essentially self-contained, the crew would be able to inhabit this station in November 1999 (once the Soyuz lifeboats were in place). In its finished configuration, the station would incorporate 72 racks of scientific equipment, and although this would represent a potent microgravity research facility, it would be very difficult for shifts of pairs of astronauts to make effective use of so much equipment. In essence, this idea derived from a study performed by Andy Petro in June 1986 in the aftermath of the loss of the Challenger, but in the originally envisaged form it would have had a truss carrying large solar panels to power all the racks of apparatus. Although Option 'C' may have been the easiest to fabricate, integrate and launch, the ascent would not have been without risk. Significantly, this configuration did not offer evolutionary growth. Furthermore, unless funding was eventually awarded to reinstate the cannibalised Columbia, it would have denied NASA an operational spacecraft.

"Each option is capable of accomplishing the mission of the space station," Goldin assured. "All of them offer significant scientific and engineering research capabilities". Where doing so would be cost-effective and schedule-enhancing, components which had passed the Space Station Freedom Critical Design Review were to be used, thereby exploiting the investment to-date.

'ALPHA'

On 17 June 1993, Clinton opted for a slightly modified form of Option 'A', revised to trim costs even more by pushing the 'permanent habitability' milestone two years further downstream than the Freedom schedule. Unimpressed, Representative Tim Roemer, a long-standing critic, argued in the House for a cancellation but his motion was defeated on 23 June by the narrowest of margins – a *single* vote. Clearly, the station was hanging by a thread.

Clinton gave NASA three months to develop a specification for his chosen option. A Transition Team was formed by O'Connor, but the day-to-day work was managed by astronaut William Shepherd. One key recommendation by the Redesign Team had been to simplify the programme's management structure. The multiple work packages had proved to be overly complicated, so Boeing was nominated as the prime contractor on 17 August and all programme implementation responsibilities were assigned to Houston. In recognition of the new reality, Reagan's evocative 'Freedom' label was discarded. Having begun the year steering Freedom into its final Critical Design Review, NASA now found itself developing a specification for a new venture called 'Alpha'. On 3 September, having supervised the station since 1989, Richard Kohrs retired. "Dick built a strong and effective program organisation which stood the test, despite the continuing annual cycles of reassessment and restructure," Aldrich noted. "With each review, it became a stronger, and more effectively focused program."

A NEW PARTNER

On 7 September, NASA submitted its 'Alpha' Programme Implementation Plan to the White House, but its technical efforts had been overtaken by international events. After two days of talks in Washington – designed to increase the scope of the existing agreement on human spaceflight cooperation – Vice President Al Gore and Russian Prime Minister Viktor Chernomyrdin had signed an accord on 2 September calling for their respective space station plans to be *merged* into a single facility in which four Russian pressurised modules would augment those from NASA, Europe and Japan. NASA promptly decided to replace the military Bus-1 of Option 'A' with a Russian manoeuvring unit. Initially dubbed the "Salyut tug", this TKS-class vehicle offered NASA the significant advantage that it would be able to provide power as well as propulsion, guidance and navigation during the early stage of assembly. Adopting proved Russian systems offered other advantages. In particular, Progress tankers would be able to resupply the tug, which would reduce the station's operating costs as Bus-1 had not been designed to be replenished. Furthermore, while Congress had obliged NASA to push the habitation module so far down the manifest that 'permanent habitability' would not be achieved until the station was operational, the Russian strategy was to launch the habitat first. Docking a Mir-style vehicle at the opposite end of the tug would enable the crew to take up residence *early* in the assembly sequence. Although it would take just as long to assemble, the station would be more productive, which was a striking unification of politics and technology. As the details were thrashed out, the tug became the 'Functional Cargo Block' (or using the Russian acronym, the FGB) and the Mir-style base block became the 'Service Module'. In acknowledging the crucial role of the Russian modules, NASA referred to them as the "core modules". The agency took advantage of this revision to reintroduce some of the systems that it had so recently elided. In this "add back" exercise the nodes and 'alpha' joints were reinstated, thereby restoring the laboratory's focus and enabling the solar panels to track the Sun. By the conclusion of this review, the 'unified' design had utilised 75 per cent of the hardware in the 1992 Freedom configuration.

When NASA had submitted its 'Alpha' plan to the White House, it had omitted the costs and schedule. On 20 September, Goldin reported that if the annual budget of $2.1 billion was assured then it would be possible to launch the first element in 1998 and complete the station by 2003, at a cost $19.4 billion. In comparison to Freedom, NASA estimated that this would save $18 billion over the 10-year life of the programme and (of crucial importance to Clinton's pledge to reduce the deficit) cut $4 billion over the Fiscal Years 1994 to 1998. However, Goldin emphasised that the costing was preliminary. Spurred on by the prospect of involving the Russians, Gore ventured that 'Alpha' was "comparable to, or better than, Space Station Freedom, but will cost significantly less". But Arkansas' Democratic Senator Dale Bumpers, a long-standing critic, was dismissive of the prediction and called for this latest plan to be killed. South Carolina's Democratic Senator Ernest Hollings was reluctant: "I hate to have to vote against this, but we can't afford it." Nevertheless, Bumpers' amendment was convincingly rejected. On 6 October the House compared the 'Alpha' Programme Implementation Plan with what was likely to result from cooperating with the Russians, and declared the "United States – International Partners – Russia" plan to be the clear winner. For once, the sheer cost was not the primary issue. As *competition* in space had symbolised the Cold War, *unification* in space expressed the new spirit of cooperation perfectly. In the larger scheme of things Clinton was keen to offer support to Yeltsin, who was eagerly 'westernising' his newly independent country's political and economic systems. Another political consideration was that by involving Russia's rocketry experts in a peaceful project, America would be able to discourage them from offering their considerable expertise to Third World countries eager to develop or upgrade their own ballistic missile programmes. In effect, therefore, the station had become a central plank in the Clinton administration's foreign policy and an important lever for the US Arms Control and Disarmament Agency. Of course, cost was still important, because reducing the deficit was still a national priority, but the programme would now serve a higher purpose.

TROUBLES FOR ESA

In mid-October 1993, ESA announced that as part of a $3.4 billion reduction in its programme running through to the end of the decade, it would discontinue the development of the Hermes space plane. It would, instead, redirect its human spaceflight effort to revive the Crew Transfer Vehicle that had been under study periodically for a decade. The four-seat vehicle would be launched by an Ariane 5 rocket and would ferry people to the space station. ESA also decided to scale down the Columbus laboratory, thus enabling it to be launched on an Ariane 5. An Automated Transfer Vehicle would serve as the upper stage of the launch vehicle, and deliver Columbus to the station. These changes represented a redirection of the agency's long-term strategy, which had previously been geared towards achieving independence. In effect, the descoped 10-tonne laboratory would be little more than a Spacelab to be permanently attached to the station.

THREE PHASES

With political commitment to the station finally in position, NASA moved rapidly. In October, it brought the Shuttle and station programmes back under a single associate administrator, Jeremiah ('Jed') Pearson,[1] O'Connor was appointed acting director of the station in Washington and Shepherd became the Station Program Manager in Houston. On 1 November the 'unified' Implementation Plan was submitted. It had three 'phases'. Phase One would build upon the agreement to fly one anothers' nationals by calling for a series of joint missions in 1995–1997 during which the Shuttle would visit Mir. In effect, NASA would sign up as one of Mir's fee-paying clients. A succession of astronauts would be able to serve tours aboard Mir in return for $400 million (subsequently raised to $482 million), and this would enable NASA to make an early start on its microgravity research programme. Furthermore, as the agency pointed out, the proximity operations and dockings would provide experience in operating the Shuttle in conjunction with a station. This agreement was historic because it was the first time that NASA had been permitted to enter into an "exchange of funds" with an international partner. Previously, it had had to trade flight opportunities in return for "donated" hardware. In Phase Two, sufficient elements of the station would be launched to enable a crew to take up residence. This milestone was, in effect, the commissioning of NASA's laboratory. Construction was to start in May 1997 with the dispatch of the FGB on a Proton rocket; NASA would add a node in July; the laboratory would be delivered by the third Shuttle, and be outfitted by the end of the year; and the Service Module would make the station habitable. Phase Three would take the assembly through to completion in October 2001, with Japan's laboratory arriving in October 1999 and ESA's Columbus (by an Ariane 5 rocket) in April 2000. "This is it," Goldin announced, when the plan was revealed. "Give us approval and we're going to go build a space station." In fact, the document was only the basis for a new round of negotiations between the international partners and it would take months to resolve the technical and political issues. Visiting Washington in early November, Koptev

---

[1] When Pearson resigned in November 1994, he was succeeded by NASA's chief engineer, Wayne Littles.

was emphatic that Russia had a great deal to offer: "Russian stations have been orbiting the Earth for 22 years." Representative George Brown of California, the Democratic chairman of the House's Science, Space and Technology committee, agreed, and was pessimistic of congressional support: "I actually have more confidence in the ability of the Russians to fulfil their part of the deal than I do in the US."

On 16 November, Goldin warned that NASA was "in chaos". With continuous reviews and budget cuts it was impossible to draw up a strategic plan. The agency was having to revise its five-year budget several times per year. "When you cannot plan, you cannot perform." Worse, the agency could not "develop a vision". At a meeting in the White House on 29 November, the congressional leadership assured Clinton of their support because it was clear that it had become the linchpin of the administration's post-Cold War policy towards the former Soviet republics and, as such, it would serve as a means of increasing business relationships with Russia. The White House's overtures to the Russians – and NASA's eager acceptance – had taken the other international partners by surprise. On 7 November they met in Montreal, Canada, to discuss the *fait accompli*. On 29 November they endorsed what had become 'International Space Station Alpha'. On 7 December Russia was formally invited to become a full programme partner, and Gore and Chernomyrdin signed the 'letter contract' in Moscow on 16 December. By the end of the year, NASA had determined that the Russian hardware would mean 25 per cent more habitable volume, 42 kW more power and (much welcomed by the international partners) an increase in the crew complement from four to six. However, if the station was to be accessible to the Russians (whose launches from the Baikonur cosmodrome in Kazakhstan were constrained by China's presence) it would have to be in an orbit inclined 51.6 degrees from the equator. NASA had planned on 28.5 degrees, as this, being due east from Cape Canaveral, was the most 'efficient' azimuth. The higher inclination, however, would significantly erode the Shuttle's lifting capacity, and in the short term it would be necessary to revise the assembly sequence. At a later stage, the development of a 'lightweight' External Tank might restore some of the capacity lost by the cancellation of the Advanced Solid Rocket Motor.[2]

With STS-61's spectacularly successful *in situ* repair of the Hubble Space Telescope in December, NASA proved to Congress and to the American people that spacewalkers were indeed capable of performing delicate work – a fact that was fairly obvious as cosmonauts had already undertaken construction work outside Mir. The plan for spacewalkers to connect elements of the space station was therefore accepted as being realistic.

---

[2] In mid-March 1994, the Marshall Space Flight Center received the go-ahead to develop the Shuttle's $172 million Lightweight External Tank. The use of aluminium–lithium alloys and ortho-grid panels would shave 3.6 tonnes off the structure, a mass which could either be assigned to additional payload to a given orbit or to lift a given payload to a more 'costly' orbit.

USING MIR

As 1994 dawned, NASA and its Russian counterpart approved the details of the Shuttle–Mir programme that would form Phase One. Docking would not be the objective of the first joint mission, however, which would simply demonstrate a close rendezvous. This task was assigned to STS-63, a mission with a free-flyer and a Spacehab that had suffered a prolonged delay while the company secured sufficient clients to justify flying its module. In the meantime, Sergei Krikalev (the veteran cosmonaut who had feared being sold to NASA as a Mir maintenance man) flew as a mission specialist on STS-60 in February 1994 under the agreement signed by Bush and Gorbachev in 1991. Vladimir Titov, Krikalev's backup, was added to STS-63 in order to supervise Mir proximity operations in February 1995. In March, an astronaut was to ride a Soyuz to Mir. If all went well he would be retrieved by STS-71 in June, when it made the first docking. To facilitate NASA's research aboard Mir, Russia agreed to finish its long-delayed modules, but in this case fitted with a substantial amount of American apparatus. At the conclusion of the Shuttle–Mir programme in 1997, Mir was to be "mothballed" to enable Russia to devote its efforts to the new station. However, not everyone welcomed the introduction of the Shuttle–Mir programme. The Space Studies Board of the National Academy of Sciences complained that three microgravity and four life sciences Spacelab flights had been commandeered for the sole purpose of ferrying missions for Mir.

In January, Spacehab Incorporated proposed using its mid-deck augmentation module as a logistics carrier for the Shuttle–Mir programme. Such a contract would boost its rather slow and "lumpy" revenue from commercial microgravity experiments which, contrary to early expectation, were proving to be scarce. Furthermore, as these missions would be unlikely to carry free-flyers, it would be possible to relocate the module over the vehicle's centre of mass instead of at the front, thus facilitating an increased payload. The prospect of a contract of up to 10 flights offered the company a 'lifeline' as this would provide *real* income. Most of the capacity so far had been used by NASA (if indirectly through its Centers for Commercial Development of Space) and the company had been operating a "zero-sum deal" in which NASA's payment for using the module exactly balanced the fee that the agency charged for carrying it. The proposal for the cargo-carrier contract demonstrated that the company was willing to adapt to commercial realities and turn its asset – a pressurised module with a flexible interior – to other uses.[3]

THE SAME OLD, SAME OLD

In his effort to reduce the federal deficit President Clinton proposed the first cut in

---

[3] Spacehab Inc. was awarded a $54 million contract in August 1995 to use its mid-deck augmentation module to haul cargo to Mir. NASA selected Spacehab, rather than Spacelab, because it would more readily accommodate bulk cargo and could be 'turned around' more rapidly. By offering to build a 'double' module, the company made its case unassailable.

NASA's budget for two decades. If inflation was taken into account, the budget of $14.3 billion for Fiscal Year 1995 was a $650 million reduction. Furthermore, this level was to be held through to the end of the decade. From now on, the Shuttle, the station, and the cost of cooperating with the Russians on Mir were to be accounted for under the umbrella of 'human spaceflight', which would be itemised in the annual budget. "This is it," Goldin warned, "we cannot cut any closer to the bone." Representative George Brown, a station supporter, was worried that if the station exceeded its annual $2.1 billion (which was likely) it would become "predatory". Representative Tim Roemer dismissed the station as a "weed in NASA's garden" that was "choking out good science projects".

In February 1994, the 'preliminary' schedule included with the Implementation Plan submitted in November was revised. The first Shuttle launch was in December 1997 instead of July 1997 and the completion date had slipped from October 2001 to June 2002. Critics in Congress were soon highlighting the erosion of NASA's early estimate that the Russian hardware would save $2 billion and advance completion by two years.[4] In defence, space station director Wilbur Trafton pointed out that the revised time-saving of 15 months was not due to problems in integrating the Russian systems but to the need to satisfy the station's $2.1 billion annual cap, whose impact was greatest in the early years. The *real* threats, warned programme manager Randy Brinkley, were both national and international politics. Representative George Brown urged NASA to develop a contingency plan to reduce the station's reliance on Russian hardware, should the Russians fail to deliver. James Sensenbrenner, the senior Republican on the House's space committee, warned that as long as there was Russian hardware in the station's 'critical path', he would oppose it. "I want us to be a partner," Brown said, "but not *dependent* on Russia." NASA responded by saying that if the funding was appropriated it would purchase the FGB directly from its manufacturer. If this was not possible, or the Russian government failed to fulfil its commitment, then the Freedom propulsion module or Lockheed's Bus-1 could be developed as a replacement. The exposure in the critical path was the Service Module, which the Russians were to develop as their principal contribution to the station. NASA had no alternative. In the longer term, the amount of hardware Russia contributed was really immaterial because the configuration of the rest of the station was fixed.

Meanwhile, NASA continued to define the contract that would specify Russia's role in the programme. It was important that the Russian Space Agency should not be seen as a subcontractor, and that the cosmonauts were not viewed as foreign guests. Although NASA regarded itself as the senior partner, it was obliged to acknowledge that its new partner had two decades of experience in assembling and operating space stations. Russian sceptics argued that once Russia had handed over its unique experience "as a free gift" the cosmonauts would be "passengers hitching rides on American Shuttles". A few months later, in response to its own funding crisis, the Russian Space Agency shed half of the cosmonaut corps. In fact, since the

[4] The $2 billion 'saving' from involving the Russians was by cutting the $19.4 billion budget to $17.4 billion, and the two-year saving was from 2003 to 2001.

collapse of the Soviet Union, Russia's space funding had declined steeply: its budget in 1994 was barely 10 per cent of the 1989 level. It was clear that Russia was unable to sustain the momentum of the massive Soviet space programme. Its space industry was in decline and the prospect for rocket manufacturers seemed bleak. Energiya's new Mir programme director, the former cosmonaut, Valeri Ryumin, warned that irrespective of Russia's commitment to the forthcoming Shuttle–Mir flights, Mir "may have to be evacuated". Little wonder that some in Congress had doubts concerning Russia's ability to meet its commitments for the 'International Space Station' (as it was now being called).

In considering the likelihood of exceeding the $17.4 billion overall budget, Sensenbrenner was ironic. "When Freedom was cancelled, we had about $20 billion to go on that; what have we gained?" In early March, Brown, who had played a pivotal role in safeguarding the station in 1993, said that this year he would vote against the appropriation. Clinton's intention was to hold the agency's budget essentially flat, at around $14.3 billion. Taking inflation into account, this was really a year-on-year cut. "If NASA's budget keeps declining," Brown insisted, "you certainly cannot have a space station." In fact, if its budget fell year-on-year, NASA would have to consider a cessation of human spaceflight. The General Accounting Office was critical of NASA's most recent five-year plan for presuming "an unrealistically high level of funding". Democratic Senator John Rockerfeller of West Virginia was the chairman of the Commerce Committee's Science, Technology and Space panel, and pointed out that "if you get below a certain funding you cannot do everything". In fact, NASA, he insisted, was different from other federal agencies and was "embedded in the psyche of this nation". It was unthinkable that the nation would "walk away" from the agency that had done so much to establish America's technological leadership.

On 24 March, the Congressional Budget Office criticised the agency for trying to maintain a programme of space exploration in the face of annual budget cuts and posed three alternatives:

- concentrate on human spaceflight, build the station and plan for future human exploration, but at the expense of cutting robotic exploration severely;
- cancel the joint missions to Mir and the station, abandon all hope of returning to the Moon to initiate an expedition to Mars, fly no more than four Shuttles per year for appropriate missions, and switch the emphasis to robotic spaceflight;
- eliminate human spaceflight in favour of using robotic vehicles.

Goldin countered that to accept *any* of these options would "destroy the dream President Kennedy began more than 30 years ago".

The Systems Design Review in March was the "most important technical milestone" of the programme to date, said Brinkley, "where we move from concepts to hardware implementation". If a major technical problem had occurred it would have been a serious embarrassment, but NASA was convinced that the programme was not exposed to any 'show stoppers'. The report was first sent to the White House, then forwarded to Charles Vest for assessment. The Congressional hearings were due in mid-April, and although NASA and Boeing had yet to agree costs, Brinkley was confident that the 20 per cent reserve incorporated into the $17.4 billion budget (excluding launch costs) would be adequate. However, by April the 'permanent habitability' milestone had slipped a year beyond the November estimate and it seemed that half of the saving would be devoured by cost overruns. Phillip Culbertson agreed that inviting Russia to become a partner had been "a good political move", but warned that "anybody who thinks that that is going to save us a lot of money, or schedule, or complexity, just doesn't understand". In fact, the General Accounting Office suggested that because hardware would have to be made to interface with and support the Russian-built core modules, the involvement of the Russians would *increase* US costs by $1.4 billion.

Despite having White House support, Goldin faced an uphill struggle to secure the requisite funding. On 13 April he told the House, "NASA cannot handle the punishment that we're going through." The station was under annual assault. It was a multi-year programme, but no matter what was agreed one year it was undermined the next. "We have got to have stable funding, and some kind of agreement about what we are going to do." A motion in the House introduced by Representatives Tim Roemer and Dick Zimmer on 29 June to cancel the station and to reallocate its funding to more deserving NASA programmes was soundly defeated. Intense lobbying by the White House had set the tone. Gore was optimistic: "The strength of the vote [278 to 155] signals the end of doubt about America's commitment to space exploration." After several months of expressing concern that the station would be pursued at the expense of other programmes which, in their own way, were just as worthy, Brown had eventually pledged his support; and Sensenbrenner had also announced that he was satisfied with contingency planning to ensure that the programme would not be delayed if the Russians failed to fulfil their commitment. "If the Russians withdraw, it will have absolutely minimal impact," assured Trafton. The agency was nevertheless considering contingencies in case the Russian-funded Service Module was delayed, and on 12 July the assembly sequence was revised to bring forward NASA's first solar power module to back-stop the FGB's 8 kW. On the other hand, the launch of the laboratory was slipped by another five months to November 1998.

On 3 August, the Senate rejected a motion by Dale Bumpers to deny funding to the station, after he had suggested that the true cost may well eventually exceed $100 billion. Senator Barbara Mikulski, the Maryland Democrat who chaired the appropriations subcommittee that provided NASA's funding, countered that $40 billion was the highest fair number conceivable, and this took into account *all* costs since the programme's inception in 1984 – *including* civil service salaries – of the development and assembly of the Baseline Configuration and then a decade of operations. Senator Paul Simon, an Illinois Democrat, opined that the station had genuine merit, but was just not worth the cost. Senator Phil Gramm, a Texas Republican and an ardent station supporter, caught the growing frustration with the annual funding battle: "Let's settle this issue once and for all!" Senator John Rockerfeller was also won over to it and mused: "We are no longer discussing a programme mired in confusing agendas." As Rockerfeller observed, Goldin "is abrasive, abrupt, fairly curt and almost always right."

The station, according to Goldin, was no longer "a design or a dream, we're building hardware", and by early September he was able to report that the scope of the work and the schedule of Boeing's prime contractorship had been agreed, and that the contract would be finalised by the end of the year.

Although the programme was at last developing the necessary momentum, the station's scientific mission was no better defined than it had been the year before. On 27 December, the General Accounting Office opined that NASA's budget had been so taxed by the cost of developing the station that it would be able to fund only one-third of the research staff that would be required to pursue the scientific programme. Of course, it was ironic that it was Congress that had imposed these financial constraints upon the agency.

As 1995 dawned, NASA was looking forward to a re-enactment of the historic 'handshake in orbit' which had marked its first joint mission with the Russians two decades previously.

11

Shuttle–Mir

History was made on 3 February 1994 when Sergei Krikalev was launched aboard STS-60 Discovery. It was fitting that "the last Soviet citizen" had more experience living and working in space than all of his astronaut colleagues combined. As a step towards having a Shuttle dock with Mir, Vladimir Titov, Krikalev's backup, was assigned to STS-63 to supervise proximity operations during a rehearsal rendezvous with the orbital complex in early 1995.

Meanwhile, 8 January 1994 saw Valeri Poliakov's return to Mir on Soyuz-TM 18, with the objective of serving with a succession of crews to establish a new endurance record. When Soyuz-TM 19 docked at the rear on 3 July, Poliakov welcomed Yuri Malenchenko and Talget Musabayev aboard. Mir's routine was disturbed on 27 August when Progress-M 24 aborted its approach just 10 metres from the front port and narrowly missed a solar panel as it drifted past. On the second attempt several days later, it turned at the last second and repeatedly bounced off the port. On 2 September Malenchenko used the newly installed TORU system to fly the errant ferry in by remote control. Without this resupply, the crew may well have had to evacuate the station, pre-empting Poliakov's marathon. One week later, Malenchenko and Musabayev inspected the forward Kurs system during a spacewalk, but found no sign of damage, prompting the belief that the fault must have been in the ferry. However, when Soyuz-TM 20 closed in on 6 October its Kurs system also failed, so Alexander Viktorenko docked it manually. The fault evidently lay in Mir's Kurs system and the engineers immediately started reanalysing the telemetry to figure out why a docking system that had worked flawlessly for years should now fail. It was important that they rectify this problem because without an operational forward Kurs system it would not be possible to resume the expansion of the complex. Soyuz-TM 20 also delivered Yelena Kondakova (the wife of Valeri Ryumin) who was to remain aboard as Mir's first female flight engineer, and Ulf Merbold, who was on the first of ESA's commercial visits, in this case lasting a month. By the turn of the year, it was found that the flight control systems of Progress-M 24 and Soyuz-TM 20 had been improperly programmed with their

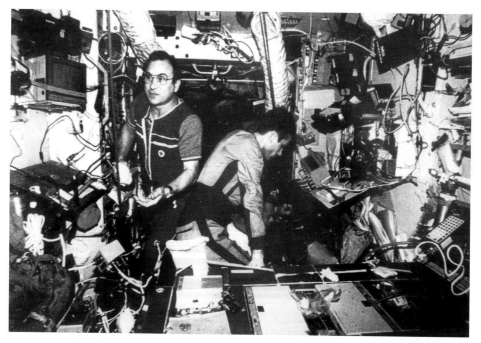

Dr Valeri Poliakov (left) during his 14-month marathon aboard Mir. Note the cluttered state of the 'base block'.

centre of gravity and this had induced pitching and yawing deviations in performing the minor attitude adjustments immediately before docking. This software had not been active when the vehicles were manoeuvred manually or by TORU. To verify the reprogramming, Soyuz-TM 20 undocked on 11 January, withdrew 160 metres and then its Kurs redocked without incident.

HISTORIC RENDEZVOUS

On 6 February 1995, STS-63 approached Mir from below and, with its nose high and its payload bay facing the station, took up position directly ahead. Mir was then reoriented so that Kristall's axis was aligned along the orbital complex's velocity vector to aim its docking port at the newcomer. Vladimir Titov on board Discovery maintained a VHF radio link with his colleagues on Mir while Jim Wetherbee slowly reduced the range to 10 metres. This 'Near-Mir' rendezvous set the scene for the first docking scheduled for the summer. It demonstrated that a Shuttle could approach Kristall's axial port within an 8-degree cone and a 2-degree tolerance in the relative orientation. "When all was said and done," Wetherbee reflected, "it turned out to be easy." Upon withdrawing, Discovery performed a slow fly-around to enable the IMAX camera in its bay to document the station's condition.

A view or Mir from Discovery during the STS-63 'Near-Mir' mission.

NASA MOVES IN

The arrival of Soyuz-TM 21 on 16 March was shown live by the main television networks in Russia and America because it carried Norman Thagard, the first NASA astronaut to visit Mir. Unlike previous visitors, Thagard's time was not restricted to the crew handover; he had already made four brief Shuttle flights and was now to serve a full tour. As a physician, his programme was predominantly biomedical: he was not only to study all the standard aspects of adaptation to weightlessness but was also to take regular blood and urine samples to monitor how his body changed on a continuous basis, to facilitate an understanding of the processes as well as the outcomes. In the base block, the Russian and American national flags were pinned to the wall to symbolise the new spirit of cooperation between the formerly competing space-faring nations. Upon departing on 22 March, Poliakov had extended the endurance to 438 days.

Vladimir Dezhurov and Gennadi Strekalov were to be relieved by cosmonauts delivered by Atlantis in the summer and return with Thagard on the Shuttle. The timing was planned so that these two resident crews would each serve a three-month tour and the Soyuz that had launched the first crew could be used to return their successors at the end of its six-month orbital life. However, as Thagard could not complete his programme until the next large module ('Spektr') was delivered, the planning had to be flexible, which was something that NASA found frustrating. If it appeared that Spektr's arrival would slip by only a few weeks, then Atlantis's launch would be delayed by that amount to give Thagard time to finish his work; but if the module was going to be significantly late they were to extend their tour to six months and return in their own ship after the next handover in August. Dezhurov and Strekalov's primary objective was to upgrade the base block's systems to accommodate the two-year Shuttle–Mir programme. Having been deemed a failure, the shower in Kvant 2 had some time previously been converted into a steam room, which was rather more successful. As part of the upgrade, they dispensed with the plumbing and commandeered the cubicle to house additional gyrodynes that were mounted internally for easier maintenance.

RECONFIGURING MIR

Progress-M 27's arrival on 12 April confirmed that the vital front port was operational. With the launch of Spektr imminent, Kristall had to be relocated, but this could not be done until its solar panels had been retracted. In the first week of May, which was devoted to preparations, Strekalov injured his arm on a metal spur. Thagard applied the available medication and the inflammation soon eased. On 12 May, their first task was to retract a panel into its container to enable the assembly to be disconnected from its drive motor. However, the retraction stalled and so Thagard, inside Mir, had to cycle the mechanism a few times to complete the process. Although they had hoped to complete the transfer on a single spacewalk, they had to cease after dismounting the retracted panel. They returned five days later and swung the bulky 500-kilogram box on the crane across to Kvant 1, but then encountered difficulties in mating it with the motor which had been installed two years earlier. On their third excursion, on 22 May, they laid out the cables and plugged the panel into Mir's power system. The panel that they moved had projected to the *left* of Kristall and *had* to be retracted to allow Kristall to be swung to the right-hand port – its assigned position in the finished orbital complex. It had been designed to be retractable and detachable specifically so that it could be moved, and the first crane had been mounted on the *left* side of the base block to facilitate its relocation. Kristall's second panel jammed with about 25 per cent of its length still exposed, but as it would not interfere with the Shuttle it was allowed to remain in this condition. It would not be able to be relocated until a crane was affixed to the mount on that side of the base block.

Progress-M 27's departure early on 23 May released the front port for Spektr, which had been launched three days earlier. Two new batteries were installed in Kristall to ensure that it would have power to operate its robotic arm for the transfer onto the axial port – a manoeuvre that was made on 27 May. The next day, Dezhurov and Strekalov sealed Kvant 2 to conduct "an internal spacewalk" (that is, they depressurised the multiple docking adaptor) and moved the second drogue from the lower port to the right-hand radial port, onto which Kristall was transferred two days later. After a long, slow, propellant-efficient rendezvous, Spektr docked on 1 June. The next day, Dezhurov and Strekalov switched the other drogue back onto the lower port, and Spektr was swung down. The final preparation for the arrival of Atlantis was to swing Kristall back onto the axial port to ensure that its androgynous port was well clear of any projections that might inhibit the Shuttle's freedom of movement. The drogue was shifted to the right-hand port on 10 June and the final transfer was made. Mir was now ready to accept its first Shuttle.

By the time that Atlantis arrived during the STS-71 mission, the Spektr module had arrived and taken Kristall's place, and Kristall had been swung onto the axis to accept NASA's Shuttle – just as had been planned for Buran.

A CASE OF OVERCROWDING

On 29 June, Robert 'Hoot' Gibson stationed Atlantis directly beneath Mir and slowly climbed to close in on Kristall's port. In making this 'R-bar' approach, the Shuttle had to fire its thrusters continually to rise. If its thrusters failed, the gravity gradient would act as a brake, so this line of approach offered a contingency against a collision. Atlantis had an external airlock set near the front of its payload bay, topped by an androgynous docking system that had been supplied by the Russians. Once the extended ring on Kristall's port had been engaged for a soft docking, Atlantis fired its thrusters to drive the collars together for a hard docking. When the Orbiter Docking System had been pressurised, Gibson made his way through its compartments to the top hatch, and then Gibson and Dezhurov shook hands in the narrow tunnel. In Kaliningrad, Yuri Koptev and Dan Goldin savoured the moment. The two crews then congregated in the base block for a 'photo opportunity' which demonstrated that Mir had not been built to accommodate *10* people. The two crews reconvened in Atlantis's

Spacelab the following morning and after a brief ceremony they began operations. Dezhurov, Strekalov and Thagard, who were well advanced with their physiological preparations for their return to Earth, were subjected to a battery of medical tests using the full panoply of the life-sciences kit. Even with an extended-duration pallet in its bay, a Shuttle is limited to two weeks in orbit. Only a permanent orbital complex could facilitate total adaptation to the space environment, Mir was the only such facility, and only by docking a Shuttle with Mir could NASA's sophisticated biomedical systems be applied to space-adapted cases. The political symbolism of this mission was rich, but it also yielded tangible scientific results. Just before Atlantis undocked on 4 July, the new crew of Anatoli Solovyov and Nikolai Budarin undocked Soyuz-TM 21 and flew around to the side of the complex to capture for posterity a photograph of Mir with an American Shuttle attached.[1]

Bonnie Dunbar, Thagard's backup, who was on Atlantis's crew, had originally hoped to stay on Mir with Solovyov and Budarin to continue NASA's programme and be relieved by the next astronaut upon Atlantis's return later in the year, but ESA's second tour had to be squeezed in before NASA could take up a continuous block of time during which other vistors would be restricted to the brief periods of the Soyuz crew exchanges. The demand for access to Mir was so great that it was a problem to fit everyone in.

Table 11.1 Soviet space station launches and re-entries

Space Station	Launched	Reentered
Salyut 1	19 Apr 1971	11 Oct 1971
Salyut 2	3 Apr 1973	28 May 1973
Cosmos 557	11 May 1973	22 May 1973
Salyut 3	25 Jun 1974	24 Jan 1975
Salyut 4	26 Dec 1974	3 Feb 1977
Salyut 5	22 Jun 1976	8 Aug 1977
Salyut 6	29 Sep 1977	29 Jul 1982
Salyut 7	19 Apr 1982	2 Feb 1991
Mir	20 Feb 1986	23 Mar 2001
Kvant 1	31 Mar 1987	23 Mar 2001
Kvant 2	26 Nov 1989	23 Mar 2001
Kristall	31 May 1990	23 Mar 2001
Spektr	20 May 1995	23 Mar 2001
Priroda	23 Apr 1996	23 Mar 2001

COMPLETING MIR

On 17 July Kristall was swung back onto the right-hand port, where it would remain. On Atlantis's next visit on 15 November, it carried a Russian-built Docking Module

---

[1] This was the only time that a Shuttle was photographed from such a perspective while docked.

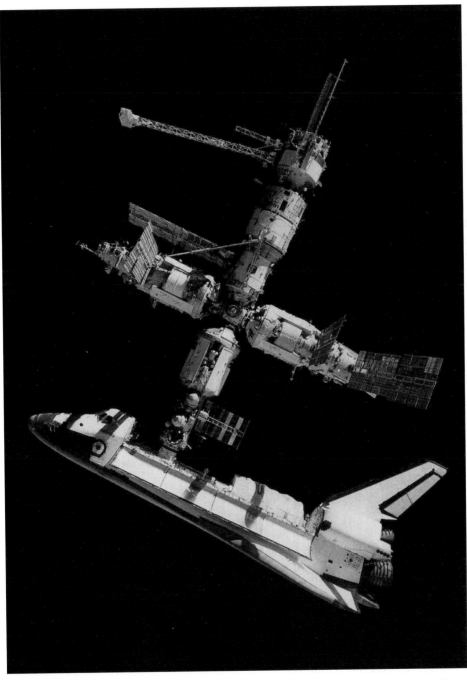

Before Atlantis departed, the crew of Soyuz-TM 21 undocked and flew around to the side to record the historic link up.

The addition of the Priroda module in April 1996 finally completed the Mir complex. The Docking Module delivered by STS-74 is on the end of Kristall. The boxy object projecting from Kvant 2 is the YMK backpack, which had been strapped outside in order to make room in the airlock.

in its bay. Once this had been mounted on top of the Orbiter Docking System, Atlantis closed in and docked with Kristall. Although the stubby module provided vital clearance, the Shuttle had to manoeuvre very close to Mir's right-hand solar panel. The docking was further complicated by the fact that Ken Cameron could not directly see the docking system, so the remote manipulator was positioned so that its TV camera provided a perspective view. This was NASA's first experience of adding a module to a space station. When it undocked, Atlantis left the Docking Module behind.

Table 11.2 Shuttle–Mir missions

Mission	Launched	
STS 63 Discovery	3 Feb 1995	Rendezvous
STS-71 Atlantis	27 Jun 1995	Docking
STS-74 Atlantis	12 Nov 1995	Docking
STS-76 Atlantis	22 Mar 1996	Docking
STS-79 Atlantis	16 Sep 1996	Docking
STS-81 Atlantis	12 Jan 1997	Docking
STS-84 Atlantis	15 May 1997	Docking
STS-86 Atlantis	25 Sep 1997	Docking
STS-89 Endeavour	22 Jan 1998	Docking
STS-91 Discovery	2 Jun 1998	Docking

LEARNING LESSONS

Starting with STS-76 in March 1996, six astronauts served a succession of tours aboard Mir, accumulating over 800 days in space. If Thagard's visit was included, then Mir enabled NASA to clock up almost 1,000 days in space, which exceeded the aggregate of all previous Shuttle missions.

The arrival on 26 April 1996 of the Proton-launched 'Priroda' module, which was transferred to the left-hand port, brought the remainder of the apparatus for NASA's programme and completed the Mir complex fully a decade after its assembly had begun.

By and large, the NASA tours aboard Mir were fairly routine, but there were several life-threating incidents. In the first, on 24 February 1997, during both Jerry Linenger's tour and the handover accompanied by ESA astronaut Reinhold Ewald, an oxygen generator in Kvant 1 malfunctioned and ignited, and because it provided its own oxidiser it burned like a blowtorch for some 15 minutes and filled the station with dense smoke. In retrospect, it was particularly concerning that the location of the inferno had blocked access to the Soyuz at the rear. Had it been necessary to abandon the station, the double crew of six would not have been able to return to Earth in the only Soyuz available. It was also of major concern to NASA that the Russians had firstly been slow to inform them of the fire, and had then argued that it had not really been a serious problem because the fire had burned itself out and the station could easily be cleaned up.

The other near disaster occurred on 25 June of the same year, during Michael Foale's tour, when, as NASA saw it, the Russians conducted an irresponsible experiment that put the station (and the lives of its crew) in great danger. On 4 March, after being discarded, Progress-M 33 had been flown back to Mir, not by its Kurs, but by the use of the TORU remote-control system that had been designed solely to fly a Progress in from its 200-metre 'holding point'. The Kurs hardware was supplied by the Ukraine, and with soaring prices the Russians were considering the alternative of having crews take control of Progress ferries much earlier. During the experiment, the Progress-M 33 test had to be abandoned early when the TORU system malfunctioned, leaving the ferry to make an uncontrolled close fly-by of the orbital complex. When Vasili Tsibiliev tried the experiment again in June using Progress-M 34, he could not gain a clear sense of 'situational awareness', lost control, and the ferry, which came in too rapidly, not only hit but punctured the Spektr module. Although the crew managed to install a hatch to isolate the damaged module, the station was thrown into a power crisis from which it took several weeks to recover. NASA was not only astonished by the risk that the Russians had taken but was also shocked that the nature of the experiment had not been explained in advance. Critics in Congress demanded that NASA terminate the Shuttle–Mir programme immediately, but Goldin insisted that Mir was inherently safe, and that the programme should continue, which it duly did.

"The Shuttle–Mir programme has been very useful in giving our astronauts good training in crisis management," wryly observed Representative James Sensenbrenner, the senior Republican on the House's space committee.

Table 11.3 Shuttle–Mir astronauts

Astronaut	Days in space	Up	Down
Norman Thagard	115	Soyuz-TM 21	STS-71
Shannon Lucid	188	STS-76	STS-79
John Blaha	128	STS-79	STS/81
Jerry Linenger	132	STS-81	STS-84
Michael Foale	145	STS-84	STS-86
David Wolf	128	STS-86	STS-89
Andrew Thomas	143	STS-89	STS-91
Total = 979			

The astronauts who served aboard Mir. Back row: Andy Thomas (left), Shannon Lucid and Michael Foale. Seated: Norm Thagard (left), John Blaha, Jerry Linenger and David Wolf.

Table 11.4 International space endurance records

Astronaut/ Cosmonaut	Launch spacecraft	Launch date	Days in space
Yuri Gagarin	Vostok 1	Apr 1961	0.07
Gherman Titov	Vostok 2	Aug 1961	1.05
Andrian Nikolayev	Vostok 3	Aug 1962	3.93
Valeri Bykovsky	Vostok 5	June 1963	4.97
Gordon Cooper	Gemini 5	Aug 1965	7.92
Pete Conrad			
Frank Borman	Gemini 7	Dec 1965	13.75
Jim Lovell			
Andrian Nikolayev	Soyuz 9	June 1970	17.71
Vitali Sevastyanov			
Georgi Dobrovolsky	Salyut 1	June 1971	23.76
Viktor Patsayev			
Vladislav Volkov			
Pete Conrad	Skylab	May 1973	28.04
Paul Weitz			
Joe Kerwin			
Al Bean	Skylab	July 1973	59.49
Jack Lousma			
Owen Garriott			
Jerry Carr	Skylab	Nov 1973	84.04
Bill Pogue			
Ed Gibson			
Yuri Romanenko	Salyut 6	Dec 1977	96.42
Georgi Grechko			
Vladimir Kovalyonok	Salyut 6	June 1978	139.60
Alexander Ivanchenkov			
Vladimir Lyakhov	Salyut 6	Feb 1979	175.06
Valeri Ryumin			
Leonid Popov	Salyut 6	Apr 1980	184.84
Valeri Ryumin			
Anatoli Berezovoi	Salyut 7	May 1982	211.38
Valentin Lebedev			
Leonid Kizim	Salyut 7	Feb 1984	236.95
Vladimir Solovyov			
Oleg Atkov			
Yuri Romanenko	Mir	Feb 1987	326.48
Vladimir Titov	Mir	Dec 1987	365.95
Musa Manarov			
Valeri Poliakov	Mir	Jan 1994	437.75

12

Building hardware

PRIME CONTRACTOR

On 13 January 1995 NASA signed the $5.63 billion 'cost plus incentive fee' deal confirming Boeing's prime contractorship. The company would subcontract as necessary to build two nodes, an airlock and laboratory and habitation modules; it would then assume responsibility for the integration of their systems. The contract was a reimbursement agreement that incorporated an award fee for management schedule and technical performance and an incentive fee for reducing costs to encourage Boeing to drive hard bargains with its top tier subcontractors. It was designed to preclude a situation in which the company received nearly all of the possible award fee for meeting a long list of performance targets, even though the product was fundamentally flawed. On 8 February, Lockheed, which had forged a strategic alliance with Khrunichev by forming International Launch Services to market the Proton, announced the contract for the FGB. Khrunichev had asked for $245 million for the spacecraft and also to be paid for the launch vehicle, but had accepted $190 million with "on-orbit delivery". At first, the contract was managed by Lockheed, but in mid-1995 Boeing took over, thereby elevating Khrunichev to the top tier. By August, when the contract was ready to be signed, the management structure had been revised and the fee raised to $210 million. Meanwhile, Energiya had started work on the Service Module, which was to be wholly funded by the Russian government.

Even as the STS-63 'Near-Mir' rendezvous mission was underway in February, Goldin was facing a White House instruction to cut $5 billion from NASA's planned programmes through to the end of the century. The obvious solution was to reduce Shuttle flights, but the trick was to do it without affecting the station. When NASA had shed jobs in previous years, Bryan O'Connor had warned that the agency had cut "too much, too fast". In March Goldin set up a panel under the chairmanship of Chris Kraft, a former director of the Johnson Space Center, to consider this, and to recommend ways of further reducing the costs of Shuttle operations. Kraft's panel said that whereas the Shuttle was now operating routinely, NASA was running it as if it were a development programme, with a complex management system that incurred high overheads, and recommended (a) that the management be simplified; (b) that the ground 'flow' processing be privatised; (c) that bids should not be solicited for the consolidated prime contractorship – a process that would in itself be time-consuming, disruptive and costly; and (d) that one of the current contractors should be upgraded. In November NASA decided to give the contract to the United Space Alliance, which was created by Rockwell (the Shuttle developer) and Lockheed Martin (the prime contractor for Shuttle operations at the Kennedy Space Center) specifically for the purpose.

Meanwhile, NASA and Boeing had formed the Integrated Product Team to manage development. Although the company had still to formalise the specific contracts with subcontractors to produce much of the hardware, in May it completed the pressure shell of the first node, which was immediately subjected to structural tests.

In early June, after the House's space subcommittee rebuffed yet another motion by Representative Tim Roemer to cancel the programme, it overwhelmingly authorised an annual $2.1 billion for Fiscal Years 1996 to 2002 to see the station to completion. Nevertheless, Representative George Brown of California, the Democratic chairman of the House's Science, Space and Technology committee, reminded Goldin that he was "not an unequivocal under-any-circumstances supporter". He wanted a "balanced" programme, and if pursuance of the station ever seemed to threaten other programmes he would vote against it. James Sensenbrenner, the senior Republican on the House's space committee, said that the degree of Republican support virtually guaranteed that the appropriation would be confirmed when it was debated by the full House. Historically, a programme with multi-year funding was more efficiently pursued than one with an annual budget. The station would be the most ambitious space construction project yet attempted. If the Senate voted the multi-year appropriation it would relieve the station of its annual ordeal. On 28 September the House set an annual cap of $2.1 billion, but approved $13.1 billion for the station's manufacture and assembly by 2002.

RETHINK IN EUROPE

The ESA Council met in Toulouse on 18 October 1995 to try to counter years of uncertainty about the nature of its commitment to the station. The delay and the escalating costs had eroded European enthusiasm and the agency had begun to fracture, with nationalistic trends. France had threatened to pull out of the development of the Columbus laboratory; Germany was frustrated by the mounting costs; Italy considered that it was not receiving contracts that were commensurate with its budget contribution, and had developed its own relationship with NASA; and Arianespace was irked that Germany had booked a payload on a Chinese rocket instead of using its Ariane.

After confirming its interest in the space station (although with Columbus scaled down) the Council decided to build France's proposed Automated Transfer Vehicle (later renamed the 'Ariane Transfer Vehicle' – ATV) as an Ariane 5 upper stage to deliver cargo and/or replenish the Service Module's propellant. The plan was thrown into doubt by Russia's announcement that it was thinking of superseding the Progress tanker with a logistics craft based on the larger TKS (only one of which would be required to be launched per year, instead of half a dozen of the smaller tankers) but a few months later this proposal was dropped. Because ESA still hoped eventually to achieve independence in human spaceflight, it ordered a study of how the ATV could be adapted to resurrect the idea of a Crew Transfer Vehicle whose initial mission would be to supersede the Soyuz as the station's lifeboat. In America, James Sensenbrenner applauded this idea as yet another means of reducing reliance on the Russians. Unfortunately on 4 June 1996 the inaugural Ariane V launch ended in disaster when a software error in the control and guidance system caused the vehicle to pitch over and the aerodynamic forces ripped it apart.

PROBLEMS IN RUSSIA

In November 1995, as STS-74 took the Docking Module to Mir, Anatoli Kiselev, Khrunichev's director-general, opined that it made no sense to abandon Mir. The final pair of science modules had just been completed and the Russian Space Agency proposed that, as an operational station, Mir ought to serve as the *basis* for the new facility. With suitable maintenance Mir should be able to be kept in service to the end of the station's assembly in 2002. On 11 December, a high-level delegation flew to Houston to propose docking the FGB on the front of Mir, enabling the old station to 'boot strap' the new one. Furthermore – extending the logic of making the best use of hardware that was currently in orbit – it was suggested that the Spektr (recently launched) and Priroda (soon to be launched) modules should be transferred from Mir to the new station as part of Russia's international contribution. Also, if the Service Module's launch was postponed until Mir was ultimately superseded, the financial pressure on the cash-starved agency would be temporarily relieved.

"We owe it to the Russians to take a look, listen to them and then go from there," admitted Wilbur Trafton. However, he would not consider the implementation of any proposals that would adversely affect the cost or the schedule. "The bottom line is that we won't redesign the space station." After six days of detailed talks, NASA refused to use Mir as the core of the International Space Station. However, it offered a deal to enable Russia to reduce its immediate costs without impacting the early assembly schedule. Specifically, the three Russian research modules and their solar power unit would be slipped to 2000, thereby providing the funding to continue the use of Mir as an independent facility for fee-paying clients. Two extra Shuttles would sustain Mir in 1998, taking Phase One to nine docking missions (within the "up to ten" specified by the original agreement). A protocol signed by Goldin and Koptev in January 1996 was designed to enable Russia to retain Mir without impeding the assembly of the International Space Station. Once the Service Module was in place, the Russians would postpone their part of the build-up until Mir was retired. While both stations were in orbit the control centre at Kaliningrad would handle Mir and Houston would deal with the new station. NASA made it clear that there should be

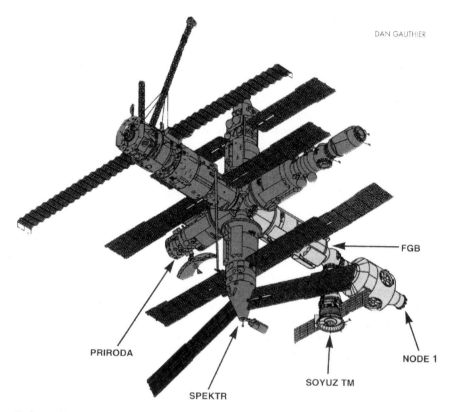

DAN GAUTHIER

PRIRODA

FGB

NODE 1

SOYUZ TM

SPEKTR

In late 1995 the Russians proposed to 'bootstrap' the International Space Station by growing it on the front of Mir, so that it would be able to be inhabited from the very beginning, and so that science apparatus could be transferred. (Courtesy of the late Dan Gauthier.)

no further revision of the terms of the bilateral agreement. "Something of this magnitude", reflected Trafton, referring to the recent change, "has to be a one-time shot. We can't do any more."

In December 1995, the Russian Space Agency said that unless the government released the promised funding it would not only be unable to meet the launch date set for the FGB (which NASA had started to refer to as the Control Module) but would also be unable to proceed with building the Service Module, so in January 1996 James Sensenbrenner flew to Moscow with California's Republican Representative Jerry Lewis to secure high-level assurance that these modules would indeed be delivered on time. On 30 January the programme took a significant step towards reality when Gore announced that William Shepherd – the station's programme manager – and Sergei Krikalev would form part of the commissioning crew. If everything went according to plan, they and the yet-to-be-named Soyuz commander would launch from Baikonur in May 1998, a month after the launch of the Service Module. When Trafton opined that the station should always be commanded by a

NASA astronaut, the Russians objected. A meeting in Houston in March decided that the responsibility should be alternated with successive crews, with Shepherd commanding the commissioning crew. Anatoli Solovyov was named as the third man, but when he grumbled about being subordinated to an American he was replaced by Yuri Gidzenko.

On 28 March, Goldin suggested to the House that the Russians should be given four to six weeks to resolve the funding problem that was delaying the fabrication of the Service Module. Sensenbrenner said that when he and Lewis had visited Russia in January, First Deputy Prime Minister Oleg Soskovets (the third highest official in the Russian administration) had assured them that funding *was* in place. A few weeks later, Koptev said that Chernomyrdin had given "strict instructions" to the finance ministry to follow a payment schedule to allow the modules to be delivered on time. However, Khrunichev was owed $20 million for work undertaken in 1995, and it would need $35 million if it was to progress with the work assigned to 1996. Despite the high-level assurances in January, the company had received only $10 million. In admitting that the Service Module had suffered a "minor" slip, Alexander Lebedev, Khrunichev's deputy director, was confident that once the funding schedule was in place, the lost time could be recovered. Gore was informed by Chernomyrdin that steps had been taken to ensure that the funds flowed to the company, and a reporting procedure was initiated to enable NASA to monitor progress. However, Russia's faltering economy undermined these honest efforts. "The problems in Russia are so much bigger than the space programme," noted Kenneth Mitchell, NASA's station liaison office in Moscow. Industry was operating in a funding vacuum. "I don't think we'd have a problem if they were more economically secure; it's just flat economics." With the presidential election due in June, the only consolation was that Gennadi Zyuganov, the leader of the Communist Party, said that he was in favour of the programme. Trafton told the House space subcommittee on 17 April he was confident that with timely funding Khrunichev would be able to catch up. On 30 May yet another motion by Representative Tim Roemer challenging the programme's funding was rejected. Notwithstanding the high-level assurances, by July Khrunichev had received only a letter to be used as a guarantee in seeking a loan to fund the work on the Service Module, which was now acknowledged to be "months" behind schedule. If the company was unable to ramp up by September it simply would not be able to deliver the module on time, in which case NASA would be forced either to slip the schedule or activate its contingency plan.

In July, Boeing bought Rockwell's space business and so made itself a major Shuttle contractor. The deal also gave Boeing a share in United Space Alliance, the company that provided ground services for the Shuttle. Nine days later, Boeing bought McDonnell Douglas, which made the company a force in the launch vehicle business. Virtually overnight, it had become NASA's leading space hardware supplier. However, in late August Boeing warned NASA that it might be $450 million over budget by Fiscal Year 1998 due to overruns. An internal assessment had earlier concluded that by the time the station was completed in 2002 the budget might be exceeded by as much as 30 per cent ($4 billion). The overruns obliged programme managers to consider options to balance the station's schedule against the $2.1 billion annual cap and the $17.4 billion total budget. One option was to delete two of the six 'utilisation flights' that were to undertake science during the assembly. Although this would release funds to counter the overrun without violating the annual cap, Congress had erected 'fire walls' between the budgets for hardware development and science activities to prevent the transfer of funds, so such a rephasing would require specific Congressional approval, which was unlikely to be forthcoming. Other rescheduling, deferments and hardware changes designed to rephase costs were also under active consideration. Richard DalBello of the White House's Office of Science and Technology Policy was "cautiously optimistic" that NASA would not have to ask for more money. Some of the overruns resulted from technical problems (for example, $100 million was incurred by the need to strengthen the node after it failed its pressure test) and others were incurred by NASA's requests for upgrades to power and communication systems. Although Boeing's interface with its subcontractors was "less than optimum", this was being addressed. "There isn't anything that I've been informed of that seems out of control to me," DalBello concluded.

By late September it had become clear that the Service Module was unlikely to be ready for launch in April 1998. It was already three months late, and could slip even more. The pressure shell was nearly finished, but the avionics would not be ready for the handover of the vehicle from mechanical fabrication to systems integration in November. Even with the cash flow, the company would require to operate a three-shift round-the-clock schedule to catch up. There were no technical problems – Khrunichev was fully capable of building it. The issue was cash flow within the Russian space bureaucracy. Although the $100 million required to complete the module had been approved by the government, the cash had not reached its destination. By this point, the programme was costing NASA $7 million per day, so a delay would soon erode the agency's margin. Accordingly, NASA started to consider an Interim Control Module as a stop-gap to minimise the impact on the assembly in the event that Khrunichev's funding hiatus continued. In 1993, when NASA had presented its trio of options to President Clinton, one variant had envisaged replacing the propulsion modules designed for Freedom with the Lockheed Titan Launch Dispenser utilised by the CIA's KH-11 imaging satellites. When the Russians had become involved, it had been decided instead to develop the FGB. However, while Lockheed's bus would provide propulsion to supersede the FGB, habitability would not be possible until the Service Module finally became available. By the end of the year, the Russian Space Agency acknowledged that the Service Module's launch would have to be delayed by a further eight months to December 1998 because the promised funds had failed to materialise. With no prospect of being paid what they were owed, Energiya's subcontractors had refused to deliver outstanding orders for components.

NASA's contingency planning was constrained by its obligation to stay within its $17.4 billion budget, and it was reluctant to push the completion date beyond 2002 because this would increase costs. Also, a switch to the Interim Control Module at this point would probably slip completion to 2004. Ironically, the simplest, and ultimately the cheapest 'solution' was to allow NASA to lend Energiya the money it required, but that was politically unacceptable to Congress. Nevertheless, NASA sought to assist its counterpart's cash flow with $20 million in 'advance payments' for Shuttles to continue to use Mir during 1998, but the damage had been done. Congress, tiring of Russian delays, began to look closely at the station.

As 1997 dawned, NASA allocated $100 million to adapt Lockheed's bus. It was able to draw this directly from the reserves in the station's budget. However, if a permanent substitute proved necessary the agency would have to return to Congress to seek additional funds. Congress concurred with the allocation of $100 million, but noted that this was only the preliminary cost estimate for the modification of the stage. "It's clear to us now that we need an Interim Control Module," Trafton admitted. Speed was of the essence. "This is a real fast-track project," said Mike Hawes, the station's senior engineer in Washington. The vehicle would require substantial modifications, however, which would consume both time and money. Its powerful engine would have to be replaced to ensure that its new 'payload' would not be overstressed. It was spin stabilised for dispensing satellites, but this would be impracticable in conjunction with the station and a three-axis stabilisation system would need to be developed. Its attitude-control thrusters were mounted on a pair of deployable booms and would need to be repositioned. Furthermore, it might prove necessary to reprogramme the avionics to enable it to control the station as its mass increased and its centre of mass evolved. "Using this control module while continuing to build the station will give us a year's worth of attitude control and reboost capability," Hawes noted. Its tanks could carry 5 tonnes of propellant, but without substantial modification it would not be capable of in-flight replenishment. Perhaps two stages could be switched periodically in service, being delivered and retrieved on an annual basis. As the 12-tonne vehicle would dominate the Shuttle's payload bay, it would be an expensive option because another flight would need to be added to the schedule – and each Shuttle launch cost $450 million. In case the Service Module was very late, NASA began to consider installing racks of life-support systems in its laboratory. Although this would reduce the number of racks available for science payloads, the laboratory could be reconfigured once the Service Module materialised.

As January yielded to February, Koptev admitted that Russia was at risk of being dropped from the programme as a result of its financial woes. As a possible 'solution', NASA hinted that the Russian companies could become Boeing subcontractors. This would guarantee the flow of funds, but would deny the Russian Space Agency any rights of access to the station.

In his State of the Union address in February, President Clinton said: "We must continue to explore the heavens, pressing on with the Mars probes and the International Space Station, both of which will have practical applications for our everyday lives." Although this was fewer than 30 words in the hour-long speech, it was nevertheless the most substantial reference to the space programme in such an address since Reagan's 1984 call to build a station.

Chernomyrdin assured Gore in early February that the Russian Space Agency would have the outstanding $100 million by the end of the month, but

Sensenbrenner, pointing out that all the earlier promises of funding had turned out to be "worthless", urged that the Russians be removed from the critical path. He was astounded to find that Chernomyrdin's pledge was in fact a decree underwriting bonds that the agency was to issue to its contractors. "The government keeps trying to get us used to working without money," wryly noted Yuri Grigoriev, Energiya's deputy general designer, but the Soviet era's mindset was dead.

Up to this point, because the fabrication of the FGB was on schedule, NASA had insisted that it would keep to the plan by launching the FGB in November, but in mid-February – a week after dismissing speculation of slippage as a "dweeb rumour" – Goldin confirmed that he was considering postponing the launch by six months. "My sense is we're going to have to slip first-element launch," Goldin told the House. It now seemed as if the Service Module's delivery could slip into 1999.

In early March, former astronaut Tom Stafford led a team of financial experts to Russia to find the reasons for the delays in releasing funding. When Goldin admitted to the House that he did not know what was happening, Republican Representative Dana Rohrabacher was flabbergasted: "It seems to me," he said, "you should know those answers without having to send someone over."

ESA UPDATE

The disastrous flight of the first Ariane 5 prompted ESA to reconsider trusting it with Columbus, and in March NASA agreed to launch Columbus on the Shuttle in return for Alenia Spazio supplying the second and third nodes for the station, thereby saving NASA the cost of building them. In fact, Boeing had intended to re-engineer its test article to serve as the second node. The second node was to host the 'international' modules. As the Baseline Configuration had only two nodes, the addition of a third would increase the scope for the station's expansion. Aerospatiale was awarded the contract for the two-layer tiles that were to be mounted on the exterior of the modules that would form the 'leading edge' of the station and take the brunt of impacts as the station swept up microscopic debris in its path.

RELIEF

It was evident that the US–Russian relationship was under strain. Chernomyrdin's order to release the funds to resume work on the Service Module had been undermined by bureaucracy. "The fact of the matter is, the Russians have to do what they say they are going to do," Goldin warned in April. "We set up a very complex programme, and the Russians have not been funding their side." Koptev pointed out that he was just as frustrated as Goldin by the state of affairs, and it was formally agreed that the FGB's launch would be postponed from November 1997 to mid-1998, and that the Interim Control Module would be sent up if the Service Module could not be launched later that year. Randy Brinkley said that although NASA would give the Russians every chance to overcome their funding problem, it would devote "equal attention" to contingency planning. On 9 April it was announced that the station's assembly would begin "no later than October 1998". This "up-to-11-month" slippage was a severe blow. Reflecting that since January 1996 the Russians had promised eight times to resolve this funding issue, Sensenbrenner said that the programme was "falling apart around us" and again urged NASA to "remove the Russian *government* from the critical path" by making the Russian companies direct subcontractors to Boeing. The House's science committee expressed its frustration by telling NASA not to fund any hardware that the Russian government should supply. It also ordered the agency to make monthly reports on Russia's status. "We knew from the outset", Goldin insisted, "that building an international space station would be tremendously challenging." It was ironic that it was the prospect of Russian participation that had saved the programme politically in 1993, yet the inability of Russia to meet its obligations had become the greatest obstacle to the station's assembly. On 11 April the Russian government arranged to release $260 million by having banks issue loans to Energiya with the proviso that the loans would be repaid with 3.5 per cent interest. With this assurance of funding flowing by the end of May, Khrunichev resumed work on the Service Module. This prompted NASA to announce that, apart from starting late, it was "cautiously optimistic" that the station's assembly was now on track.

On 14 May the Space Station Control Board published a new schedule in which the FGB would be launched in June 1998, the node in July and the Service Module in December. This new schedule had "much more flexibility" Brinkley observed. A Shuttle flight had been added to provide a "contingency opportunity", giving time to send up the Interim Control Module in December 1998 if the delivery of the Service Module slipped in 1999. If the Service Module was launched on time, this 'extra' mission would deliver logistics.

Table 12.1 Revised manifest (circa May 1997)

1998	June	1A/R	FGB
1998	July	2A	Node 1 and two PMAs
1998	December	–	SM
1998	December	2A.1	Either SM logistics or the ICM
1999	January	3A	Z-1 truss, etc
1999	January	–	Crew on Soyuz

In order to provide "defence in depth" against an escalation of the Russian financial crisis, NASA brought the Z-1 truss forward to early January in order to make a start towards providing a power module. Also, in case the virtual shutdown of the Semyorka rocket's production line meant that there would be no Progress tankers to replenish the station, NASA added a mission in October 1999 to mount the Interim Control Module on the rear of the Service Module to preserve its fuel.

By October 1997, NASA had to acknowledge that the station could not be completed before December 2003, some 18 months after the previous target of June 2002. Congress, having been asked to augment the Fiscal Year 1998 budget with

$430 million to cover overruns was sceptical. In November, after suggesting that NASA should terminate work on the Interim Control Module and reassign the cash towards the overrun, Trafton resigned and was superseded by Joseph Rothenberg. In February 1998, in his Capitol Hill debut, Rothenberg was "sceptical" about the Service Module being ready by December 1998. A disbursement from Moscow which had been due that month had not materialised. With the prospect of the Service Module slipping into 1999, it was decided to postpone the launch of the FGB rather than 'waste' its service life sitting in orbit. As long as the Service Module was ready by April 1999, it would not seriously disrupt NASA's schedule. Yeltsin fired Chernomyrdin in March – a development that Congress took to be further bad news for the station – so Sensenbrenner personally flew to Moscow to assess the situation.

In March, NASA admitted to the House that cost overruns and schedule slips meant that the station would have cost $21 billion by completion in December 2003, which was some $4 billion more than the $17.4 billion bugdet. In April, Goldin released the independent assessment by consultant Jay Chabrow, whose report, 'Cost Assessment and Validation report on the Space Station', indicated that there was significant scope for delay during assembly and that a slippage of up to three years in the completion date was likely. Furthermore, NASA would require $130–250 million per annum more than it had estimated. Between them, he warned, this additional funding and the cost of such a slippage could easily push the total budget to $25 billion, exceeding the target by some 40 per cent. "We are all concerned about the cost overruns and schedule slips," admitted Goldin, who had commissioned the report the previous September, "and I'm not going to sugar coat them." It was clear that the $2.1 billion annual cap could not accommodate the combination of the cost of overruns and the Russian crisis. Each month the Service Module slipped was costing NASA $120 million. If Congress insisted upon imposing the cap, then the inevitable result would be to slip the completion date by "a year or two", he admitted.

According to the schedule announced by the Space Station Control Board on 30 May 1998, the FGB (which NASA referred to as the Control Module and the Russians had recently named 'Zarya', for 'dawn') would be launched "no earlier than 20 November", the node (which NASA had recently named 'Unity') would be delivered by STS-88 "no earlier than 3 December", and the Service Module would follow "no earlier than 28 March 1999". Despite the late start, it was hoped that the lost time would be recoverd as the assembly progressed in order to complete the station by January 2004.

Reminding the White House that Chabrow's report had said that the programme was underfunded by $130–250 million per year, Sensenbrenner urged Clinton to find the shortfall because it was his administration that had introduced the Russians who were now causing NASA so much angst. However, Jacob Lew, director of the Office of Management and Budget, was insistent that "additional funding" was not available. In effect, the administration could not abandon its foreign policy and admit that Russia had failed in its obligations, and Congress did not wish to appear to sanction "foreign aid" to Russia. The White House was hoping that NASA would find the money that it required by raiding other programmes and Congress was determined not to let this happen. In September, the agency proposed a two-tiered strategy. The immediate priority was to alleviate Russia's financing shortfall by "investing" to guarantee the timely production of Russian items required to put the station on track; that is, the Service Module, Progress tankers and Soyuz lifeboats. Secondly, NASA should seek long-term "independence" from Russia. It should build a replenishable propulsion module and consider the offers by ESA and Japan to provide logistics vehicles. It proposed a $1.2-billion one-time injection of funding to "fix" the programme[1] and urged the immediate sanction of a $60-million "advance" to ensure that the Service Module was delivered on time. To preclude criticism that it was just another bail-out, NASA described it as "prepayment" for the two Soyuz lifeboats. It also proposed four annual $150-million instalments to guarantee three Progress tankers for orbital reboost as the station was being assembled, and a pair of Soyuz spacecraft per year, each of which would serve a six-month tour. The funding would be provided on the strict understanding that it would be used to produce crucial station hardware. Senator John McCain, the chairman of the Commerce Committee, was concerned that this would simply establish a precedent that the Russian government might exploit. Sensenbrenner said that it showed that putting Russia in the station's critical path had been an expensive mistake, and that NASA ought to have seen the problem coming.

As John Glenn launched aboard STS-95 on 29 September, former anchorman Walter Cronkite asked Bill Clinton, live on CNN, if he was willing to spend "lots of money" to keep the International Space Station on track. "If it were required," Clinton replied, "I'd be supportive of it." He noted that NASA had had "hardly any increase in funding" since his election to office. In fact, taking inflation into account, the agency had had a 20 per cent cut under the Clinton administration. "If it were required now to help the Russians through this difficult period, which will not last forever, so they could continue to participate, I'd be in favour of it. I think we're doing the right thing with this space station, and we need to stay with it." John Logsdon of the Space Policy Institute at George Washington University was very encouraged by Clinton's candid support for the programme. "What Clinton said at the launch was clearly scripted," he observed. "No President goes on television and makes up commitments like that." However, Clinton's support appeared to have been directed specifically towards supporting the part of the station that was sitting in the Russian critical path – the $660 million. This represented only half of the $1.2-billion plan proposed by NASA to eliminate reliance on the Russians, and a mere 28 per cent of the projected $2.4-billion shortfall by the end of the assembly period. By limiting his support to the immediate issue, Clinton had failed to address the real issue: *the station could not be built within the projected budget.*

---

[1] The $1.2-billion 'fix' comprised $660 million to ensure that crucial Russian hardware was produced, $90 million to modify the Shuttles so that they could provide orbital reboost, and the rest was to develop and launch the new propulsion module in 2002 as a long-term solution to the reboost issue.

GO, GO, STOP!

The flawless launch on 20 November 1998 of a Proton carrying Zarya marked the long-awaited start of the International Space Station's assembly. Koptev took advantage of the occasion to say that he was actively "working with external investors" to try to secure funding to extend Mir operations into the new millennium. Goldin, however, was blunt: "We expect the Russian government to live up to its ISS commitments." In fact, the basis of a compromise was becoming evident. The Russian government could satisfy NASA by denying funds to Mir and releasing Energiya to continue to run Mir as a commercial venture with fee-paying clients, but to run both programmes would create a competition for the limited supply of Soyuz and Progress spacecraft and Semyorka rockets. To keep Mir operational would require the production line to triple its current low-level output, which seemed unlikely.

Viktor Blagov, deputy director at Kaliningrad, reported that Mir's physical state was "fully satisfactory". He dismissed the claim that Mir would draw resources from the International Space Station. "Enough Soyuz and Progress are already built to ensure Mir's use through 1999," he insisted. To continue further would require only two additional Progress tankers and one Soyuz ferry. In fact, Mir had enough propellant to see it through 2000. Fee-paying international missions had recently contributed $120 million per year, he noted. Although true, this was mostly due to the Shuttle–Mir contract. On the other hand, the fact that NASA had booked Mir's 'research seat' for so long had prevented others from utilising its facilities. As soon as it was clear that Mir would not be de-orbited, France booked a 100-day mission starting early in the new year.

In January 1999 NASA announced that the launch of the Service Module would have to be postponed for six months until the autumn. The vehicle was effectively complete, but post-manufacture testing was proving to be more time-consuming than expected, which had delayed its shipment to Baikonur. On 1 February Congress refused NASA's proposed $600 million "investment" in the Russian Space Agency, and on 25 February Rothenberg admitted to the House's space subcommittee that although the official date was September, "we have our reservations" and the Service Module might not be ready until November or December. A progress meeting with the Russians was planned for 15 April, but NASA would not commit itself to a specific launch date until June or July. NASA's deputy programme manager for space station operations, Frank Culbertson, was more upbeat: "It'll launch; it's just a matter of when." The Russians were simply "not going to ship it before it's ready".

In early March some of the senior staff changed hats. Randy Brinkley resigned and was superseded as the station's programme manager in Houston by Tommy Holloway. Veteran flight director Ronald Dittemore filled the vacated slot of Shuttle programme manager, and when Gretchen McClain left to take up an industrial post, Mike Hawes took over as the station's chief in Washington.

On 8 March, it became clear that the Russians were assessing their options in the event that the Service Module launched but was unable to dock. Although Zarya's TORU system could be directed from Kaliningrad, it would be preferable to launch a crew to dock with the Service Module, so that they could steer the Zarya–Unity combination in manually. In effect, bringing the commissioning crew forward would reinstate the earlier schedule for occupancy. This option was added to the agenda for the next management meeting in Moscow in April. An 'early' launch might also calm Congressional critics who were upset about the Russian delays. At the April meeting, the Russians confirmed that the Service Module would be sent to Baikonur in May, and would be launched on 20 November. It was also decided not to pursue the option of occupying it immediately. "We'll send a crew up in early 2000," Goldin announced. The vehicle was 'rolled out' by Energiya with considerable fanfare on 26 April. "As far as I can tell," said Goldin when it had been safely despatched, "we're on track."

The new schedule was:

- STS-96   24 May 1999
- Service Module   20 November
- Progress   3 December
- STS-101   12 December
- Progress   22 December
- 2R/Soyuz   25 January 2000
- STS-92   24 February
- Progress   6 March
- STS-97   23 March
- STS-98   20 April
- Soyuz   5 May
- Progress   19 May
- STS-102   29 June

Meanwhile, the Russians were considering what to do with the FGB's backup vehicle. One option was to reconfigure it as the Docking and Stowage Module that Russia was scheduled to add to the station in 2003. As yet, no progress had been made towards defining the science modules, so the Ukraine offered to supply one, subcontracting the work in the same way as Zarya had been financed by Boeing. Of course, this module would be dispatched using the Ukranian-built Zenit launcher. Spacehab Incorporated signed a deal with Energiya to jointly build and operate their own commerical module for microgravity research. Both of these modules, like Italy's life sciences module, were intended to expand the station beyond its Baseline Configuration.

On 27 May, after an almost six-month hiatus, STS-96 launched with a Spacehab module and an Integrated Cargo Carrier in its payload bay to ferry logistics to the station to 'preposition' internal stores and external apparatus for later use. After a short-circuit prompted one of Columbia's main engines to shut down on 23 July while carrying the Chandra X-ray observatory, NASA found similarly eroded wires in the remainder of its vehicles. The wiring repairs delayed the Hubble-servicing and topographic radar mapping missions and slipped the next station logistics mission into the new year. After the loss of Protons on 5 July and 27 October owing to faults in the second stage, the Russians indefinitely postponed the Service Module's

November launch. The problem was found to be spurious material in the engines by poor workmanship in manufacturing. To provide time to recertify the modified stage, the Service Module was rescheduled in December for "no earlier than May 2000". In response, NASA investigated the possibility of launching Atlantis on two 'back to back' flights with the same Spacehab configuration for a fast turnaround (as it had done the previous year with a microgravity research Spacelab), one just prior to the Service Module's launch and the other just afterwards. In January 2000, Koptev committed to a launch in July and planning for the 'back to back' mission was deleted. "It's frustrating," Goldin admitted, "but we've had our own problems." An integrated test of the hardware and software of NASA's laboratory had found defects in its command and control system. If it were not for Proton problems having delayed the Service Module, NASA's laboratory would by this point have become the 'pacing' item.

Meanwhile, upon the completion of the French fee-paying tour on 28 August 1999, Mir had been vacated for the first time in a decade. It was briefly reoccupied between April and June 2000 and then 'mothballed'. When it was de-orbited in March 2001 the challenge of human spaceflight was passed to the International Space Station. After STS-101 delivered logistics in April, the Service Module was launched on 12 July. When it completed the station's core by docking with Zarya on 25 July, it opened the floodgate for NASA's Shuttles to begin the real assembly work.

13

Orbital assembly

Prior to the fitting of the shroud which would protect its multiple docking adapter during the ascent through the atmosphere, the 'Zarya' (the Russian word for dawn) module had been loaded with equipment, disinfected, pressure-tested and fuelled. On 15 November 1998, at the Baikonur Cosmodrome in Kazakhstan, it was mated with its Proton launch vehicle and driven horizontally on a train to the launch pad, where it was raised to the vertical position.

Over the next three days all of the lines between the launch vehicle and its payload were connected, as the combination was prepared for flight. A commission representing all of the major ISS partners met on 19 November and confirmed the intention to proceed. The final preparations began at 1630 American Eastern Standard Time on that day, when the engineers began to load the hypergolic propellant into the Proton. Then at 2030 the gantry embracing the vehicle was retracted and final pre-lift-off tests were completed. The Proton's first stage ignited at 2040, flame belched down into the trench, which was designed to channel it away from the base of the rocket. Clamps held the Proton in place as its six first-stage engines built up to the required combined thrust of 167 tonnes. At mainstage, the clamps released and the 54.86-metre-tall vehicle climbed into the overcast Kazakh sky, and the first ISS element was finally on its way into space.

After 2 minutes and 6 seconds, the first stage shut down and was jettisoned to fall into the Kyzyl-Kum desert. One minute after staging, with Zarya already above the thickest portion of the Earth's atmosphere, the shroud separated into two halves and was jettisoned. The four second-stage engines then ignited, and Zarya continued to climb. After a 3-minute 28-second burn, the second stage shut down and it too was jettisoned. The final thrust to attain orbit was provided by the single-engined third stage, which was jettisoned 9 minutes 47 seconds after launch.

Zarya's initial elliptical orbit had an apogee (the point in its orbit farthest from Earth) of 354 kilometres and a perigee (the point nearest to Earth) of 185 kilometres. The trajectory was inclined at 51.6 degrees to the equator. The orbital period was 90 minutes. In orbit, Zarya was 12.5 metres long and had a maximum diameter of 4.1

The Zarya module is prepared and mated with its Proton launch vehicle.

metres. The onboard computer deployed the Kurs and Compares antennae, extended the docking probe and then deployed the solar panels, each of which was 10.6 metres long and 3.3 metres wide. The two photovoltaic arrays produced electricity from sunlight and provided it to 6 nickel–cadmium batteries, from which up to 3 kW of electrical power would be available to the ISS's systems. Zarya's 24 large and 12 small manoeuvring rockets drew propellants from 16 tanks providing a total capacity of 5,446 kilograms. Two primary rocket engines were available for major orbital manoeuvres and for countering orbital decay. The tanks were capable of being refilled by an automatic Progress-M tanker. Inside Zarya, pale-yellow walls lined the main pressurised compartment, and the panels could be unbolted to access stored items. The air ducts and cables around the walls gave the interior the appearance of the Mir base block when it was first placed into orbit; that is to say, when it was clean and tidy. Given Mir's longevity, Zarya was designed for a minimum life in space of 15 years.

Zarya is launched.

Three hours after launch, the Russian flight controllers in the Korolev control centre in Kaliningrad radioed a command to set Zarya slowly rolling about its primary axis. This passive thermal roll ensured that energy from the Sun was distributed evenly across the exterior. If any one area was illuminated too long it might overheat. Alternatively, any area that remained in shadow risked the freezing solid of any fluids contained in that area, with the possibility of burst plumbing. Over the next two weeks, Zarya would be returned to this roll whenever it was not performing a task requiring it to adopt another orientation. Later on the first day, Russian controllers radioed commands to assess the articulation of the solar panels, and both were able to follow the Sun faultlessly.

In a launch-day press conference, NASA's administrator, Dan Goldin, emphasised the scale of the programme, "We only have 44 more launches to go, and about a thousand hours of spacewalks, and countless problems, and countless issues."

The next day, the Korolev flight controllers tested the two black-and-white television cameras which would be needed during the docking with the Service Module, due sometime in the summer of 1999; the cameras presented no problems. They also test fired one of Zarya's two primary manoeuvring engines for 10 seconds, then again for 1 minute 40 seconds in the first of a series of manoeuvres in preparation for the rendezvous of STS-88 in a few weeks time. This initial burn resulted in an orbit ranging between 363 and 251 kilometres.

On 22 November, the suspicion developed that two 1.2-metre-long antennae of the Teleoperator Control System (more commonly referred to by its Russian initials TORU) remote-control docking system may have failed to deploy. Following the firing of the bolts that held the antennae in place they ought to have unrolled like a child's party favour. Testing to establish their status continued throughout the week.

Korolev's controllers commanded the same engine that had been used earlier in the week to fire twice on 23 November. It first performed a 31-second burn to increase Zarya's velocity by 6.4 metres per second. The main burn lasted nearly 2 minutes and increased Zarya's velocity by a further 23.7 metres per second. The overall effect was to manoeuvre Zarya into an orbit ranging between 339 and 312 kilometres with a period of 91 minutes. As the check-out proceeded, the motion control system was verified in the free-flight mode. The atmospheric measuring equipment and the fire detection and suppression equipment were functioning as expected, but the gas analyser indicated a higher than expected internal humidity level which was initially thought to indicate a defect in the instrument's calibration. Also, for some reason Battery 1 indicated a higher charge than its five companions. A 1-minute 56-second burn on 24 November raised the orbit to 403 x 386 kilometres. Over the next two weeks this orbit would decay to a nearly circular 389 kilometres, as required for the STS-88 rendezvous, which was scheduled for 6 December. The following day was taken up by checks on the two multiplexers/demultiplexers that formed part of Zarya's command and data handling system in that they formed the interface between the computers and most of the module's major systems. One device was turned on, and operated for 3 minutes before being turned off again as the backup. The second was activated and left on for the coming rendezvous. During the day, the earlier indication of high humidity inside the module was traced to faulty software on the ground. Tests confirmed that there had been no leakage of any fluids onboard the spacecraft. Also, Battery 1 was functioning as planned but its associated electronic apparatus appeared to be at fault. By 26 November, the indications were that one TORU antenna was at least partially deployed. With no further testing planned, the STS-88 crew would be charged with observing and photographing the antennae to confirm their status. Although the Kurs automatic primary docking system continued to function according to its specification, the TORU would provide a manual backup for the crucial docking with the Service Module.

Zarya was put through a final series of systems tests on 27 November, simulating the module's preparations for the STS-88 rendezvous. The activities included:

- locking the solar panels in the berthing position;
- conducting an electrical checkout of the RMS grapple fixture;
- inhibiting specific thrusters from firing during the rendezvous;
- manoeuvring Zarya to the position from which STS-88 would capture it;
- activating and testing three external cameras; and
- turning on Zarya's external navigation lights.

During the day, a software update rectified the high humidity reading. Humidity within Zarya was monitored over the next three orbits and confirmed to be within normal parameters. By this time, managers were considering sending new power distribution equipment for Battery 1 into orbit on STS-88 and having Russian cosmonaut Sergei Krikalev replace the faulty equipment. He had performed similar work onboard Mir. No further major activities were planned prior to the launch of STS-88 on 3 December, so the controllers simply monitored Zarya's status over the weekend. A study of the telemetry indicated that the Battery 1 problem lay with the Storage Battery Current Regulator System (known by its Russian initials PTAB). Although a replacement current converter unit and PTAB would be loaded on board STS-88, the decision as to whether Krikalev should replace the units would be exercised in real time, depending on the original PTAB's performance in the meantime.

STS-88 ADDS UNITY

STS-88 – Endeavour – SSAF-2A1

Commander:	Robert Cabana
Pilot:	Frederick Sturckow
Mission Specialists:	Nancy Currie; Jerry Ross; James Newman; Sergei Krikalev (Russia)

Endeavour was rolled from the Orbiter Processing Facility at the Kennedy Space Center (KSC) on Merritt Island, Florida, to the nearby Vehicle Assembly Building (VAB) on 15 October 1998. It was left in the transfer isle overnight, and then mated with its External Tank (ET) and Solid Rocket Boosters (SRB) inside the VAB's High Bay the following day to complete the STS-88 stack.

Following a Shuttle Interface Test lasting several days, the STS-88 launch vehicle began its agonisingly slow 5.4-kilometre-long journey on the Crawler Transporter to Launch Complex 39's Pad A on 21 October. Once in position over the flame trench, the Mobile Launch Structure was locked down and the crawler withdrew to its parking bay alongside the VAB. The day continued with a series of pad validation tests, and the engineers on the night shift hot-fired Endeavour's Auxiliary Power Units 1 and 3. The Rotating Service Structure was moved in early the following morning to allow engineers preparing the vehicle access to all levels as well as to protect the vehicle from the weather. On 23 October, the three Space Shuttle Main Engines (SSMEs) passed their flight readiness test.

On 5 November, Robert Cabana's flight crew participated in briefings on the emergency escape procedure from a Shuttle standing on the launch pad. The next day they took part in the Terminal Countdown Demonstration Test, during which the crew performed exactly as they would do on launch day. The test ended with a simulated SSME cut-off at the point in a real count where they would be ignited, at which time the crew exercised the full emergency evacuation procedure. At the end of the day, they flew back to Houston in order to spend the weekend with their families. Meanwhile, work proceeded in preparation for installing Unity into Endeavour's payload bay.

Unity had arrived at KSC on 23 June 1998. It was offloaded from the delivery aircraft that same evening and taken to the Space Station Processing Facility (SSPF) for final preparation and testing. Its hatch was shut for the final time on 20 September. It was subjected to a number of pressurisation and leak tests before finally being purged with clean dry air in preparation for launch. It was transferred from the SSPF to the launch pad payload changeout room on 27 October, and was installed in the rear part of Endeavour's payload bay on 13 November. Ten days later, Shuttle and ISS managers gathered at KSC for the Flight Readiness Review, and confirmed 3 December as the launch date. Cabana's crew returned to KSC on the night of 29 November. The countdown commenced at 0700 the following morning, as the work in Endeavour's bay was completed. The Air Force meteorologists were predicting a 60 per cent chance that low cloud ceiling and showers would force a postponement. On 1 December, Endeavour's bay doors were closed, the orbiter's navigation system and the SSME final preparations were completed, the inertial measuring units were tested, the communications systems were tested, the cryogenic reactants were loaded and the mid-body umbilical was retracted into the Fixed Service Structure.

The stowage of the final flight crew equipment began at 0600 on 2 December. Safety personnel completed an inspection of the two service structures to verify that all equipment and debris had been removed. The Rotating Service Structure was moved away from STS-88 at 0800 and placed in its parked position. Prior to propellant loading, mission managers met to review the chances of launching during the 10-minute window. The earlier estimate of a 60 per cent chance of a launch cancellation was reduced to 30 per cent. Loading of the ET's propellant was started at 1840, and by 2140 the liquid oxygen (LOX) and liquid hydrogen (LH2) had been pumped into its two insulated tanks from the storage facilities on the periphery of the launch complex.

In the Crew Quarters of the Operations and Checkout Building, Cabana's crew was awakened and allowed to shower and dress in casual clothing. They then had their final pre-flight medical examination. After eating a traditional pre-launch breakfast, they moved on to don their orange launch and re-entry partial-pressure suits. After the walk-out for the television cameras, they boarded the transfer van for the short drive to Pad A of Launch Complex 39. An elevator carried them to the 59-metre-level of the Fixed Service Structure, where they waited until called forward individually to enter Endeavour. The orbiter's side hatch was closed for flight at 0156.

The countdown proceeded normally until the scheduled hold at T−9 minutes, during which the final review determined that all the weather constraints were satisfactory except those for the Return to Launch Site (RTLS) abort. A 'go' was given to come out of the T−9 minute hold on schedule and proceed to T−5 minutes, at which point the countdown would be held again while everything waited for the local weather to clear. The 'go' for the RTLS constraint was given at 0342. The Orbiter Access Arm was retracted at 0351. After the count was resumed, it passed through T−5 minutes without pause. However, at T−4 minutes and 24 seconds, a Master Alarm sounded on Endeavour's flight deck for the Shuttle's Hydraulic System 1. The count was automatically held at T−4 minutes. An investigation found that the hydraulic system had temporarily dropped below 196,886 grams per square centimetre at the moment that the Auxiliary Power Units which drive the orbiter's hydraulics switched from low to high power. The fact that the readings had then returned to normal and were holding steady prompted a 'go' to continue the countdown to T−31 seconds, at which point there would be an opportunity for a further hold of 42 seconds. Unfortunately, the count was unable to be restarted to reach zero within the launch window, so the launch attempt was scrubbed. After STS-88 had been made safe, the crew disembarked. The launch was rescheduled for the following day.

The countdown for the 4 December launch attempt was started at T−11 hours. There was a slight delay in loading the propellants into the ET, while a piece of ground equipment was changed. The crew left the Operations and Checkout Building just after midnight and were transported to the pad, whereupon they repeated the same ingress procedures as they had the previous day. At 0020, when Cabana made a communications check, he told the conductor in the Launch Control Center, "Let's go do this tonight". The hatch was closed for flight at 0110. The last 20 minutes of the terminal countdown began at 0229. It reached the 45-minute-long hold at T−9 minutes at 0241. The Orbiter Access Arm was retracted at 0333, and the command was sent for the crew to close and lock their visors. This time, everything proceeded as planned. The three SSMEs were ignited and run up to full power, and then the two SRBs ignited and the clamps holding them in place were released at 0335. STS-88 climbed slowly into the night sky, lighting up the Florida coast as she went. The crew called back the readings on their instrument panel to Houston, which had assumed control as soon as the vehicle had cleared the tower. They also recorded their personal feelings about the launch over the internal voice loop.

Two minutes after launch, at a height of 45 kilometres, the spent SRBs were jettisoned. They continued to rise by their own momentum until overcome by gravitational attraction, then fell Earthward along ballistic trajectories, until their parachutes deployed to lower them into the ocean, where ships were waiting to recover and return them to Port Canaveral. After being transported back to the manufacturer, if they were not damaged beyond repair they would be refurbished for re-use on a later flight.

Endeavour continued to climb under the thrust of the three SSMEs consuming the propellant in the ET. As the acceleration built up, the engines were throttled to control the structural loads on the vehicle. Now clear of the thick lower regions of the atmosphere, Endeavour arched over as it headed out across the Atlantic. As the trajectory was slowly brought parallel to the Earth's horizon, it left the crew flying in a 'heads down' position. Eight and one half minutes into the flight the SSMEs also shut down and the empty ET was jettisoned and left to burn up as it re-entered the atmosphere. The Orbital Attitude Manoeuvring System (OAMS) thrusters were used to ease Endeavour to orbital velocity. Meanwhile, Zarya was completing its 222nd orbit and, with the exception of Battery 1's power distribution unit, its systems were continuing to perform well.

After 30 minutes in orbit Endeavour's crew opened the payload bay doors to enable the radiators mounted on their interiors to transfer the heat from the vehicle's electrical systems directly to the vacuum of space. It would have been obliged to return to Earth within hours if its doors had failed to open. The Ku-Band antenna was extended, and the Hitchhiker package in the payload bay was activated. After troubleshooting a problem in the Orbiter Communications Adapter (OCA) system, which would be used for transmitting software files between the spacecraft and stations on the ground, Cabana's crew started its first sleep period two hours late, at 0921.

Flight Day 2 started with a wake-up call at 1636, when Chris Hadfield, a fellow astronaut serving as CapCom, radioed up the message that, "It's time to get ready to build the International Space Station".

While Shuttle commander Cabana and pilot Frederick Sturckow monitored Endeavour's systems, the crew went about their own tasks. Jerry Ross and James Newman, who were scheduled to undertake three spacewalks, or in NASA-speak extravehicular activities (EVAs), inspected Endeavour's airlock and prepared their Extravehicular Mobility Units. They also checked the small thruster pack that would fit around the EMU's Portable Life Support System (PLSS), the Simplified Aid for EVA Rescue (SAFER) which would provide emergency thrust to help an astronaut to return to the Shuttle in the event that a safety tether became detached. In preparation for the EVAs, Endeavour's crew compartment atmospheric pressure was reduced from 998 to 717 grams per square centimetre, which would reduce the time that Ross and Newman would need to spend pre-breathing pure oxygen in order to flush the nitrogen from their bloodstreams prior to switching over to their partial-pressure, pure oxygen EMUs.

As day turned to night in Houston, Ross and Newman set up the Orbiter Space Vision System (OSVS). This combination of cameras, mirrors and marks on the payload bay and ISS modules would allow Nancy Currie to precisely control the Remote Manipulator System (RMS) even when she could not see its end-effector. The OSVS had been thoroughly tested on earlier missions in preparation for the assembly of the International Space Station. Currie powered up the RMS and used the video camera mounted alongside its end-effector to complete a thorough inspection of the payload bay, the Unity node, its attached Pressurised Mating Adapters (PMAs), the Orbiter Docking System (ODS) and the Hitchhikers. The RMS performed these tasks without any significant problems. One of the cameras temporarily jammed at the extreme left of its pan movement, but its range of motion was later restored. Finally, the outer androgynous ring of the ODS was extended into its 'ready' position.

After breakfast on Flight Day 3, Currie powered up the RMS again, and then engaged Unity's grapple fixture and hoisted the node 4 metres straight out of the bay. The action was so slow and deliberate because there was only a 2.54-centimetre gap on either side of the node between it and the bay wall. Currie then turned Unity to a vertical position, with PMA-2 facing towards the payload bay. Once she had the PMA a few centimetres above the extended ring on the ODS she relaxed the RMS and Cabana fired Endeavour's thrusters to engage the two units at 1845 on 5 December, and the latches mated and locked on the first attempt. "We've got it firmly attached to Endeavour, and we're off to a great start on building the International Space Station", Cabana reported delightedly.

During the STS-88 mission, the RMS lifted the Unity node (with a PMA on each end) out of Endeavour's bay, rotated and mated it with the Orbiter Docking System.

The tunnel between Endeavour's crew compartment and the PMA was then pressurised to match the reduced pressure in the Shuttle. Cabana and Ross then opened the hatch and entered the tunnel to place caps on vent valves in preparation for the crew's entry into Unity later in the mission. They also cycled the docking system on PMA-1, at the other end of Unity, in preparation for mating it with Zarya. With this done, Cabana had to manoeuvre Endeavour to avoid flying close to a spent rocket stage from a Delta II launch on 6 November. The 6-second burn was added to the flight plan after US Space Command, which tracked debris in orbit, informed Houston that if nothing was done the stage would pass uncomfortably close to Endeavour. It was decided to make the burn immediately, rather than to interrupt the sleep period, between 0436 and 1136 on 6 December. As a result, when they were awakened for Flight Day 4 Endeavour was 32 kilometres further behind Zarya than scheduled on the flight plan, but this would easily be able to be made up during the final rendezvous, which began with Endeavour 88.5 kilometres in trail of Zarya.

By virtue of being in a lower orbit, Endeavour was orbiting slightly faster than Zarya, so it was progressively closing in. At 1430, Cabana fired the thrusters to climb to slow the rate at which he was catching up. At 1615, when some 14.5 kilometres behind Zarya, Endeavour performed the Terminal Initiation (TI) burn to place Endeavour 183 metres below Zarya one orbit later. At 1745, now on the aft flight deck, Cabana took manual control of Endeavour to complete the rendezvous by flying a trajectory that first took Endeavour 106 metres in front of Zarya and then 76 metres directly above it. Relying on television images from the OSVS, Cabana flew Endeavour down towards Zarya, coming to a stop just 3 metres from it. Throughout this final phase, Newman and Krikalev had used handheld laser range-finders to provide a steady flow of range and range-rate data.

During the STS-88 mission, the RMS grappled the Zarya module and mated it with PMA-1, on top of the Unity node.

Currie began her attempt to capture Zarya using Endeavour's RMS. Unable to see past Unity, she had to rely on the OSVS to position the RMS's end-effector over the grapple fixture on the exterior of Zarya. After engaging it at 1847, she started to manoeuvre the most massive piece of equipment ever manipulated by the RMS. When she had positioned Zarya so that its docking system was 4 centimetres directly above PMA-1's extended ring, she once more relaxed the RMS, then Cabana fired Endeavour's thrusters at 1948 to achieve a soft docking. On the ground, an indicator suggested that the two modules were misaligned, but this was attributed to the fact that the RMS was pulling Zarya to one side. Once Currie had disengaged the RMS, PMA-1's ring was retracted to achieve a hard docking between the first two ISS modules. This was what the Shuttle had been built for. The ODS, Unity and Zarya combination projected some 23 metres from the forward end of its bay. Currie used the cameras on the RMS to conduct a thorough external inspection of Zarya and determined that the pins holding the two TORU antennae in their retracted positions had indeed fired, but the antennae had not unfurled. In Houston, consideration was given to having Ross and Newman visually check them during their initial EVA, but no decision was made as to whether the astronauts should attempt to unfurl the two antennae manually.

Ross and Newman began Flight Day 5 with the preparations for their first EVA. The outer hatch of Endeavour's airlock was opened at 1710 on 7 December. In contrast to previous excursions, each man had two safety tethers. Each man also wore the SAFER rocket system on his back. In the event that one of them did become detached from Endeavour he would be able to fire the SAFER thrusters to return to a position in the payload bay where he could at least hold on to something until his tethers could be re-secured by his partner.

Currie used the RMS to move her two colleagues from place to place with the same agility as she had shown when manipulating Unity and Zarya. Ross worked from a mobile platform mounted on the RMS's end-effector. Newman had his own mobile work platform which he moved by hand, and installed in special receptacles at each work position. Working from the aft flight deck, Sturckow used the flight plan to remind the two EVA astronauts what they should be doing and kept an eye on the timeline.

The two astronauts began their EVA by running eight cables between Unity and PMA-2. These cables had to be installed with the stack erected because there had been insufficient room in Endeavour's payload bay for them to be fitted to Unity when it was stowed in its launch position. For each cable, Ross had to uncover the socket on the spacecraft that the cable was to be plugged into. Newman then handed him the 6- and 9-metre-long cables, which had been stored in a box in Endeavour's payload bay. Ross visually checked the numerous pins of the plugs at either end of the cable to ensure that they were straight. He had a tool to straighten bent pins. Having plugged one end of the cable into the socket, he then turned a locking device to keep it in place and covered the whole plug and socket with a thermal blanket. The procedure had to be repeated at the other end of the cable. Installing the cabling externally was deemed preferable to running them through the open internal hatches. When their work on PMA-2 had been completed, Currie swung the two men up the

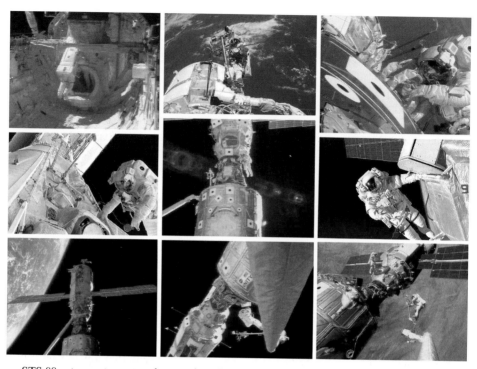

STS-88 astronauts ventured out to install cables to link the systems between the various parts of the International Space Station.

stack so that they could install another eight cables between Unity and PMA-1. Currie then relocated them so that they could connect 24 cables between PMA-1 and Zarya.

While passing over Russia, a radio signal was transmitted to activate a pair of American–Russian voltage converters so that the electricity produced by Zarya's solar panels and stored in its six batteries would be routed to the electrical systems in Unity. When Houston next received telemetry, it confirmed that the module was receiving power and various systems, including a pair of data relay boxes on the module's exterior that would serve as a central control for the node, were activated. Once Unity's heaters were confirmed to be active, Ross removed the thermal blankets from the data relay boxes. Currie positioned Newman where he could see the two TORU antennae on the outside of Zarya. He verified that the deployment bolts appeared to have fired but that, for some reason, the antennae had failed to unfurl. Perhaps their deployment had been impeded by cabling which had stiffened upon being exposed to the extreme cold of space, or perhaps it was due to interference from a thermal blanket. When they returned to Endeavour's airlock, Ross and Newman had been out for 7 hours 21 minutes. Ross, who had just set a 30-hour 8-minute cumulative spacewalking record for an American astronaut, reported that he had lost a wire carrier and a tool socket. The tools were meant to be tethered to the astronaut's EMU or to his workstation at all times. He could not account for how he had lost them. They had become part of the growing cloud of debris which circles the Earth and poses a collision risk to spacecraft. During the EVA, managers had decided that Krikalev would install the new PTAB once Zarya had been opened up on 10 December, so that the electrical power from Battery 1 would be distributed properly.

After a successful but hectic day, the crew stayed up past the start of their sleep period relaxing and admiring the view through Endeavour's windows. About one hour after they should have begun their sleep period, Cabana asked that their wake up call be delayed to give his crew members more time to themselves. The request was granted, and so Flight Day 6 began about half an hour later than called for by the flight plan.

Ross and Newman relaxed during the day. Aided by Currie and Krikalev, they replaced filters in the EMUs and replenished their PLSSs for their second EVA. Meanwhile, Houston verified a remote link that would enable it to command Unity via Korolev. It was used to switch on fans and filters in Unity, to warm the node prior to the crew's entry. At 1530 on 8 December, Cabana started a 22-minute series of thruster firings to raise the apogee by 8.85 kilometres, to 399 kilometres. During the manoeuvres the 8-storey-tall Unity–Zarya combination swayed gently on its ODS base. Throughout its life, the ISS would have to be repeatedly boosted to counter the orbital decay induced by the frictional drag of the tenuous gas of the Earth's outer atmosphere. Initially, this would be done either by Zarya's two primary engines, or by visiting Shuttles. Once the Service Module was in position, it would take over the burden from Zarya.

At 1741, Cabana, Sturckow and Currie were interviewed for the American ABC, Discovery and MSNBC television channels, and then at 2030 they began a half-day off, during which they could do as they wished. They were all able to talk to their families over a video link. Just before they started their sleep period, Houston reported that PMA-2 had been pressurised and warmed sufficiently for the crew to enter the two ISS modules as planned.

Flight Day 7 began at 1136 on 9 December, and Ross and Newman started pre-breathing pure oxygen and all six crew members began preparing for their part in supporting the second EVA. At 1533 Ross and Newman exited Endeavour's airlock feet first, pulling two large box-shaped antennae behind them. Currie used the RMS to lift Newman and the antennae half-way up Unity, while Ross made his way up the handrails on the outside of the module. The video downlink showed the Sun reflecting from the aluminium node, its 'UNITY' nameplate clearly visible. Once they were in position, they started work on installing the two antennae, one on either side. These were the antennae for the S-Band Early Communications System. Once the system was active, Houston would be able to send commands to, and receive data from, Unity via NASA's Tracking and Data Relay Satellite (TDRS) network. This would offer the ISS almost 24-hour communications, rather than the sporadic communications offered by the limited Russian network of ground stations. Ross and Newman then moved on to removing the launch restraint pins from the four exposed hatches on which other ISS elements would be installed by later missions. They next installed sunshades over Unity's two data relay boxes, and put thermal covers on three of the four metal pins which had supported Unity while it had sat on the payload bay cradle; Ross, who was rapidly acquiring a reputation as a litterbug, had inadvertently let go of the other thermal cover and it had drifted away.

As this second EVA drew to a close, Currie lifted Newman on the RMS to a point where he could use a grappling hook on the end of a 3-metre-long pole to nudge the TORU antenna which was next to Zarya's nadir docking port. After being prodded a few times, the antenna unfurled to its limit. The two spacewalkers then removed the cable by which PMA-1's androgynous system had been commanded while docking Unity with Zarya. This connection was to be permanent, so that particular command path would not be required again. The cable, which ran from PMA-2 along the outside of Unity to PMA-1, had been installed prior to launch because the command path would be required prior to the first EVA. Its removal now was a 'get ahead task' to save time on a later mission when PMA-2 would be relocated. Finally, Ross and Newman bundled up a number of the exterior cables and verified that they did not obstruct the OSVS alignment marks. They then returned to the airlock and began re-pressurisation procedures at 2235, thereby bringing this 7-hour 2-minute EVA to a close. With the EVA complete, Currie used the RMS cameras to inspect Endeavour's bay and to document the OSVS alignment targets on both Unity and Zarya.

Flight Day 8 saw a change of pace. There were to be no EVAs today. Instead they were to open six hatches:

- from Endeavour to the ODS;
- from the ODS to PMA-2;
- from PMA-2 to Unity;
- from Unity to PMA-1;
- from PMA-1 to Zarya's Pressurised Adapter (PA); and
- from Zarya's PA to the Instrument Module (Zarya's main compartment)

in order to gain successive access to all of the components of the nascent space station. As a preliminary, Endeavour's cabin pressure was first raised back to sea-level pressure and then equalised with that beyond the first hatch. At 1454 Cabana and Krikalev, representing America and Russia, opened the hatch to gain access to Unity. Wearing matching blue-and-white striped sweatshirts, they squeezed through the hatch together. Unity's lights had to be switched on manually, because the module had accidentally been launched with their power supply turned off. Krikalev turned somersaults to celebrate. He was due to return as a member of the Expedition 1 crew, and serve a 5-month tour of duty. The other crew members were hot on their heels. Cabana, Ross and Newman turned on Unity's fans and hooked up air ducts from Endeavour to remove carbon dioxide and humidity from the new module. Sturckow removed some wall panels to retrieve stored apparatus and installed it ready for future visiting crews. Ross and Newman made the internal connections so that the TDRS satellite link could be used. Cabana, Ross and Newman checked out the video-conferencing equipment on board Zarya. They spoke to the six American astronauts and five Russian cosmonauts at Houston comprising the first four resident crews, along with Krikalev.

Cabana and Krikalev shook hands and hugged one another, then moved deeper

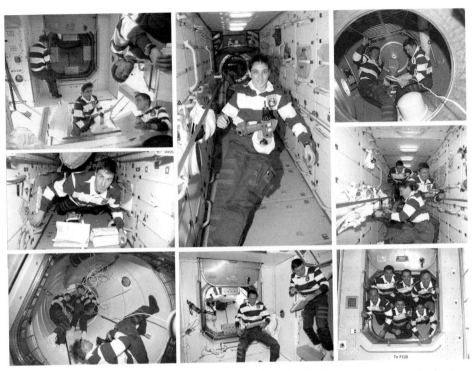

Once they boarded the International Space Station, the STS-88 astronauts checked out its systems and removed materials required only during launch.

into the station. At 1612 on 10 December they opened hatch from to Zarya together, and once again the others followed them in. Currie helped Krikalev to replace the balky power distribution system for Battery 1. Sturckow and Currie removed restraining bolts from some of the panels inside Zarya. The bolts had been installed prior to launch so that that none of the panels was shaken open by the vibration of launch. This was another 'get-ahead task' designed to assist later crews. At 2145 the entire crew gathered inside the ISS modules for radio and television interviews, before returning to Endeavour to sleep.

Back in Unity on Flight Day 9, Cabana, Currie and Sturckow resumed removing access panels and installing equipment, although they lost an hour searching for a lost pivot pin from one of the revolving racks. Following the removal of the launch restraint bolts the pin slipped behind the panel during an attempt to install it and was never retrieved.

Endeavour's crew increased the atmospheric pressure in Zarya to 1033 grams per square centimetre. As they closed each hatch, they set the pressure at a slightly lower rate than that in the previous compartment. The positive pressure on the 'inside' of each hatch would help to push it into its seal. Dessicant bags were installed in each of Unity's fans to remove humidity and the portable, battery-operated fans were left running. After a few final somersaults in Unity's spacious interior they turned off the lights and ventilation. The hatch to Unity, which had been open for some 28 hours 32 minutes, was closed at 1926 on 11 December. Cabana admitted that the crew had left behind a gift for the next crew to venture into the ISS, but he refused to say what this was.

In Endeavour once more, they again lowered the cabin pressure to 710 grams per square centimetre to shorten Ross and Newman's pre-breathing time before their third and final EVA. After being interviewed for American television at 2036, Ross and Newman spent the evening testing their EMUs and SAFERs and prepared the tools that they were to use. In an effort not to lose any further apparatus they fitted a second tether to each item. Later, in Houston, Greg Harbaugh, the chief of EVA assignments, explained to the press that the astronauts viewed the 162 EVAs which would be required to assemble the ISS as such a daunting prospect that they referred to this programme as 'The Wall'.

The final EVA took place on Flight Day 10. Ross and Newman left Endeavour's airlock at 1533 on 12 December. The first task was to release the ties on four of the cables that they had connected during their first EVA. During a television survey, Houston had identified excessive tension in these cables. If the crew did not relax the tension by removing the ties, the cables might not have enough 'play' to endure the constant expansion and contraction of the modules as they were warmed in sunlight and frozen in shadow during each orbit. They then inspected the cover on a cable connector that they had installed during their first EVA, confirming that it was fully insulated. After a box of wrenches, power tools and various EVA aids had been mounted on the exterior of PMA-2, Ross and Newman inspected some of the OSVS targets where television inspection had shown what appeared to be paint bubbling in the unfiltered sunlight. Currie used the RMS to lift Ross to a position where he could deploy Zarya's other TORU antenna, as before using the grappling hook. Once an EVA handrail had been mounted on the far end of Zarya, a photographic survey of the entire exterior of the two ISS modules was conducted.

Having packed their tools away, Ross and Newman both tested their SAFER units, while tethered to Endeavour. The telemetry indicated that the units used more nitrogen than expected, but the brief test firings were considered a success. The two men returned to Endeavour after an EVA lasting 6 hours 59 minutes, bringing their STS-88 total to 21 hours 22 minutes, and Ross's NASA record accumulated over seven spacewalks to 44 hours 9 minutes. Once they were back inside, Sturckow depressurised the tunnel between Endeavour's ODS and PMA-2 as a preliminary to undocking.

The wake-up call at 1136 on 13 December, for Flight Day 11, was the first time that the new ISS Control Room in Houston had called an ISS crew into action. After the final preparations, Endeavour undocked at 1525. Sturckow fired the thrusters to slowly back his spacecraft to a position 137 metres above the nascent ISS, then, beginning at 1545, took Endeavour on a nose-forward fly-around during which his colleagues used a variety of still, video and IMAX cameras to document the ISS. After the separation burn at 1649, the ISS slowly faded into Earth's horizon. Six hours later, the range was 112 kilometres, and was increasing by 30.5 kilometres per orbit. Commands were radioed up to turn the ISS so that Unity faced directly

After undocking, STS-88 performed a fly-around to document Zarya connected to the Unity node by PMA-1, with PMA-2 at the far end.

downward. Another command put the station into a passive thermal control roll around its primary axis, completing one rotation every 30 minutes (three times per orbit). Over the coming months, both American and Russian flight controllers would check the two modules. Zarya's motion control system would be activated once per week. Endeavour had left the ISS in a nearly circular 395-kilometre orbit, but this would fall. Zarya's guidance system would be periodically updated for the latest orbital parameters.

After a few hours of free time, the STS-88 crew set up a number of non-ISS-related tasks. Cabana fired one of Endeavour's OAMS thrusters at 2115 as part of a Department of Defense test called SIMPLEX, in which an Earth station in Peru would observe the engine firing. At 2230 the Shuttle's crew gathered together to be interviewed for American radio.

At 2331 Cabana, Sturckow and Ross worked together to deploy the SAC-A satellite from the payload bay for Argentina. As well as five technology experiments, this cube-shaped package had a device to track whales in the South Atlantic. The day ended with Ross, Newman and Krikalev storing much of the equipment used in the final EVA. Meanwhile, Currie returned the cabin pressure to 1,033 grams per square centimetre preparatory to the return to Earth. By wake-up for Flight Day 12, SAC-A was 56 kilometres in trail of Endeavour.

Cabana and Sturckow spent the day preparing for re-entry and landing. At 1530 they began checking the flight control systems and aerodynamic surfaces. One hour later they fired Endeavour's thrusters to verify their status. At 1800 they all gathered for the traditional end-of-flight press conference, which this time included the Canadian Space Agency Headquarters at St Hubert in Quebec. After three hours of free time Cabana, Sturckow and Ross worked together to deploy MightySat at 2109 on 14 December, a US Air Force and Phillips Laboratory satellite designed to demonstrate advanced technologies, including the composites used in its construction and photovoltaic arrays. As their last full day drew to a close, bad weather in Florida threatened to cancel the first re-entry opportunity, in which case they would land at Edwards Air Force Base in California.

Flight Day 13 saw the final preparations for return to Earth. The Ku-Band antenna was retracted and then the payload bay doors were closed at 1907 on 15 December, and the crew strapped into their seats. Retrofire at 2147 involved the two large engines in the OAMS pods being fired for 3 minutes 8 seconds to slow the vehicle and lower its trajectory to intersect the atmosphere an hour later. Endeavour was reoriented into a nose-forward attitude so that it could present its belly to the ever-thickening atmosphere following the entry interface, which occurred at 2222. Ionised molecules surrounded the spacecraft, causing a radio blackout lasting several minutes. Endeavour's computers managed the transition from a spacecraft to a heavy glider and the aerodynamic surfaces superseded her reaction control system as the main form of in-flight control. A series of large 'S' turns in the atmosphere helped Endeavour to shed energy. NASA Lockheed T-38 chase planes climbed up to meet the descending spacecraft. One plane flew around Endeavour while its pilot made a visual check of its exterior. Cabana read off his altitude and velocity, and the chase plane pilot confirmed that Endeavour's instruments were reading accurately.

Crossing the Gulf of Mexico, Endeavour fell through the air as it approached Florida from the southwest. Flying across the Space Center, it made a sweeping 299-degree turn out over the Atlantic Ocean. This was the Heading Alignment Circle. At the end of the turn, Endeavour lined up with the Shuttle Landing Facility, Runway 15 at KSC, and pursued the characteristic steep straight-in approach. The landing site's microwave landing system constantly updated the computers. As all of his predecessors had done before, Cabana assumed manual control for the landing itself.

The chase plane pilot continued to talk Cabana down, reading off the distance to touchdown. To those waiting on the ground, Endeavour emerged suddenly out of the darkness into the area of sky lit by the runway lights. Cabana's professionalism brought Endeavour down to a perfect two-point landing at 2253. This was only the tenth night landing in the Shuttle programme. Endeavour rolled down the runway on its rear undercarriage. Cabana lowered the nose wheel to meet the runway. As Endeavour rolled out at high speed, Cabana opened the tail-mounted airbrake and deployed the red and white drag parachute. Then he gingerly applied mechanical breaking to the wheels and Endeavour slowed to wheel-stop at 2254. A convoy of recovery vehicles gathered around the waiting orbiter and men in pressurised suits plugged hoses into its propulsion system to draw the toxic fumes safely away to a closed tank, while others completed a series of close-out tasks on the vehicle. An hour later, the crew were granted permission to exit by the side hatch. They were greeted on the runway by a crowd which included Dan Goldin. The first ISS assembly misson was over, but the task had barely begun.

ROUTINE OPERATIONS

With Endeavour back on the ground, controllers in Houston and Korolev continued to monitor the ISS's systems by telemetry. On 16 December, in a routine that would be undertaken more or less on a weekly basis, Zarya's computer was updated with the latest orbital parameters. On this occasion, both of the main engines were fired together for the first time. The manoeuvre raised the orbit to 411 × 397 kilometres with a period of 92 minutes. Two more engine firings on 21 December boosted the ISS into an orbit from which it would slowly drift down until it was at the proper altitude for the arrival of the Service Module, whose launch was then expected mid-1999. These two 5-second burns occurred 45 minutes – half an orbit – apart, and raised the station's perigee to 399 kilometres. The ISS was returned to its passive roll attitude. No further major manoeuvres were scheduled prior to the Service Module's launch. During the week the performance of the Kurs and TORU systems were verified and all six of Zarya's batteries were 'deep cycled' (that is, given a full discharge followed by a full recharge) in what was intended to become a fortnightly routine.

On 28 December, the ISS was commanded into an attitude that maximised power generation via the solar panels. The test was intended to demonstrate that the station could adopt a 'power-friendly attitude' in the case of low battery levels. After the successful test the ISS resumed its passive thermal roll, where it remained for the next week. In the week 7 to 13 January 1999 a decrease in the voltage provided by Zarya's six batteries was noticed. On 10 January some of the heaters in PMA-1 were turned off to reduce the load, and then several of Zarya's smoke detectors switched off automatically to further reduce the drain on the power supply. Housekeeping tasks were commanded to improve the battery performance. In the routine deep-cycling activity, the batteries were discharged and recharged one by one, and it was decided to reduce the time between cycles from once every two weeks to once "every few days". This returned the batteries to their optimum power output. Weekly cycling became the routine while flight controllers continued to monitor their condition.

On 27 January the American Early Communications System which had been installed in Unity by the crew of STS-88 was used to modify the passive thermal roll, to demonstrate the ability to relay commands to Zarya's systems via Unity, as a contingency for a failure in the Russian communications network. Meanwhile, everything continued to go well in orbit. Throughout February, flight controllers continued to test their ability to command Zarya via the Early Communications System. The tests included demonstrating general command sequences by way of geostationary the TDRS relays. The passive thermal roll was adjusted to conserve propellant by minimising thruster firings. On Friday 5 March, Houston noted a degradation of communications via one of two Early Communications System antennae. In given attitudes there was a 15 per cent drop in the starboard antenna's performance. Controllers immediately turned off the starboard antenna and switched communications through the port antenna. Troubleshooting took place throughout the week. On 16 March, Houston turned on an external television camera mounted on Zarya in an effort to inspect the faulty antenna on Unity, but this inspection was inconclusive. Possible causes under consideration ranged from loose insulation to possible damage by orbiting debris. During the following week, it was determined that the antenna could be used without causing further damage, and it was reselected and used for communications later in the week. During that same week, Korolev completed a routine test of Zarya's Kurs automatic rendezvous and docking system.

In the first week of April, flight controllers turned their attention to systems tests of the Zarya–Unity combination, to prepare for STS-96's launch in May. The first of three tests confirmed that Zarya could support a high call for electrical power by Unity. Turning on several heaters in Unity steadily increased the power drawn from Zarya's storage batteries. The test showed that Zarya could continuously feed almost 1 kW to Unity in orbital conditions similar to those that would apply during the flight of STS-96. Meanwhile, Houston had drawn up plans for the Shuttle's RMS end-effector television camera to inspect the malfunctioning omni-antenna on Unity, although no EVA repair activity was scheduled at this time. The second power test commenced on 14 April, and it involved repositioning the station using Zarya's motion control system to place the ISS side on, rather than vertical with respect to the Earth. The test involved proving Zarya's capability to provide 1.5 kW to Unity upon demand, in the attitude required during a Shuttle docking. Power levels were maintained until the evening, then Zarya resumed its earlier vertical attitude with Unity aimed downward. The following week an erroneous command was sent to the

ISS by Korolev. When a single digit was misunderstood and transmitted incorrectly, it was received as a command to power up one of Zarya's solar panel retraction motors. Although the power was initially sent to the motor, the computer recognised the command as invalid and turned off the power supply before the motor started to retract the panel. Later that week, the link via the Early Communications System's right-hand high-gain antenna was lost. It was decided that a transmitter was at fault, but a replacement was already onboard the ISS. To overcome the problem, all future communications were sent via Korolev and the left-hand high-gain antenna. As a backup, the option remained of utilising the omni-antennae (one of which was operating in a degraded state) at a much lower data rate. The third power test was completed later in that week. Korolev up-linked and verified a patch to Zarya's software which would inhibit Zarya's main engines from firing while a Shuttle was in close proximity or docked to the ISS; during such times it would be able to use only its small attitude control thrusters. In the first week of May, Korolev noticed a false indication from one of Zarya's eight internal smoke detectors; the faulty detector was turned off.

On 12 May, the ISS was manoeuvred to the orientation parallel to the horizon, with Zarya facing the direction of travel. The guidance system was calibrated using the horizon as a reference. After 3 hours the station resumed its vertical orientation, but this time with Zarya pointing down and without reinstating the passive thermal control roll, which was how it would be during STS-96's rendezvous and docking. After 4 hours in this attitude, it rotated 180 degrees so that Unity once again faced Earth and resumed its passive thermal control roll. On 24 May controllers began activating the heaters on the ISS in preparation for the next Shuttle.

STS-96 LOGISTICS FLOW

Having inserted STS-96 into the schedule as a cargo-hauling mission in advance of the launch of the much-delayed Service Module (which the Russians had decided to name 'Zvezda', meaning 'star') the Shuttle now faced a complication, because it would be required to dock with the ISS without the Service Module's sophisticated attitude control system. In order to do this, consideration was given to Discovery pulling alongside the ISS and grabbing Zarya with its RMS to make the docking, more or less as STS-88 had done, but this time having to manipulate a much heavier mass. However, simulations showed that the Zarya–Unity stack was too unwieldy. Instead, Zarya's more basic attitude control system would have to hold the ISS steady while the orbiter made its final approach. A software patch had already been up-linked to the module to inhibit its main engines from firing during these manoeuvres, and Korolev would deactivate its other thrusters immediately upon docking, after which the Shuttle would take control of the complex. The requirement to manually send the command to deactivate Zarya's thrusters meant that the Shuttle's approach had to be changed: rather than approaching from beneath the station, until it was in position directly in front of the ISS, it would now have to make its final approach from above to ensure that it would not block the antenna on

Zarya's exterior which would receive the vital command; and, of course, the docking would now have occur over a Russian ground station.

On 8 May 1999, STS-96 Discovery was in place on Launch Complex 39's Pad B, surrounded by the Rotating Service Structure, with just the top of its ET protruding from the Weather Protection System. After a hail storm on that date, pad workers found almost 650 pits ranging from 1 to 5 centimetres in diameter etched in the ET's thermal insulation. Water lodged in the insulation would freeze when the ET was loaded with cryogenic propellant. There was concern that if the small globs of ice fell out during the ascent at supersonic velocities they might strike Discovery's windows, so a scaffolding was erected to allow workers to repair most of the damage with the vehicle *in situ*. This scaffolding also facilitated an inspection of areas that otherwise were not accessible with the vehicle on the pad to determine whether it would be necessary to roll Discovery back to the VAB to effect repairs. A rollback threatened to delay the launch by a minimum of seven days. The decision to return STS-96 to the VAB was made on 13 May. On 15 May the SRB/ET stack for STS-93 was removed from the VAB's High Bay A and stored in the Mobile Launcher Platform Refurbishment Site. STS-96 was then returned to Bay A the next day. Consideration was given to transporting the STS-93 stack to the pad so that it could utilise the Weather Protection System, but it was decided not to do so. In the VAB, STS-96's ET was inspected to determine the scale of the repair task, which would involve sanding down the insulation so as to take out the hail damage, filling the resulting hole with new insulation and then resanding for a smooth finish. The repair was complete by 18 May. It was initially thought that predicted bad weather would delay Discovery's return on 20 May (which had been the original launch date) but the weather was kind and the STS-96 stack was returned to the pad on that date.

In fact, the Cape was not a very happy place in May 1999. The 174-day gap since STS-88 in December 1998 was the longest since the Challenger's loss in January 1986 had grounded the fleet for 975 days. That accident had been surrounded by the loss of four satellite launch vehicles. The superstitious Cape workers drew ominous comparisons between 1986 and 1998/9. Starting in August 1998, the Air Force had lost four Titan IV launch vehicles, one after the other. The failure of two Delta III launch vehicles to place their satellites into their correct orbits did little to improve the mood. Only on 19 May did a fifth Titan IV break the run of bad luck. The Flight Operations Division, acknowledging that a prolonged delay between launches would be detrimental to efficiency, had been running a five-month series of launch and mission simulations. The situation was not helped by Russia's continued public disagreement with America over the bombing of Yugoslavia. Nor was it helped by the continuing delays in the preparation of Zvezda, which had reached Baikonur on 20 May but was still several months away from launch. The Russians said that it would be ready by the end of September, but most Americans thought that it would be nearer the end of November. Some even suggested that it would not launch before the New Year. Any further delays would throw the 1999 Shuttle schedule into disarray. There was a real prospect that STS-96 would be the only flight to the ISS of the year, and that the launch of the Expedition 1 crew would have to be delayed into 2000.

STS-96 – Discovery – SSAF-2A.1

Commander:	Kent Rominger
Pilot:	Rick Husband
Mission Specialists:	Ellen Ochoa; Tamara Jernigan; Daniel Barry; Julie Payette (Canada); Valeri Tokarev (Russia)

Kent Rominger's crew arrived on 21 May, and the countdown was initiated on 24 May, with a view to launching on 27 May. On that morning, the crew awoke, underwent pre-flight medicals and then ate the traditional launch day breakfast, after which they went to the suiting-up room where they donned their orange suits. On time, they emerged from the Operations and Checkout Building and were driven out to Pad B. One by one, the seven astronauts strapped into their couches. Launch was on time at 0649 and the vehicle climbed into a hazy morning sky. At the time, the ISS was over the Atlantic Ocean, northwest of Bermuda. Less than 9 minutes later, Discovery was in orbit.

The principal payloads were a double Spacehab module loaded with supplies, spare parts and tools for the ISS, and an unpressurised Integrated Cargo Carrier with a pair of cranes, one American and one Russian, which would be mounted on the ISS's exterior by spacewalkers. The other payloads included the Student Tracked Atmospheric Research Satellite for Heuristic International Networking Equipment (STARSHINE) which was just a sphere with a highly polished surface to reflect sunlight which would allow students to visually track the satellite in orbit; the Shuttle Vibration Forces Experiment to assess the in-flight forces acting upon equipment; and the Orbiter Integrated Vehicle Health Monitoring Technology Demonstration, which would record the performance of various Shuttle systems to provide a basis for enhancing future safety and efficiency.

Having spent several days before launch adapting their body clocks to their intended schedule, Rominger's crew had actually launched late in their working day, so they began their sleep period a few hours after achieving orbit. By the start of Flight Day 2, Discovery was trailing the ISS by 1,435 kilometres, but this range was reducing by 111 kilometres per orbit. The only significant fault on this first full day was one of the four corner cameras in the payload bay, but this malfunction would not impair the mission. Rominger and his pilot Rick Husband spent the day checking out the vehicle's systems in preparation for the rendezvous and docking, planned for late the following day. At 2130 on 27 May, Rominger fired Discovery's thrusters in the first of two manoeuvres to place the orbiter on a looping approach to the ISS. Meanwhile, Ellen Ochoa and Julie Payette opened the hatch and inspected the Spacehab module. The three EMU's (two primes and one backup) were tested by Tamara Jernigan, Daniel Barry, Payette and Husband. The checks of the EMUs were carried out early in the flight so that there would be time to attempt to overcome any major problems, but no problems were identified. During the latter part of the workday Ochoa and Payette powered up the RMS for a series of tests in preparation for its use later in the flight. They made an inspection of the payload bay using the end-effector's camera. Jernigan and Ochoa then extended the outer ring of the Orbiter Docking System. As the workday came to an end, Discovery's internal pressure was lowered to shorten Jernigan and Barry's pre-breathing prior to their 29 May EVA. As part of the pre-sleep activities, Valeri Tokarev transferred some items from the flight deck to the Spacehab module in order to relieve the rather cramped conditions.

By the start of Flight Day 3 Discovery was 222 kilometres in trail of the ISS and closing in at 76 kilometres per orbit. At 2000 on 28 May, the ISS was manoeuvred so that Zarya was aimed towards Earth and Unity on top, for a docking. Zarya's revised computer software deactivated the main engines. At 2153, when Discovery was 14.8 kilometres from the ISS, Rominger performed the TI burn. This put the Shuttle on a trajectory that took it directly towards the ISS during the next orbit. He took manual control as Discovery closed to within a kilometre, flying it from the aft flight deck. At 2305, when 182 metres directly beneath the ISS, Rominger initiated a semicircular manoeuvre that took Discovery through a position 106 metres ahead of the ISS to a stable position 76.2 metres above it, half an hour later. From there, he began the slow descent towards docking, pausing at a distance of 51.8 metres for 15 minutes until within range of a Russian ground station, then he descended to a distance of 9.1 metres and paused once again. Ochoa was using the range-finder to monitor the range and range-rate. Jernigan was standing by with the docking mechanism. With everything verified, Rominger resumed the approach. Soft docking occurred at 0024 on 29 May, and Korolev promptly ordered Zarya to shut down its attitude control thrusters. As the two spacecraft came together for the first time, Rominger reported "We have contact!" The outer ring was retracted for a hard docking at 0039. Jernigan and Ochoa performed leak and pressurisation tests, then opened the outer hatch and stored the docking targets, video camera and lights that had been mounted within the tunnel for the docking. Then Rominger and Husband removed the CBM controllers. These four boxes of electronics supplied power to Unity's berthing ports. Removing them cleared the tunnel for the transshipment of bulky items from Discovery to the ISS. The units were put in storage in Unity until restored by a later crew in preparation for the arrival of the American Laboratory Module. Finally, they checked the hatch seals in the tunnel before closing the hatch again, because the flight plan did not require the ISS to be entered until 31 May. In any case, the hatch had to be closed during the coming EVA. Jernigan, Barry and Payette devoted the latter part of the day to testing the SAFER packs and the tools that they were to use while spacewalking.

Jernigan had trained to make two EVA's from Columbia in 1996, but these had had to be cancelled when a loose screw fouled the mechanism of the airlock's outer hatch. On Flight Day 4, there were no such difficulties. The hatch was opened at midnight and Jernigan shouted, "Okay, the hatch is unlatched. Unbelievable!" She was manoeuvred around the payload bay on the end of the RMS by Ochoa. Payette kept track of her activities and checked them off against the flight plan. Meanwhile, Barry scrambled around the ISS's exterior using a number of hand and foot holds and a slide-wire attachment. The first task was to remove the 1.52-metre-long US-built Orbital Transfer Device (OTD) from the carrier in the bay and affix it to PMA-1. As the ISS neared completion, this crane, which could be extended to 5.48 metres, would be installed on a cart that would run along rails on the truss structure and would be able to transfer loads of up to 272 kilograms. The second task was to affix the first elements of the Russian-built 'Strela' (Arrow) crane on PMA-2. Jernigan had to unscrew the bolts that held the various elements to the cargo pallet, a task that took an hour. The first two elements comprised the circular mounting plate and the 2.2-metre-tall operator's post. The 13.7-metre-long boom would be delivered by a subsequent mission. Eventually, the crane was to be assembled on Russia's Science Power Platform. After installing a pair of portable foot restraints on PMA-1, Jernigan and Barry opened a storage box in the payload bay and transferred three equipment bags into storage compartments on the ISS's exterior. This was another 'get ahead task' that would enable the tools to be available for a later crew. The next item was to install a thermal cover on the metal pin on Unity which could not be covered during STS-88 because Jerry Ross 'lost' that cover. Before retreating, they made a photographic inspection of the hulls of both Unity and Zarya. Finally, they inspected an area of paint on Zarya's exterior which appeared to be discoloured, and Unity's faulty starboard Early Communications System omni-antenna. Earlier in the flight, they had noticed a brownish stain in its vicinity, possibly the result of impingement by the exhaust from Zarya's thrusters. When the camera they were using to document these items jammed they had to fetch another camera before they could continue. They returned to the airlock without testing the OTD, because they had run out of time.

Back inside the airlock, Jernigan's bad luck appeared to come back to haunt her. With the airlock still depressurised, she had trouble plugging an airlock umbilical into her suit. Nearly two hours later, with the problem finally resolved, the airlock was pressurised and the two EVA crew members returned to the mid-deck. Including this delay, the EVA officially lasted some 7 hours 55 minutes, and it bought the ISS EVA total to 29 hours 17 minutes.

After breakfasting on Flight Day 5, the crew prepared to open the hatches and enter the ISS. This involved equalising the pressures at each of the six hatches in turn. Houston activated the lights in each compartment as they proceeded. Jernigan and Tokarev entered Zarya at 2207 on 30 May. There, they uplifted the 'floor' panels to replace 12 of the module's 18 MIRTS,[1] completing work on four of the module's six batteries. Russian ground stations commanded each battery system off and on as required. Later in the day, Barry and Tokarev placed insulation 'mufflers' around some of Zarya's noisiest equipment in order to minimise noise levels in the Russian-designed module from over 72 decibels down to 60 decibels. This done, they replaced a power distributor and a transceiver in Unity's Early Communications System. Meanwhile, Jernigan and Husband had mounted shelving in two stowage racks inside Unity. In the afternoon, the entire crew participated in the transfer from the Spacehab which contained 1,815 kilograms of equipment, most of which was stowed

---

[1] The Russian acronym for the Battery Charge Controller on Zarya.

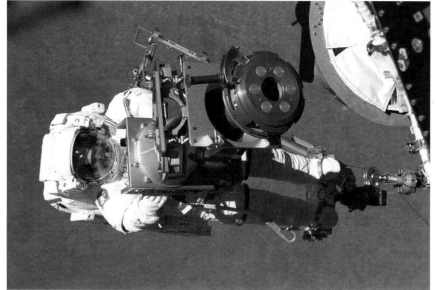

On STS-96, Tamara Jernigan, standing on a work station on the RMS, installed the mounts for Russian (left) and American (upper right) cranes.

in 120 large canvas bags. As mission loadmaster, Ochoa provided direction and checked off each item as it was transferred to the ISS. Overall, three days had been set aside for crossloading. As the day drew to a close Rominger, Jernigan and Barry discussed the mission so far with a number of American television networks.

Husband, Ochoa, Jernigan and Barry devoted most of Flight Day 6 to the logistics transfer. Meanwhile, Payette and Tokarev replaced the MIRTS on Zarya's other two batteries and reset the panelling. While Barry and Tokarev resumed installing acoustic insulation, Rominger relayed video of the mufflers on portions of the air ducting that circulated the air, to indicate to the engineers that they were causing some of the flexible ducts to collapse, restricting the airflow through the ISS and raising temperature to 78 °F and the humidity to 60 per cent. Although this was within acceptable limits, it made life uncomfortable when performing physically intensive work such as moving large canvas bags of stores throughout the complex. When the worst point of constriction was identified, that panel of insulation was removed. With the insulation in place, Barry made a series of readings of the noise level at various positions. It turned out that the mufflers had reduced the noise level within Zarya by only 3 decibels to levels between 65 and 69 decibels. If this could not be reduced further by some means, then long-duration crews might be obliged to wear earplugs. The day ended with a briefing by Payette on how well the transfer was progressing: the task was more or less 50 per cent complete, and many of the larger items had been loaded on board.

Flight Day 7 was devoted to the continuing cargo transfer to the ISS. Payette and Rominger exchanged greetings with Canada's Prime Minister Jean Cretien and Science Minister John Manley, and they answered questions put by children from Canada. Rominger, Husband and Ochoa were interviewed by the US television networks. Ochoa ended the working day with a second transfer briefing.

Once the cargo transfer had been completed on Flight Day 8, Ochoa used the RMS to conduct an external video survey of the starboard Early Communications System antenna on Unity. Some extra tasks were performed, including installing elements of a strain gauge system to measure the structural loads that were imposed as more modules were added, checking smoke alarms and cleaning filters. The final task was to verify that each of the stowage bags was secure and to photograph it. The concern was that if these bags came free they might alter the complex's centre of mass, which would complicate the task of controlling its orientation during the forthcoming rendezvous and docking with Zvezda. Leaving the ISS for the final time, Rominger's crew switched off the lights and shut each hatch in turn as they retreated to Discovery. By the time they shut the final hatch, at 0444 on 3 June, they had spent a total of 79 hours 30 minutes inside the ISS. Shortly after 0530, Discovery's thrusters were fired a total of 17 times during a 37-minute interval in order to raise the station's altitude by 9.6 kilometres. The ISS was now in a 395.8 × 387.8-kilometre orbit. By the time Zvezda arrived, this would have decayed to an ideal 357.2 kilometres. Husband undocked Discovery at 1839 on 3 June, on Flight Day 9, then withdrew 122 metres to make a 2.5-circuit fly-around to photograph the ISS's exterior, before manoeuvring clear at 2053.

Jernigan and Barry spent their evening stowing the EMUs in the airlock, and

Ellen Ochoa in STS-96's heavily-loaded Spacehab cargo module (upper right), whose contents she and Julie Payette (left) stored in Zarya.

Payette and Tokarev helped to stow equipment. Meanwhile, Rominger and Husband spent some time flying simulated landings on their laptop computers to upgrade ther proficiency after a week in weightlessness. The afternoon was declared time off. Flight Day 10 saw the initial preparation for re-entry. Ochoa assisted Rominger and Husband on the flight deck, as they test fired Discovery's thrusters. They then activated one of the Shuttle's three Auxiliary Power Units and tested the orbiter's aerodynamic surfaces. At 0321 on 5 June Payette ejected the beachball-size STARSHINE from its canister in the payload bay.

Flight Day 11 began at 1650 on 5 June with the music from the classic film *The Longest Day* to commemorate the imminent 55th anniversary of the Allied landings in Normandy, France, on 6 June 1944. With the preliminaries complete, Rominger manoeuvred Discovery to a heads-down tail-first attitude at 0020 on 6 June. The first retrofire opportunity was just 34 minutes away and meteorologists were still monitoring the weather at the Cape carefully. At 0036, Flight Director Linda Ham gave the 'go' for re-entry at the first landing opportunity. At 0054 Rominger fired the OAMS engines in a 3-minute 36-second de-orbit burn and then turned heads-up nose-forward for re-entry. To observers on the ground Discovery emerged from the darkness at the end of Runway 15. Rominger held the orbiter's nose up as the chase plane pilot counted down the last few metres. Finally, the two main landing gear touched down and Rominger lowered the nose. STS-96 was home. As the wheels stopped, the 94th Shuttle mission came to a close after 9 days 19 hours 13 minutes and 57 seconds.

BAD AIR: THE TRUTH ABOUT LIFE ON BOARD THE ISS

Two weeks after landing, the STS-96 astronauts admitted that they had all suffered headaches, nausea and vomiting, and had a burning sensation in their eyes and on their skin while working onboard the ISS. Finally, they periodically had to return to the Shuttle to relieve the symptoms, and even then it took as much as two hours to recover. They had not reported this at the time in order to preclude the flight being cut short.

NASA's engineers put the problems down to there having been too much carbon dioxide in the station's atmosphere as a result of inadequate air circulation. The carbon dioxide exhaled by an astronaut working in a given location would remain nearby, with the result that successive inhalations would become enriched in carbon dioxide. The situation was worst when several astronauts worked inside Zarya at the same time. It was also uncomfortable when the entire crew held their end-of-flight press conference in Unity. Another post-flight complaint was of the strong odour of solvent. When they removed wall panels to install the sound insulation material in Zarya the problem was made worse but no additional difficulties were experienced when they removed the floor panels to repair Zarya's batteries. They thought that it might have been due to the presence of large amounts of Velcro. Upon retreating to Discovery's cabin, they were able to recover in an atmosphere that was efficiently recirculating to remove carbon dioxide. NASA set up a 'Tiger Team' of various specialists to resolve this issue as a matter of priority. A rapid correction was vital because once the first resident crew moved in, they would not have the option of retreating until their symptoms cleared. Their docked Soyuz would be in a powered down state, so its life support systems would not be on. As an expediency, it was decided that each crew member on STS-101 – the next mission, scheduled for December – would be issued a personal air monitor to indicate the quality of the air in which they were working.

RETURN TO AUTOMATED FLIGHT REGIME

With Discovery gone, flight controllers commanded the ISS to assume a vertical attitude with Unity at the bottom, and reset the passive thermal control roll. When US Space Command warned that an inert rocket stage would come within a few kilometres of the ISS, American and Russian flight controllers worked to draw up a command to manoeuvre to open the range. It was sent to the ISS by Korolev on 12 June, but possibly because controllers had forgotten to factor in the centre of mass following the STS-96 cargo shipment, the command was incorrect and Zarya's computer rejected it and shut down its guidance system. No longer under positive control, the ISS began to slowly tumble end over end. The station made one orbit of Earth before the guidance system was ordered back on line. The sequence of events was subjected to a thorough review which confirmed that the relevant data had indeed been correctly entered into the station's computer after the STS-96 flight. It was decided that in similar circumstances in the future the command sequence would be fully tested prior to being transmitted. In the event, on this occasion the spent rocket came no nearer the ISS than 7.24 kilometres.

The six-monthly service procedure for Zarya's batteries was started the following week. Although discharging and recharging Battery 6 was straightforward, the next one, Battery 1, failed to fully discharge. It was ordered off-line while a 'training' procedure was devised prior to trying the other four batteries. This maintenance was in addition to the regular cycling of the batteries. In the week ending 8 July, a new problem arose with communications via the Early Communications System when using Unity's left-hand antenna. While troubleshooting continued, the possibility remained open for having STS-101 replace the balky antenna. Maintenance of Zarya's batteries resumed on 17 July, when Battery 2 was fully discharged and charged up again. The onboard software began to cycle Battery 1 in parallel. While this was not as intended, a review showed that two batteries could be safely cycled at once, so it was allowed to continue. Battery 3 was next, followed by Battery 5, but Battery 4 would not be cycled until the week ending 12 August. With Battery 1 having a different voltage output to the others, the Russians raised the option of STS-101 delivering a replacement battery and associated electronics in the hope that this would solve the problem once and for all.

During the week ending 19 August, a test was made of Zarya's Kurs system preparatory to the launch of Zvezda. For this test, the ISS manoeuvred into the standard docking attitude. Once the test was complete, the ISS resumed its vertical orientation and reinitiated the passive thermal roll. The station continued to perform well, despite one of Zarya's six batteries having been disconnected. On the ground, plans continued to be refined for the battery's replacement during STS-101's visit. In the final week of August, Korolev ordered Battery 1 on line for a single orbit in order to monitor its performance. Although it operated normally, its performance remained slightly below average. The engineers attributed the slight degradation to one of its cells not operating properly. Suspecting that the larger performance problem that had prompted its disconnection in August may have been an isolated event in the ancilliary electronics that would not recur, it was reconnected on 3 September for three orbits (some four and a half hours) to gather more data. This revealed that the battery was charging and discharging in an unexpected manner, so it was once again disconnected so that the engineers could consider its status. In the week ending 16 September, Zarya's propellant pumps were tested by pumping nitrogen gas from tank to tank to simulate the flow of fuel and oxidiser. The transfer of propellants was planned for October, in preparation for the docking of Zvezda. One week later, while one of the five functional batteries was being routinely cycled, the remaining four batteries became overloaded so some non-essential systems that were being operated at the time were switched off and the ISS was reoriented so that its photovoltaic arrays faced the Sun at a more favourable angle in order to supplement the output from the batteries. The week ending October 7 was spent verifying backup command links through the Early Communications System, as commands were sent to Zarya via TDRS and Unity's antenna. The system was verified independently by both Houston and Korolev. Zarya's motion control system was also activated to allow the passive thermal roll to be monitored. This action was repeated each Sunday for the next month. The following week, the Early Communications System was switched from Unity's left-hand to its right-hand antenna.

On 24 October, the US Space Command alerted NASA to an inert Pegasus rocket stage which would pass within 1.4 kilometres of the ISS in the early hours of 27 October. As a result, Zarya's motion control system reoriented the ISS for a manoeuvre which would move it clear of this space debris. Thirty-three minutes later, Zarya's two main engines were fired for 5 seconds to raise the station's orbit by 1.5 kilometres, thereby slowing it so that by the time of the encounter it would be 25 kilometres from the spent stage. After the manoeuvre, the ISS resumed its standard orientation. On 28 October, tests verified that each control centre could communicate with its own module using the communication systems in its counterpart's module. In the week ending 4 November, Zarya's Batteries 2 to 5 started their regular cycling process. Battery 1 remained disconnected. The second phase of 'deep cycle' maintenance (a year into Zarya's mission) was initiated on 18 November. Although Battery 2 was fully discharged and recharged, it then failed to discharge on the second round; it was disconnected from the electrical bus and some of Unity's heaters were turned off to minimise power usage. The ISS could run in this semi-dormant state on Zarya's four batteries. Once Battery 3 had been deep cycled without incident, Battery 1 was brought on line on 28 November while a review considered whether to proceed with deep cycling batteries 4, 5 and 6. Battery 1 functioned properly for one and a half weeks, and then failed to discharge correctly, so it was disconnected once again. These battery problems were surprising.

On 1 December, Zarya performed two manoeuvres to put the ISS in a 392 x 380-kilometre orbit. Both burns utilised both of Zarya's main engines. During the week ending 9 December, Zarya's Kurs was tested, and this time its relative onboard velocity readings differed from those shown by radar tracking, so the test was ordered to be repeated to determine whether interference from other ISS systems was responsible for this discrepancy. A software patch which was uplinked to Zarya on 8 December meant that an additional 68 electrical power system parameters could be transmitted via the Early Communications System so that it could be monitored more effectively.

In early January 2000, with Zvezda's preparation running far behind schedule, Houston began to consider dispatching STS-101 prior to Zvezda's launch in order to replace Zarya's malfunctioning batteries and their ancilliary equipment. It would also provide an opportunity to deliver a load of equipment ready for the Expedition 1 crew.

On 15 January one of Unity's Remote Power Controller Modules (RPCMs) which routed electrical power to the node's systems suffered a momentary problem while performing its self-test. While Unity was in its semi-dormant state, only two RPCMs were running. The other two were brought on line only when a Shuttle was docked. The RPCM recovered almost instantly and the other on-line unit ran normally, but Houston started an investigation to find out what had happened. Meanwhile, the backup channel for the drive for Zarya's photovoltaic array systems was tested for three hours, and when it was confirmed to be working satisfactorily, the primary was reinstated.

On 27 January, ISS managers elected to "protect the option to fly" STS-101 to perform maintenance work on Zarya in advance of the delayed Zvezda launch. A decision would not be made until a Joint Programme Review meeting in February had assessed the status of the Proton rocket, which had recently suffered several failures. If it was decided to proceed, this mission would be "not earlier than 13 April". In effect, STS-101 was 'half' of the originally manifested STS-106 mission. The commander, pilot and two of the mission specialists from STS-106 were to be reassigned to STS-101 and the crew rounded out by the three Expedition 2 crew members. The other STS-106 crew members, whose work assignments required Zvezda to be in place, were to remain on that mission and fly with the rest of the original STS-101 crew, and the mission would be flown approximately 30 days after Zvezda docked, whenever that was.

Meanwhile, analysis of photographs taken during the STS-96 fly-around had revealed that Jernigan and Barry had installed the Orbital Transfer Device (OTD) in the wrong position. Investigations on the ground showed that the OTD was loose in its interface socket. Planning continued to determine whether it would be able to carry out its assignments from its current position, or whether it would have to be relocated. At that time, it was not known if the next crew to visit the ISS would perform any EVA tasks, or whether the task would have to be passed downstream.

In Washington, on 3 February, Dan Goldin announced that if the Russians failed to launch Zvezda in July, NASA would launch the Interim Control Module (ICM) to the ISS on a redesignated STS-106 in December. Furthermore, if Zvezda was not in place by the end of August it would become necessary to reschedule the launch of the Expedition 1 crew. Meanwhile, rumours were rife that the Russians had found private investment for the Mir space station and were planning to reopen it. Goldin told reporters "The Russians have got to decide if they want to work peacefully with other countries or do they have issues so important to their national interests they want to break their commitment. We are at the moment of truth. It is up to the Russians to demonstrate to America and the international partners that they are committed." The fact was, however, that even if the Russians had launched Zvezda on time, the assembly of the ISS would then have stalled because, as Goldin had to admit, the development of his agency's laboratory module was almost 18 months behind schedule.

On 16 February, Zarya's Kurs system was subject to two tests, with the Early Communications System's antenna set on high and low gain. The fact that it performed perfectly implied that the anomalous velocity readings experienced in December 1999 were not due to electromagnetic interference from this system. Meanwhile, Battery 5 began to show the same symptoms as Battery 1. It was kept on-line, so that its performance could be monitored.

After the Joint Programme Review meeting, plans for STS-101 were finalised as an additional logistics flight. The tasking was still under review, but it was clear that the crew would have to replace some of Zarya's batteries and an EVA by James Voss and Jeffrey Williams would ensure that the OTD was securely mounted. A meeting of the Space Station Control Board agreed yet another new flight schedule when they met in March:

Table 13.1 ISS assembly sequence (circa March 2000)

Date	Manifest	Launch vehicle	Payload
17 Apr 2000	2A.2a	STS-101	Double Spacehab
– Jul 2000	1R	Proton	Zvezda
19 Aug 2000	2A.2b	STS-106	Double Spacehab
21 Sep 2000	3A	STS-92	Z-1 Truss Structure, PMA-3
30 Oct 2000	2R	Soyuz	Expedition 1
30 Nov 2000	4A	STS-97	P-6 photovoltaic array unit
18 Jan 2001	5A	STS-98	American Laboratory
9 Feb 2001	4R	Soyuz	Russian Docking Compartment
15 Feb 2001	5A.1	STS-102	MPLM Leonardo
19 Apr 2001	6A	STS-100	MPLM Raffaello, SSRMS
30 Apr 2001	2S	Soyuz	Lifeboat replacement
17 May 2001	7A	STS-104	Joint Airlock Module
21 Jun 2001	7A.1	STS-105	MPLM Donnatello
23 Oct 2001	UF-1	STS-109	Utilisation flight

GOOD NEWS AND BAD NEWS

In early 2000, the Russian Space Agency announced that plans were underway to send another Soyuz crew to Mir to prepare that venerable station for a privately-funded mission lasting up to 45 days. Meanwhile, there were conflicting reports regarding whether Mir would then finally be de-orbited. Despite concerns in some quarters that Mir's reoccupation would draw vital Russian resources away from the ISS, NASA released information that had been presented to the Joint Programme Review that 6 Soyuz and 14 Progress vehicles were in production and so there was no real conflict – one Soyuz could be used to launch a crew to Mir and the second could launch the Expedition 1 crew to the ISS, while the third could serve as a backup to be used to launch a crew to the ISS in the event that it proved necessary to dock Zvezda manually using TORU system. A Progress tanker would de-orbit Mir following this final period of occupancy, three would be needed by the ISS between Zvezda's docking and the delivery of NASA's laboratory, with six more following at two-monthly intervals in the year after that. At about this same time, Boeing completed the preliminary design of an American Propulsion Module. But even if Congress allocated $300 million for its construction, it would not be ready until early 2003. Not only would a Shuttle flight have to be set aside to ferry it up, but Shuttles would have to be configured to carry propellant to replenish it while it was docked at the station. As a contingency against the Russians failing to supply the requisite number of Progress tankers, it was decided to spend an additional $240 million modifying the Shuttles so that their forward RCS could make the ISS reboost burns.

On a rather less positive note, 16 March saw representatives of the American Government Accounting Office (GAO) testify before the House of Representatives Space and Aeronautics Subcommittee and criticise the Russian Zarya and Zvezda elements of the ISS for failing to meet NASA's published safety standards in four key areas. The most serious flaw was the noise level in Zarya (and most likely Zvezda, also). Fans, valves and motors performing their normal operations produced noise levels well in excess of standards set for NASA's modules by the US Occupational Safety and Health Administration. The levels were high enough to make normal communication difficult, and even to mask a caution and warning alarm. Cosmonauts and astronauts serving tours of duty on board Mir had complained about the noise level in that space station, and some had even suffered sufficient hearing loss to preclude them from future missions. NASA's initial upper level was 55 decibels, and when it became obvious that the Russian modules would not meet this the requirement was raised to 60, but the first crew to enter Zarya measured levels in excess of 70. The second crew had installed mufflers behind the wall and floor panels, but this had reduced the level only to about 65 decibels, so NASA had asked the Expedition 1 crew to wear ear defenders when working in Zarya for long periods, which was a move hardly likely to improve their ability to hear a caution and warning alarm. The second cause for concern was the lack of protection against impact from space debris. The Russians had argued that installing the required shielding on Zvezda before it was launched would make it too heavy for the Proton to lift, so NASA had issued waivers following a Russian declaration that extra shielding would be installed in orbit to bring Zarya and Zvezda up to standard within four years. The third item on the list was the capability of the Russian portholes to withstand long-term exposure to the space environment (a strange matter for the GAO to raise, considering that Mir's portholes had not developed any problems during its extended lifetime). Finally, the GAO expressed concern that the Russian apparatus would be able to function in the event of a decompression accident. The computers that controlled the station's attitude were positioned inside Zarya and were air cooled, so the circuitry would overheat and fail if that module was exposed to vacuum, leaving the station without positive control of its attitude.

NASA blamed "shortfalls in Russian funding, designs based on existing Russian hardware, and technical disagreements between Russian engineers" for all these deficiencies. Allen Li, the associate director of GAO's National Security and International Affairs Division, told the hearing, "the lack of approval currently stands in the way of [Zvezda's] launch". NASA signed the appropriate waivers and in the final week in March Zvezda was cleared for launch in July 2000.

STS-101 'EXTRA' MISSION

STS-101 – Atlantis – SSAF-2A.2a

Commander:	James Halsell
Pilot:	Scott Horowitz
Mission Specialists:	Mary Ellen Weber; Jeffery Williams; James Voss; Susan Helms; Yuri Usachev (Russia)

The STS-101 crew was originally named as James Halsell, Scott Horowitz, Mary Ellen Weber and Jeffery Williams, with Edward Lu and Russian cosmonauts Yuri Malenchenko and Boris Morukov as mission specialists. It was to have followed the launch of Zvezda with a double Spacehab of equipment for the Expedition 1 crew, who would fly up in a Soyuz later in the year. When it was decided to split this manifest over two flights, Lu, Malenchenko and Morukov were reassigned to STS-106, which would fly the second 'half', and their places on STS-101 were taken by James Voss, Susan Helms and Yuri Usachev, the Expedition 2 crew, who would get an early look at the ISS.

With Zvezda's launch stalled by the requirement to recertify the Proton launch vehicle, Zarya was approaching the end of its 'guaranteed' life as a stand-alone module with responsibility for attitude control and other crucial functions. By this point, two of Zarya's batteries had been switched off due to problems with their ancilliary systems. Also, one of two communications antennae associated with the American Early Communications System had failed and the other was degraded by 60 per cent. The STS-101 crew were tasked with these repairs.

This would be Atlantis's first mission after a refurbishment lasting 18-months. Its flight deck was now a 'glass cockpit' of eleven flat-screen Multifunction Displays which replaced a large number of the original electro-mechanical instruments.

STS-101's launch had originally been scheduled for 13 April, but when Halsell sprained his ankle on 15 March the launch date was rescheduled for "not earlier than 24 April". The stack was rolled out to Launch Complex 39's Pad A on 25 March. When testing revealed an anomalously high pressure in the Power Drive Unit (PDU) of the rudder/speed brake on the orbiter's tail, it was decided to 'borrow' the PDU off Columbia, which was undergoing refit at the Rockwell plant in California. Columbia's PDU was removed on 8/9 April and flown across country on 11 April. The Rotating Service Structure was moved clear of the vehicle, to facilitate access. The hydraulic lines leading to the PDU were frozen using liquid nitrogen to prevent air penetration. With the new PDU in place, the Rotating Service Structure was reinstated on 13 April. Engineers took advantage of this cycle to replace a box of electronics in Atlantis's aft compartment and a fluid discharge hose of Auxiliary Power Unit 1. System tests were run throughout the weekend and the projected launch date of 24 April was confirmed, with lift-off set for 1615, at the start of a five-minute window.

The crew flew to Florida on 21 April and the countdown began at 1900 that evening. At 1100 on 23 April the countdown entered a 13-hour 22-minute planned hold. It resumed at 0022 on what would hopefully be launch day. The weather forecasters predicted a 30 per cent chance of a thunderstorms. At $T-9$ minutes the count entered a planned 40-minute hold. During the hold the lift-off time was reassigned to 1617:17, with the launch window closing at 1622:19. The crosswinds at the Shuttle Landing Facility rose above the 17-knot limit so at 1606 the count, still holding at $T-9$ minutes, was scrubbed. A 24-hour recycle was ordered, with the window opening at 1552. On 25 April, the weather offered a less than 20 per cent chance of launch. By 1400, the winds were at 34 knots. At 1420 Launch Director David King scrubbed the launch and a second 24-hour recycle was ordered for 1526 on 26 April. This time, although Atlantis stood under a clear sky, two of three Trans-Atlantic Abort sites were heavily clouded and a weather front was approaching the third site. With no technical issues on the spacecraft, the count continued to the $T-9$ minute hold, which was planned to last 40 minutes. With both Spanish abort sites still overcast and excessive crosswinds at the site in Morocco, the third scrub in as many days was ordered at 1521. With the Eastern Test Range reserved for other vehicles, the next launch opportunity would not be until 18 May, so the next launch was tentatively set for 0600 on that day.

Meanwhile, increased solar activity had prompted the Earth's atmosphere to inflate. At the end of April the ISS's orbit had decayed to 365×339 kilometres, and the drag was eroding this by 2.7 kilometres per week. In the absence of Zvezda, Atlantis was to have performed a reboost manoeuvre. Following the scubs, the ISS controllers had to decide whether to await launch or to fire Zarya's main engines and so consume more of that module's diminishing propellant supply; they decided to give STS-101 one more chance. On 15 May, the US Air Force attempted to launch a European satellite on the first Atlas III commercial launcher, but the launch was cancelled when equipment at the Bermuda tracking station malfunctioned. A second attempt to launch the Atlas III on 16 May was cancelled due to high winds. STS-101 was slipped from 18 May to 19 May in order to give the Atlas a third attempt, but when that attempt failed the Cape's radars were recalibrated for the Shuttle and STS-101 finally left the pad at 0611 on 19 May. The launch was silhouetted against the breaking dawn, producing some spectacular pictures. Ten minutes later Atlantis was safely in orbit. Five hours into the mission, Halsell's crew retired for the 'night'.

Flight Day 2 began shortly after 1800. Halsell and Horowitz tested Atlantis's navigation equipment, and made a series of manoeuvres to set up the rendezvous. During one burn, telemetry suggested that the primary system had failed to close one of the propellant valves on the left-hand OAMS. The backup system promptly closed the valve. Houston recommended that they use only the right-hand OAMS system in future manoeuvres, although both OAMS would be needed for the de-orbit burn. Williams and Voss inspected the EMUs to be used on the flight's one scheduled EVA. Horowitz assisted Weber when she tested the RMS and used its end-effector camera to survey the payload bay. Later in the day she also activated and extended the outer ring on the ODS. By the end of the shift, Atlantis was some 650 kilometres in trail of the ISS and closing in at a rate of 48 kilometres per orbit. On Flight Day 3, Halsell commenced the final manoeuvres at 2139 on 20 May. Just prior to 2300, he took manual control. At 2309 he commenced a large semi-circle which took the Shuttle from beneath the ISS through a position directly in front, and on above it at 2334. At 2339, 52 metres above Unity, which was on top of the vertical stack, he began a 20-minute stationkeeping pause until they flew into range of the Russian ground station that was to command Zarya's thrusters off, as on the previous mission, and docking was over the Ukraine, at 0031 on 21 May. Weber promptly retracted the docking ring and triggered the latches. The rendezvous and docking procedure had become routine. Meanwhile, Voss and Williams continued to prepare for the EVA and the air pressure in Atlantis's cabin was reduced from 1,033 to 717 grams per square centimetre.

Flight Day 4 was to be devoted to Voss and Williams's EVA. They began donning their EMUs at 1911. The EVA was intended to begin at 2231, but they were ready far ahead of schedule and Houston approved an early egress at 2148. Their first task was to make their way to PMA-1 to inspect the improperly secured Orbital Transfer Device, which they removed and reinstalled. When they tested their handiwork the crane no longer wobbled back and forth as it had done previously. Next, they completed the assembly of the Russian Strela crane, which had also been begun by the STS-96 crew. Williams replaced the malfunctioning antennae of Unity's Early Communications System and then ran a camera cable across the station's hull. Finally, they affixed eight EVA handrails for use by future astronauts. Throughout the EVA, Weber controlled the RMS while Voss rode a work platform mounted on the end-effector. Horowitz served as coordinator. In all, the two men attached 148 kilograms of apparatus. The 6-hour 44-minute EVA ended at 0432 on 22 May.

On Flight Day 5 Atlantis's crew turned their attention to the preparations to venture into the ISS. Usachev and Helms entered Unity at 2033 on 22 May, followed closely by Voss, to take a look at what would be their home upon their return as the Expedition 2 crew. They entered Zarya at 2058. Their initial tasks included taking air

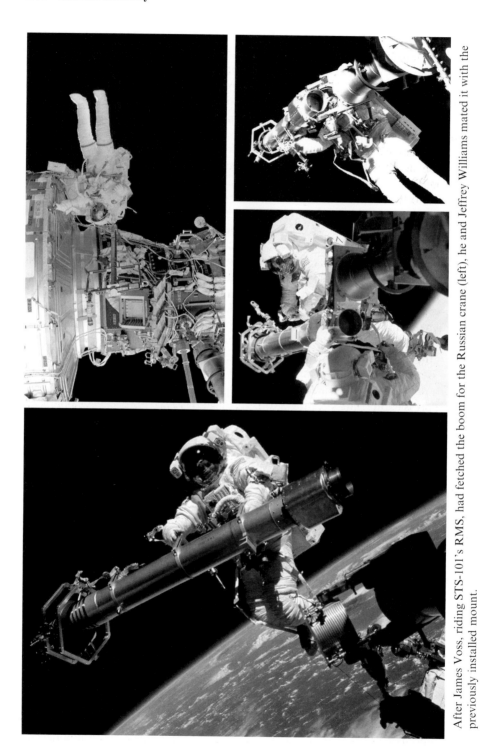

After James Voss, riding STS-101's RMS, had fetched the boom for the Russian crane (left), he and Jeffrey Williams mated it with the previously installed mount.

samples from both ISS modules and checking air circulation efficiency and carbon dioxide concentration. After they revised some of Zarya's ducts to try to improve the circulation in that module, they installed additional acoustic mufflers around some of its noisier equipment. Helms and Usachev opened floor panels in Zarya and replaced the two malfunctioning batteries together with their associated charging equipment. While in Zarya, the astronauts wore earplugs to dampen the noise from its equipment, and they set up portable fans to ensure that exhaled air did not linger in front of their faces. Meanwhile, in Unity, Horowitz and Voss replaced a power distribution unit for the Early Communications System. In the double Spacehab module in Atlantis's payload bay, Weber checked out the mass of logistics to be transferred to the ISS the coming day. As they slept, controllers started charging the first of the two batteries that had been installed in Zarya. It was fully charged and then monitored for a period before being pronounced in "excellent condition". The charging of the second new battery had to be halted for 90 minutes on Flight Day 6 while Helms and Usachev replaced a third battery and its related systems. Ten smoke alarms which were nearing the end of their life were replaced. At 2001 on 23 May Halsell made the first of 27 firings over a 58-minute period of Atlantis's thrusters to raise the orbit by 14.5 kilometres in the first of three such sessions designed to raise the station's orbit by 43.5 kilometres overall, as a preparation for the rendezvous with Zvezda in July.

When the astronauts had first entered the ISS, the interior temperature had been 86 °F. Twenty-four hours later, they reported that it was down to a comfortable 70 °F. Horowitz said that conditions had improved "probably about 100 per cent". During a press conference he told a journalist: "We have had no operational impacts at all as a result of poor air quality; the air quality has been just fine." Likewise, in an interview with a CNN correspondent, Horowitz put to rest fears about the noise level in Zarya. "The noise level is a lot less than I thought it would be. It is like motors running in the background. It is not very loud. All the insulation wrap [the STS-96 crew] put up has helped a lot."

On Flight Day 7, Helms and Usachev replaced a fourth battery in Zarya. This time their work was scheduled so that it did not conflict with the charging of the battery that they had replaced the previous day. They then replaced Zarya's radio telemetry system, which had reached the end of its design life; this stored telemetry data when there was no tracking station beneath the ISS, and 'dumped' it to the next site when the ISS came into range. Helms and Usachev also filled four large bags with potable water from Atlantis's fuel cells and transferred them to the ISS, where they were stored with seven bags left by the STS-96 crew. Meanwhile, Williams and Voss constructed new storage compartments behind Zarya's panels. During the day, Shuttle managers decided to delay the undocking by one orbit on 26 May in order to give the crew a full 8 hours of sleep prior to the critical manoeuvre. As a result, the post-undocking fly-around would be reduced to just half of one circuit. This change would have no impact on the scheduled landing time on 29 May.

The wake-up call for Flight Day 8 was placed at 1611 on 25 May. This was to be their final day of offloading stores from Spacehab. The third battery was sharing the load of supporting station activities alongside the first two replacements and the two

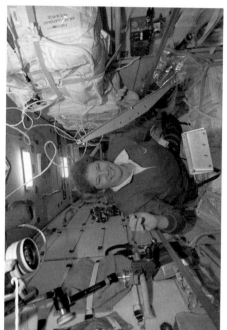

On STS-101, Susan Helms performs maintenance on the batteries in the decidedly cramped Zarya module.

remaining original batteries. As the fourth battery was being charged – an action that was due to be complete before Atlantis undocked from the ISS – one of the new batteries caused a power fluctuation as it was brought on line, and the engineers suspected that its charger had not been installed properly. Helms and Usachev corrected the problem. When one of the new smoke alarms in Zarya went off, Voss was dispatched to inspect behind the relevant wall panels to verify that there was no fire. The alarm was subsequently logged as a false activation. The last repair task was the replacement of a broken fan. By this time, they had transferred a total of 1,497 kilograms of equipment into the ISS which, with canvas bags strapped along its walls, resembled a warehouse. At 0123 on 26 May, the astronauts began the lengthy procedure of evacuating the ISS, securing and checking for an airtight seal on each hatch as they left. Fittingly, Usachev, Voss and Helms were the last three to leave, closing the final hatch at 0404. At 1903, on Flight Day 9, Halsell undocked. The two spacecraft had been mated for 5 days 18 hours 32 minutes. Horowitz backed Atlantis away and then flew a semicircular photodocumentation fly-around, followed by the separation manoeuvre at 1941. Having completed their housekeeping chores in Atlantis, the crew relaxed.

On the ground there were other problems to be dealt with. At Cape Canaveral Wild Life Reserve a fire had been burning out of control since 25 May, and the local winds had blown thick smoke across the Shuttle Landing Facility, so there was the prospect of postponing Atlantis's landing. On the other hand, the forecasted weather conditions for 29 May threatened shower conditions and crosswinds which might also preclude a landing.

The final full day in orbit was spent preparing Atlantis for re-entry, testing its systems and stowing material on the flight deck, mid-deck and in Spacehab. As Atlantis passed over the respective control centres the crew answered questions from reporters in Houston and Korolev, then retired for their final sleep period. The predicted bad weather did not materialise so Atlantis was cleared to return. The de-orbit burn occurred at 0113 on 29 May, with the first contact with Runway 15 at 0220 followed by a smooth roll out to wheel-stop. With STS-101 back on the ground, all eyes were turned to the Baikonur Cosmodrome, where Zvezda was being made ready for launch in July.

PREPARING TO MOVE ON

Once STS-101 had departed from the ISS, the station was put in the correct attitude to support the planned rendezvous and docking with Zvezda. Over the next few days, it was confirmed that all six of Zarya's batteries were now performing according to specifications, so Russian controllers recommenced the regular 'cycling' routine. Within days, however, three of the ten new smoke alarms in Zarya went off individually at irregular times. The fact that only one alarm activated at any given time indicated that there was no fire, so power to these alarms was switched off. During the week ending 8 June, Zarya's TORU was tested to verify that its external television and angular rate sensors were functioning. The Kurs system was similarly verified a week later.

The state of the International Space Station after the STS-101 visit.

At about this time, an internal NASA report highlighted the fact that if Zvezda suffered a depressurisation accident in its first year or so the resident crew would be unable to carry out many key functions required to keep the rest of the station in operation, which raised the possibility of the station being lost. This risk would be eliminated when SSAF-8A delivered hardware that could assume control of such systems in an emergency. Furthermore, the loss of Zvezda during the vital early months could conceivably deny the resident crew access to their Soyuz ferry, if this was docked to its rear (wake) port.

On 20 June, Korolev began to transfer Zarya's propellant to the tanks required for the rendezvous and docking with Zvezda; some 750 kilograms of propellant was transferred over the next few days

'EXPEDITION ZERO'

If the automated Kurs system prevented Zarya/Unity from docking with Zvezda, the Russians had developed a contingency involving an additional flight, 'Expedition Zero', in which Gennadi Padalka and Nikolai Budarin would be launched by Soyuz on 10 August to dock with Zvezda so that Padalka could employ the TORU system to dock Zarya/Unity manually. Although all of Mir's expansion modules docked automatically, some of the later Soyuz commanders earned 'bonus payments' by docking their spacecraft manually when the Kurs system reportedly malfunctioned. When this contingency was proposed for docking Zvezda, there was criticism by the American media that the Russians were simply seeking an excuse to occupy the ISS ahead of the 'official' NASA-commanded crew. The Russians insisted that after ten days of checking out Zvezda's systems, the contingency crew would return to Earth.

ZVEZDA IS ADDED

The Proton launch vehicle assigned to Zvezda arrived at the Baikonur Cosmodrome on 2 June. A number of recent launches had recertified the second stage, after several failures in 1999. Zvezda had been delivered some months before. On 23 May, while the STS-101 crew had been working in the ISS, Gennadi Padalka, Nikolai Budarin and Kenneth Bowersox had taken the opportunity of the Service Module's presence to familiarise themselves with its major systems.

Following a Joint Programme Review and a General Designers' Review in late June, Zvezda's launch was set for 12 July. The module was fuelled on 5 July, mated to its launch vehicle on 6 July, and transported by rail to the pad on 7 July for commencement of final preparations. Twenty-six months behind the original ISS schedule, Zvezda finally lifted off at 0056 American Eastern Daylight Time on 12 July. The Proton performed flawlessly and put its payload into a 172 × 355-kilometre orbit. All of Zvezda's solar panels, antennae and other external fittings successfully deployed. The Kurs systems on Zvezda and Zarya were both tested to ensure that they were capable of supplying valid range and range-rate data. When telemetry failed to verify that two of Zvezda's TORU docking targets had deployed as programmed. The solar panels were locked into position for two test firings of Zvezda's rocket motors. At 2327 on 12 July, the first manoeuvre added 1 metre per second to lift its apogee slightly to 339 kilometres. Half an orbit later, a 2-metre-per-second manoeuvre raised the initial perigee to 175.5 kilometres. Later, on 13 July, the Korolev flight controllers confirmed that the solar panel drive motors were maintaining the photovoltaic arrays facing the Sun. The electrical power was being directed to the four batteries that had been installed for launch (four more batteries would be installed by the crew of STS-106). The inertial and star-tracker navigation systems were also verified to confirm that Zvezda would be able to undertake the forthcoming rendezvous. The panels were again locked into position so that Zvezda could manoeuvre into a 361 × 269-kilometre orbit. The remainder of the day was spent reviewing telemetry to confirm that the solar array drive motors were working satisfactorily.

Most of 15 July was spent verifying the telemetry system by commanding Zvezda's computer through various software modes so that the resulting telemetry could be studied. The next day's schedule had been left 'free' to provide time to troubleshoot any outstanding issues, but everything was working correctly. Testing resumed on 17 July. Zvezda's motion control system and navigation system were tested using Sun sensors, proving that the software could control the guidance system. The routine of 'cycling' the batteries was initiated. Meanwhile, it had been noted that Zvezda's engines had been over-efficient, so Zarya was ordered to make two 5-second burns at 2259 on 17 July and 0025 on 18 July to raise its orbit by 3.3 kilometres so as to restore conditions for rendezvous. At 2247 on 19 July, Zvezda made a 15-second perigee-raising burn resulting in a 361 × 290-kilometre orbit. Later in the day, the tests on the telemetry system's secondary computer were completed.

On 25 July, while Zvezda held its attitude, Zarya completed the rendezvous using its Kurs automatic system. Zvezda's solar panels were then turned edge-on to the approaching vehicle in order to minimise the impingement by the exhaust from Zarya's thrusters on the surface of the photovoltaic cells. The docking occurred at 1945. Zvezda's solar panels were instructed to resume tracking the Sun. The ISS now comprised Unity, Zarya and Zvezda connected in line. A few days later, Zvezda's computers were integrated with Zarya's and Unity's, then Zarya's motion control system was switched off and Zvezda took responsibility for attitude control, orbital reboost and manoeuvring. At this time, Unity's Early Communications System was routed into Zvezda so that Korolev could send commands via Houston.

With Zvezda integrated into the ISS, the station now had its main control centre and living quarters so, after two years of frustrating delays, the programme could finally kick into high gear. But now that the torch had been passed to America, NASA began to encounter problems.

Preparation and launch of the Zvezda module.

UNSAFE SAFER UNITS

In July 2000, when Carleton Technologies in Buffalo, New York, was undertaking routine maintenance on one of the dozen SAFER units, it discovered an oily substance in an oxygen regulator which, in an emergency in which an astronaut required the SAFER's oxygen, would regulate the flow rate from the high-pressure tank. If the flow had been started, the residue could easily have caused a fire in the unit. Investigation revealed that the same regulator in every unit was similarly contaminated. This danger had undoubtedly been present all along. Although the regulators in the primary oxygen systems of the EMUs were uncontaminated, no EVAs would be possible until this problem had been rectified. In fact, the issue threatened to ground the Shuttle, because even if no EVAs were planned, two EMUs with SAFER units were carried on each mission as a contingency in case the payload bay doors failed to close. Although it was unlikely that the cause of the contamination would be identified, the SAFERs were modified to pass oxygen through a new cold-trap to flush out any traces of contaminants which had not been caught by the in-built filters. Fortunately, this upgrade became available in time for STS-106, on which an EVA was planned.

RUSSIA CALLS FOR ADDITIONAL AMERICAN EXPENDITURE

In the weeks leading up to the Zvezda's launch, Congress had authorised NASA to purchase $14 million of hardware from Russia, including a docking system to enable the Interim Control Module to dock with Zarya in the event of yet another long postponement of Zvezda's launch, or if it launched and was lost. At the press conference after the successful launch, Yuri Semenov called for more American expenditure on Russian hardware and services. "I wish Goldin would add to his moral support other commitments," he said. As a joint effort, the ISS should be "funded jointly". Goldin was observed to be shaking his head in disbelief as he listened. In fact, NASA was awaiting approval to spend a further $21 million in Russia to undertake work that American contractors could not complete. Included in that package were five SAFERs similar to the NASA's, but to be used by the Russian Orlan suit; an electrical system simulator to ensure that American equipment was compatible with Russian hardware on the ISS; software testing to ensure that American and Russian modules would fully integrate; and a condensor to extract water from the air-conditioning system and recycle it, thereby exploiting systems proven on Mir.

At the same press conference, Yuri Grigoriev noted that the Russian Docking Compartment, which was to start Russia's expansion of the ISS, had slipped a year behind schedule and so would not be ready until 2001. He also said that work on other hardware had stalled due to lack of funding by the Russian government.

PROGRESS-M1 3

As Zvezda closed in on the ISS, the first M1 variant of the Progress tanker was being made ready at Baikonur. It was launched at 1426 on 6 August. Following the standard two-day rendezvous, its Kurs system steered it to Zvezda and it docked at the wake at 1614 on 8 August. The STS-106 crew would unload the ferry's dry cargo in September, to complete the preparations in advance of the arrival of William Shepherd's Expedition 1 crew. The 615-kilogram load in the orbital module included food, clothes and various computers. However, the primary payload was propellant for Zvezda. This was automatically pumped aboard through pipes in the docking system, just as on Mir. The only problem was an instrumentation fault which temporarily halted the oxidiser transfer. During the fuel transfer, a ground-command error deactivated Zvezda's motion control system, but positive attitude control was soon re-established. With one set of fuel and oxidiser tanks in Zvezda refilled, the tanker's engine was fired on 15 and 17 August to raise the ISS's orbit slightly.

On 21 August, controllers in Korolev noticed an irregularity when 'cycling' Battery 4 in Zvezda, so it was decided to have STS-106 replace the battery. When the problem was traced to the PTAB recharging system, it was decided to replace this as well. The other batteries were working normally. In the weeks before the STS-106 launch, one of Zvezda's three computers failed and was taken off line. Its memory was downloaded for analysis. In the meantime, tests found that it functioned correctly while disconnected from the Zvezda flight control system, so the relationship was investigated.

At this point, Yuri Semenov admitted to a press conference that Energiya had no spacecraft for the ISS that were being funded by his government. The company was owed $40 million. It accepted its responsibilities to the ISS, "but could no longer stake everything". He referred to the Proton/Zvezda and Soyuz-U/Progress-M1 3 launches as "good will gestures on the part of Energiya". He believed Progress-M1 4 would be launched on time in September, with the Expedition 1 crew following by Soyuz in October, but thereafter the flow of vehicles would be dependent upon the government meeting its commitment.

NEW TRAINING ROOM OPENED

On 16 August, NASA announced the opening of a new training Flight Control Room in the Mission Control Building at the Johnson Space Center in Houston. It accurately mimicked the Flight Control Rooms and could be linked to the various astronauts' training devices in order to undertake integrated training. The requirement for a new training control room had been identified two years earlier, when it became obvious that once the long-duration crews were on board the ISS both the Space Station Flight Control Room and the Space Shuttle Flight Control Room would be occupied with flight operations and so would not be available for training purposes.

ISS RESCHEDULED AGAIN

In the wake of the Progress-M1 3 docking with the ISS, programme managers agreed to reschedule the latter part of the ISS assembly. The missions assigned through to the end of 2001 would be flown as planned, but rephasing the others would have the effect of slipping the Baseline Configuration's completion into 2006.

Table 13.2 ISS assembly sequence (circa August 2000)

Date	Manifest	Launch vehicle	Payload
12 Jul 2000	1R	Proton	Zvezda
8 Sep 2000	2A.2b	STS-106	Double Spacehab
5 Oct 2000	3A	STS-92	Z-1 Truss Structure, PMA-3
30 Oct 2000	2R	Soyuz	Expedition 1
30 Nov 2000	4A	STS-97	P-6 photovoltaic array unit
18 Jan 2001	5A	STS-98	American Laboratory
15 Feb 2001	5A.1	STS-102	MPLM Leonardo
– Mar 2001	4R	Soyuz	Russian Docking Compartment
19 Apr 2001	6A	STS-100	MPLM Leonardo
17 May 2001	7A	STS-104	Joint Airlock Module
21 Jun 2001	7A.1	STS-105	MPLM Donnatello
4 Oct 2001	UF-1	STS-109	Utilisation flight

STS-106 UPGRADES ZVEZDA

STS-106 – Atlantis – SSAF-2A.2b

Commander:	Terrance Wilcutt
Pilot:	Scott Altman
Mission Specialists:	Edward Lu; Daniel Burbank; Richard Mastracchio; Yuri Malenchenko (Russia); Boris Morukov (Russia)

In the event that Zvezda had failed to launch, or failed to dock, STS-106 was to have been redesignated SSAF-2A.3 and have delivered the ICM, which would have been attached to Zarya in order to maintain the ISS's orbit until Zvezda (or its replacement) appeared. With Zvezda safely in place, STS-106 was loaded with stores. In addition to a spacewalk, this crew would be called upon to unload a total of 2,723 kilograms of cargo from their Spacehab and from the docked Progress, effectively making them space stevedores.

Wilcutt's crew entered Atlantis at 0600 on 8 September 2000. Despite fears over the weather, the countdown was flawless and STS-106 launched on time at 0845. Two hours later, the mission specialists began activating the Spacehab. Meanwhile, Wilcutt and Altman initiated the rendezvous sequence.

From the beginning, the crew set out to conserve electrical power in the hope of being granted a one-day extension. After the first sleep period, they were awakened at 2145 for Flight Day 2. While the rendezvous proceeded, Lu, Malenchenko and

The launch of Atlantis for mission STS-106.

Burbank prepared the ODS for the docking and then checked out the EMUs and the tools that Lu and Malenchenko would use during their EVA. Mastracchio activated the RMS to confirm its full functionality and then he used its end-effector camera to survey the payload bay. During the day, it was discovered that the upward-looking star tracker was not working, so during the rendezvous Wilcutt would have to reorient Atlantis from time to time to utilise the sideward-viewing star tracker. Wilcutt and Altman fired Atlantis's manoeuvring thrusters just after 2300 on 9 September, on Flight Day 3, for the Terminal Initiation burn. During the final approach to the ISS, Wilcutt paused just a few metres short of the station until they flew into range of a Soviet ground station so that it could monitor the docking at PMA-2 at 0152 on 10 September. Immediately thereafter, the crew opened the hatches between Atlantis's airlock and the PMA to draw a sample of air, then they resealed the hatches. The atmospheric pressure in Atlantis's cabin was then reduced in preparation for the forthcoming EVA. An inspection confirmed that at least one of Zvezda's docking targets had not fully deployed immediately following the module's launch.

Lu and Malenchenko began Flight Day 4 with the preparations for their EVA. Upon finishing their pre-breathing they donned their EMUs. Meanwhile, Burbank went over the timeline and the activities they were to perform, and Mastracchio confirmed that the RMS was fully functional. Three EVAs had previously been performed outside Mir in which Russian cosmonauts and American astronauts had worked together. During those excursions they had all worn Russian Orlan suits, had spoken Russian, and had employed Russian procedures. During this first international EVA on the ISS, operations were to be carried out using American EMUs and Russian safety protocols that called for each man to employ two safety tethers. As he moved about on the station's surface he would disconnect one tether, move to his next position, and connect the tether at its new location. By alternating tethers it would be possible to progressively translate to any part of the station that was accessible to spacewalkers. The operating language would be English. This would be Malenchenko's third EVA, the previous excursions having been made from Mir. It was to be Lu's first EVA. Lu exited the airlock at 0045 on 11 September. Malenchenko followed a quarter of an hour later. At that time, they were in Earth's shadow. They collected their equipment and prepared for the tasks ahead of them. Once they were ready, they mounted the platform that they had installed on the RMS's end-effector and Mastracchio lifted them to the limit of the arm's reach – some 15 metres above the payload bay. There, each man hooked his tethers to attachments on Zarya's exterior and set off on a hand-over-hand 'climb' towards Zvezda. Burbank had a route avoiding the numerous communication antennae and other exterior hardware detailed on the flight plan and he provided directions as the spacewalkers maintained a running commentary on their movements, specifying their positions in terms of the many location-identifiers on the surface of the vehicles.

Once at the interface between the Zarya and Zvezda modules, Lu and Malenchenko set about installing nine cables that they had carried with them on their backs. One end of each cable had to be connected to an outlet on Zvezda, and the other end to an outlet on Zarya. Two 4.8-metre-long cables would carry data between Zarya and

Zvezda, and enable Zvezda to control Zarya's solar panels in concert with its own. Two other 4.8-metre-long cables would provide video transmission between the two modules. Four 8-metre-long cables would carry electrical power produced by the solar panels that STS-97 was to mount on the Z-1 Truss Structure on top of Unity. From time to time, the shadow cast by the very large American arrays would tend to reduce the output from Zarya's and Zvezda's arrays. These power cables would enable the Russian modules to make up this shortfall by drawing power from the American arrays. The ninth cable that they fitted was a fibre-optic link that would be used once the ISS's airlock had been berthed on Unity's starboard CBM. It would route data from Russian Orlan EVA suits to Zvezda through the American modules. With the cables installed, Lu and Malenchenko worked their way up Zvezda to a point 33.5 metres above Atlantis's payload bay (the farthest any EVA astronaut without an independent Manned Manoeuvring Unit had been away from a Shuttle) to retrieve a magnetometer that the Russians had mounted temporarily on its exterior. The instrument was to report the station's orientation with reference to the Earth's magnetic field; however, it could not provide reliable readings while resting on the metal hull. Budget restrictions had prevented it being mounted on a self-deploying boom, so Lu and Malenchenko installed a 2-metre-long boom in a socket and then mounted the magnetometer on its end, in yet another demonstration of the cost-benefit to be derived from using spacewalkers for orbital assembly. They returned to the airlock at 0701, having been out for 6 hours 14 minutes.

Atlantis's crew cleared up the cabin after the EVA. Wilcutt and Altman fired the Shuttle's RCS thrusters 36 times over the following hour in order to raise the docked combination's orbit by 7.2 kilometres. They began the preparations for their entry into the ISS on Flight Day 5. The first hatch was opened at 2240 on 11 September. Breathing masks and goggles were worn to protect against particulates in the newly launched Zvezda. Even as some crew members worked their way into the station's compartments, testing the pressure and air quality behind each hatch before opening it, the others began to transfer equipment from the Spacehab module into Unity. Wilcutt and Malenchenko finally entered Zvezda at 0120 on 12 September and reported it to be in excellent condition. Zvezda had been too heavy to be launched on a Proton with all of its systems in place, so its life support system had been brought by Atlantis and the crew were to install it. In the meantime, Atlantis was to provide it with oxygen. Flexible hoses were extended from Atlantis and fans were turned on to circulate the flow of oxygen within it. The last hatch to Progress-M1 3, at the far end of the stack, was opened at 0222. With the Progress's orbital module open and no particulates discovered, the seven crew members removed their masks and goggles and began immediately to offload the cargo from the ferry craft. Equipment was now entering the ISS from both ends. The frames which had restrained apparatus in Zvezda against the vibration of launch, Zarya's now superfluous TORU system, and the docking probe from the Progress were to be returned to Earth on Atlantis. Some of these items would be able to be re-used on future missions, so their retrieval was a cost-saving measure. At 0945 Houston relayed the decision that, as had been hoped, the mission had been extended by one day in order to provide more time to unload the cargo.

Immediately after their breakfast on Flight Day 6, Burbank and Morukov began work on replacing one of the two batteries in Zarya that STS-101 had not replaced. However, this time there was a complication. Four metal brackets had been riveted over the top of pairs of bolts prior to launch and these would have to be removed. On the advice of the Russian engineers, Burbank and Morukov took a hammer and chisel to the rivets holding the brackets in place. Once the brackets had been removed they were able to replace the battery and the equipment that controlled the flow of current through it without incident. Meanwhile, in another upgrade of Zvezda to overcome its initial weight-limit, Lu and Malenchenko installed two more batteries and Wilcutt, Altman and Mastracchio loaded new equipment into the ISS from both ends. Despite the frantic pace, Wilcutt managed to take a break to communicate with television stations in his home state of Kentucky as he flew overhead. Overnight, one of Zvezda's two new batteries began to malfunction, and the Russian controllers set up an investigation to identify the fault. During Flight Day 7, while Lu and Malenchenko fitted two more batteries in Zvezda to bring its total up to eight, Burbank and Morukov replaced the final battery in Zarya. Mastracchio served as the ISS's loadmaster and coordinated the unloading of equipment from the Shuttle and from the Progress. The ferry craft had delivered a varied cargo. Two Orlan EVA suits were stored on the station for later transfer to the airlock module. The station's toilet was installed in Zvezda, but it was reserved for the Expedition 1 crew, who were scheduled to take up residence in November 2000. At 0045 on 14 September STS-106 passed the halfway point of its docked operations. At 0213 the reboost operation was resumed by firing Atlantis's thrusters 36 times over one hour to raise the station's orbit another 7.2 kilometres. As the day was winding down, Wilcutt and Burbank were interviewed by reporters from a number of American news outlets.

On Flight Day 8, Lu and Malenchenko spent their third day on the ISS installing the power converters in Zvezda to enable the Russian modules to draw power generated by the American solar panels (they had installed the cables for this transfer during their EVA). Later in the day, they installed components of the Elektron unit that formed part of Zvezda's life support system. Based on a system used on Mir, this would electrolyse molecules of water into hydrogen and oxygen. The hydrogen would be vented overboard as waste, but the oxygen would be pumped into tanks. It would be activated by the Expedition 1 crew. At the request of the Russian engineers, they then disconnected Zvezda's malfunctioning battery. Mastracchio, Burbank and Morukov spent the day unloading cargo and bags of water from Atlantis. The accumulated packaging materials were dumped into the now-empty Progress, to be discarded when that spacecraft undocked. Wilcutt and Altman made the third series of 36 manoeuvres to boost the orbit by a further 5.6 kilometres, and then television interviews brought the day to a conclusion.

Altman, Lu, Burbank and Morukov spent Flight Day 9 installing the exercise treadmill in Zvezda. This went by the name of the Treadmill with Vibration Isolation System (TVIS) because the treadmill contained a vibration-dampening system which would allow an astronaut to run on it without disturbing the numerous microgravity experiments that would be ruined by the vibrations sent through the hull of the

The STS-106 crew were the first to board the Zvezda module (upper left). While Boris Morukov works in Zarya (right) Scott Altman transfers cargo (lower left).

station by a running machine that was mechanically coupled to the vehicle. Once the new treadmill was in place, each crewman was to increase his exercise regime to 90 minutes twice a day. Burbank and Mastracchio installed the CBM controllers which had been removed from PMA-2 so as to allow the crews of STS-96, STS-101 and STS-106 to pass large objects through the hatch. Their reinstatement was a preparatory step for the attachment of the American Laboratory Module 'Destiny', scheduled for January 2001, at which time PMA-2 would have to be relocated. Throughout the day, the transfer of cargo from the Spacehab module continued. They loaded the final items of rubbish into the Progress. When flying over Australia, Wilcutt televised his encouragement to the athletes at the opening ceremony of the Sydney Olympics.

The wake-up call for Flight Day 10, their final full day of working inside the ISS, came at 1946 on 16 September. Starting with the Progress hatch at 2300, they worked their way back to Atlantis, sealing the final hatch at 0830 on 17 September. In all, they had spent 5 days 9 hours 21 minutes working on board the station. Wilcutt and Altman performed the fourth series of 36 thruster firings to boost the ISS's altitude by a further 5.6 kilometres, so they would leave it in an orbit ranging between 388 and 375 kilometres.

As they slept, Hurricane Gordon was raging off the Florida coast and lashing the Kennedy Space Center with strong winds and rain. Flight Day 11 began with the preparations for undocking, which occurred at 2344 on 17 September. Once the spring-loaded mechanism had pushed Atlantis 0.6 metre clear, Altman fired its thrusters to open the range to 137 metres, then manoeuvred Atlantis around the station while the mission specialists photo-documented the state of the ISS. He made the separation manoeuvre as they passed directly over the station during the second circuit. At 0211 on 18 September, all seven astronauts gathered on the mid-deck for a live video press conference with reporters in Houston and Korolev, then, after a six-hour 'free' period in which they relaxed, they retired.

On Flight Day 12, while Wilcutt and Altman prepared Atlantis for the return to Earth, the mission specialists closed up the double Spacehab module and stowed away all unwanted equipment. The final wake-up call was at 1846 on 19 September. The de-orbit preparations began just before 2300. The bay doors were closed at 0010 on 20 September, the de-orbit burn was performed at 0250, and Willcutt took over manual control and guided his spacecraft to a perfect landing on Runway 15 at 0356. They had left the ISS in an excellent state of repair and well stocked ready for the Expedition 1 crew.

BUDGETS FOR FISCAL YEARS 2001 AND 2002

On 14 September, Congress set NASA's budget expenditure on the Shuttle and the ISS at $14.1 billion for Fiscal Year 2001 and $14.6 billion for Fiscal Year 2002. The expenditure on the Shuttle flights to the ISS was capped at $17.7 billion, and the total development cost for the ISS was ordered not to exceed $25 billion. Emergency funds for the two years were set at $3.5 billion for the Shuttle and $5 billion for the

The state of the International Space Station after the STS-106 visit. Top to bottom: PMA-2, Unity, PMA-1, Zarya, Zvezda and Progress-M1 3.

ISS. This was the first NASA Reauthorisation Bill on which both Houses had agreed since 1992. Since that time, the budget had been set in the annual Appropriations Bill. The Bill required NASA to obtain assurances from Russia that it would place a higher priority on its commitments to the ISS. The Bill also relaxed the restrictions that the previous year's budget had imposed on the development and use of TransHab, the inflatable module for testing on the ISS with a view to future use on deep space missions. Although the restriction banning NASA from using its budget to develop inflatable structures for use in space remained, the new Bill allowed NASA to lease such a module in the event that one was developed by a private company.

STS-92 ADDS THE Z-1 TRUSS

STS-92 – Discovery – SSAF-3A

Commander:	Brian Duffy
Pilot:	Pamela Melroy
Mission Specialists:	Koichi Wakata (Japan); Leroy Chiao; Peter 'Jeff' Wisoff; Michael Lopez-Alegria; William McArthur

---

Cosmonaut Nikolai Budarin had been originally assigned as a mission specialist on the STS-92 crew, but in September 1999 he informed the Press that his flight had been cancelled, offering no explanation. It was speculated that this was an expression of American dissatisfaction with the Russian government's ongoing failure to pay its way in the ISS.

Towards the end of the STS-106 mission Hurricane Gordon was off the Florida coast and looking likely to approach the Cape, so short-notice plans were set in place to roll STS-92 back to the VAB out of the weather, but on 13 September the hurricane turned north and started to lose power. Although Discovery was left on the pad, the facilities were secured against the winds and rain. The STS-92 launch was set for 2138 on 5 October. Even though a tropical weather system hung over the Cape on 4 October, the countdown was continued in the expectation that conditions would improve overnight, but then low clouds and atmospheric moisture threatened to violate the RTLS abort constraints. The loading of propellants into the External Tank was held up by a problem with an explosive bolt that connected Discovery to the ET. The launch attempt was cancelled at 1325 and rescheduled for 2116 on 6 October, but the weather predictions for the new date remained poor. The problem with the explosive bolt had come to light during analysis of data from STS-106. On that flight the bolt in question had not fully retracted following ET jettison. This raised the spectre of drag on an extended bolt causing the ET to manoeuvre unpredictably and possibly strike the underside of the orbiter. Three teams were assigned to study the problem. A review of films from previous launches revealed that the bolt had failed fully to retract on several flights without causing the ET to tumble. A new problem arose during the turnaround for the new countdown. As Discovery's main propellant system valves were cycled a pogo-suppression valve to control oscillations in the turbo-pumps performed sluggishly. The valve was on the LOX side of the propellant system, so Discovery's cryogenic storage tanks were drained between 0200 and 0800 and later that morning a management meeting decided to replace the valve, a task which would take three days, and the launch was reset for 9 October. Once the Rotating Service Structure was back in place, workers entered Discovery's aft compartment and, after further tests of the valve, began the 12-hour removal and replacement process on 7 October. With the new valve installed and verified, the aft compartment was resealed on 8 October and the cryogenics were reloaded. Discovery was declared ready for launch at 2005, but the wind still posed a threat. The Rotating Service Structure was retracted at 0100 on 9 October and the countdown was picked up at T−11 hours at 0510. Preparations for ET loading were due to start at 1040, but at 0200 high winds had prevented the launch team from affixing the gaseous oxygen vent hood (more commonly known as the 'beanie cap') to the apex of the ET. Based on the weather predictions, a decision was made to try again at 0800. That attempt also failed due to the 45-knot winds being 3 knots over the acceptable limit for the procedure, and as loading could not take place without the vent hood in place there was no option but to scrub the count and reschedule it for the next day.

The STS-92 countdown was picked up at T−11 hours, at 0445 on 10 October with a view to launch at 1940. The gaseous oxygen vent hood was secured in place just before 0500. Launch Complex 39's Pad A was cleared, and propellant loading was due to begin at 1044. The day remained windy, but wind strengths were not as high as the previous day. Nevertheless, the chances of another scrub due to weather remained high. The countdown was held and propellant loading delayed in order to investigate a fault with the Pyrotechnic Initiator Controller (PIC) which controlled the firing of the tie-down bolts that held the base of the SRBs to the launch platform. The team dispatched to resolve the issue evacuated the pad once more at 1200. The countdown was resumed at 1115, at T−6 hours, and propellant loading procedures began with the system being chilled preparatory to loading cryogenic fluids. A planned two-hour hold at T−3 hours began at 1340, and loading of the ET commenced. During the hold an Ice Inspection Team made a check of the vehicle. Jorge Revera, a member of the team, was using a pair of binoculars to view the Shuttle from the Rotating Service Structure when he noticed a foreign object apparently wedged between the lower strut from the right-hand SRB to the ET and a liquid oxygen pipe running between the ET and the SSMEs. Further investigation found this to be a locking pin from one of the Rotating Service Structure work platforms. The pin should really have been accounted for when the platform was lifted prior to moving the Rotating Service Structure away from the Shuttle. A suggestion to use a high-pressure water hose to dislodge the pin was ruled unlikely to succeed due to the height of the strut above the pad, so the count was reluctantly scrubbed – a scrub after ET-loading cost NASA $600,000 in lost propellant – and rescheduled for the next day. Discovery finally climbed into the moonlit Florida sky at 1917 on 11 October 2000, as the 100th Shuttle mission. The separation of the ET occurred without undue tumbling. After configuring their vehicle for orbital flight, the seven-person crew retired for the night.

Early on Flight Day 2, the signal from Discovery's Ku-Band antenna failed. This reduced the voice and video communications to mission control to a garble and blurred images. Despite fault-finding efforts by the crew, the signal remained indistinct, which meant that the near-continuous coverage of in-flight activities received from the earlier flights would be absent from STS-92. Houston and Korolev would have to make do with sequential still-image video coverage and brief periods of moving video coverage when the ISS was in the correct position for alternative communications systems to be used to carry the pictures. More seriously, the fault meant that the antenna would not be able to serve in its other role as a radar for the rendezvous, but other apparatus would perform the function. Wakata put the RMS through a series of tests to verify that it was performing correctly, then made a video survey of the Z-1 Truss Structure, the primary payload on this flight.[2]

The Z-1 Truss Structure was to be mounted on Unity's zenith CBM to serve as a temporary mount for the Port 6 Photovoltaic Array, which would supply power to supplement that produced by Zvezda and Zarya. Four Control Moment Gyros (CMGs) were held in the Z-1 as the ISS's primary attitude control system, allowing it to maintain its orientation in orbit without the need for thruster firings. It also had a number of antennae for communication with TDRS satellites.

Mission specialists Chiao, McArthur, Wisoff and Lopez-Alegria made routine checks on the EMUs that they would use on the mission's four scheduled EVAs. As Duffy and Melroy pursued the rendezvous, Houston activated Unity's heaters to prepare it for the transfer of supplies and some hardware associated with the Z-1 Truss Structure.

After breakfast on Flight Day 3, Duffy, flying Discovery from the aft flight deck, began the final approach. Once lined up, he waiting until they were in range of the chain of Russian ground stations and then moved in and docked with PMA-2 on the front (ram) end of Unity at 1345 on 13 October. After pressure tests, Lopez-Alegria opened the hatch giving access to Unity at 1715. Duffy, as the mission commander, was first inside, but he was closely followed by Lopez-Alegria, Chiao and Melroy, and they were soon all busy transferring items of cargo from Discovery to Unity.

Preparations to transfer the Z-1 Truss Structure to the ISS began after breakfast on Flight Day 4. As they worked through the check list, an electrical short-circuit disabled the Orbiter Interface Units through which the astronauts' laptop computers communicated with the ISS, a camera mounted on the keel of the payload bay and the Orbiter Space Vision System which was to coordinate the RMS during the transfer. The crew made repairs and restored all functions except the payload bay camera. With the repairs completed, Wakata, now running some two hours late, secured the RMS's end-effector to the lifting pin on the Z-1 Truss Structure, the restraint latches were released, then the structure was hoisted from the bay and positioned directly over Unity's zenith CBM. Melroy used her laptop to command

---

[2] The video from the end-effector's camera could be viewed at the RMS's control station, even though it could not be relayed to Houston.

Standing on STS-92's RMS, Michael Lopez-Alegria works with a 'reactionless' power tool.

the CBM's 16 latches to open. Supplementing the information provided by the Orbiter Space Vision System with verbal instructions from Lopez-Alegria, who was viewing the scene through a porthole in Unity, Wakata carefully lowered the Z-1 Truss Structure into position and Melroy closed the CBM's latches to lock the unit into place at 1320 on 14 October. In Unity, Melroy and Wisoff were to open the hatch between Unity and the pressurised vestibule beneath the Z-1 to make electrical connections. This was postponed when it appeared that there was an air leak between the two units, but once this had been traced to a faulty hose they opened the hatch to make the electrical connections. This done, they retreated to Discovery and sealed Unity's hatch at 1857. The day's schedule was revised to accommodate the delays. In particular, the plan to continue transferring cargo to the ISS was put off until the end of the flight. Duffy and Melroy reduced the cabin pressure, preparatory to the EVAs scheduled over the next four days. Chiao and McArthur spent the rest of the day preparing their tools.

The main activity of Flight Day 5 was to be Chiao and McArthur's EVA. On Discovery's aft flight deck, Wakata prepared the RMS. After emerging into the bay at 1027 on 15 October, Chiao and McArthur spent an hour deploying their various tools and EVA aids. Once McArthur had installed a work platform on the RMS's end-effector and climbed on board, Wakata elevated him alongside Unity's zenith CBM. Once Houston had switched off power to the Russian–American Conversion Unit 5 (RACU) inside Unity, McArthur connected six external electrical cables between Unity and the Z-1. McArthur, who was making his first EVA, said, "I was wondering what it was going to be like being out on the end of the arm – not being able to see the Shuttle. It's a strange feeling. My toes are curling right up." His last comment was not totally facetious. Research on long-duration space flights had shown that astronauts suffered from lower back pain, caused by the fact that they curled their toes up in microgravity. It was an innate reaction from our ape heritage designed to counteract loss of balance. For the next task, Chiao took his place on the work platform and Wakata lifted him towards the top of the Z-1. They jointly unstowed the S-Band Antenna Subassembly (SASA) from its launch position and relocated it to a temporary mounting to be deployed and activated by a later crew. This cleared the way for two DC–DC Converter Units to be installed on the third EVA. RACU-5 was then powered up and RACU-6 was powered down to allow McArthur to connect a further four cables. With the connections made, power was restored to RACU-6. These ten cables delivered power to heaters on the Z-1. This done, they removed the Space to Ground Antenna (SGANT) dish and put it on its boom assembly. Their final task was to relocate a tool stowage box, which was located on the structure that was holding PMA-3 in Discovery's bay. They closed the airlock's outer hatch at 1655, having been out for 6 hours 28 minutes. The rest of the day was spent 'resizing' the EMUs for Wisoff and Lopez-Alegria.

On Flight Day 6 Wisoff and Lopez-Alegria entered the payload bay at 1030 on 16 October and set about preparing their tools. Meanwhile, Wakata mated the RMS end-effector with PMA-3. After some initial difficulties, the spacewalkers released the latches which held PMA-3 on its payload bay support structure. While Wakata lifted PMA-3 to its new location on Unity's nadir CBM, Wisoff and Lopez-Alegria climbed up to the top of the Z-1 to release the latches that would engage the P-6 structure when it was emplaced by STS-97. Melroy used her laptop to open the 16 latches on the CBM to accept PMA-3. With no direct line of sight and no camera, Wakata relied upon verbal cues by Wisoff and Lopez-Alegria to locate the PMA correctly. Once it was in position, Melroy activated four of the latches to bring the seals of the CBM and PMA-3 together in order to allow the thermal differences between the two seals to equalise. Wisoff and Lopez-Alegria retreated to the airlock at 1722, having been out for 7 hours 7 minutes. As the spacewalkers ran through the airlock procedures, Duffy and Melroy fired the thrusters 18 times over a 30-minute period to boost the combination's orbit. Twelve hours after PMA-3 had been loosely secured, Houston commanded the remaining latches to secure and complete its installation.

Flight Day 7 was devoted to the third EVA. After leaving the airlock at 1030 on 17 October, Chiao and McArthur retrieved two DC–DC Converter Units from the payload bay and then Wakata used the RMS to transfer them up to the Z-1. The installation procedure took approximately two hours. Once the P-6 photovoltaic array was in place, these units would convert their 160-volt feed to 124 volts to power the ISS's equipment. Chiao and McArthur then completed the final electrical cable connections between Unity and the Z-1, as well as the cables to allow power to flow from PMA-2 to PMA-3. After mounting a second toolbox on the exterior of the ISS, they recovered the bag of tools which Jernigan and Barry had positioned during STS-96 and placed them in the newly installed toolbox. They re-entered the airlock at 1718, having been out for 6 hours 48 minutes. Immediately afterwards, Duffy and Melroy made a second set of thruster firings to raise the station's orbit.

Wisoff and Lopez-Alegria began the mission's fourth EVA at 1100 on 18 October, on Flight Day 8. After removing the grapple fixture by which the RMS had held the Z-1 Truss Structure, they deployed a utility tray that would provide power to the American Laboratory Module, Destiny. Lopez-Alegria held the tray in place while Wisoff unlatched the four bolts that had held it in its launch position. Lopez-Alegria then placed the tray in its new position and held it in place while Wisoff secured it. Wakata then used the RMS to lift Wisoff to the top of the Z-1, where he used a tool to open, close and reopen the latches which would engage the P-6 unit, to verify their functionality. He also verified the operation of the temporary berthing mechanism which would be used during Destiny's installation. These tasks over, the two men mounted the RMS and Wakata lifted them high above Discovery's payload bay. Each took turns to push himself away from the RMS to verify his SAFER unit, then flew down to the airlock with Wakata manoeuvring his colleague down beside him, ready to intervene in the event that the SAFER malfunctioned – because Discovery would not be able to undock and give chase to a loose spacewalker. With time running out, the final test had to be cancelled. It was have assessed the ability of a healthy astronaut to help an incapacitated colleague to return to the relative safety of his spacecraft by pulling him along, much as a lifeguard assists a drowning swimmer.[3] Wisoff and Lopez-Alegria re-entered the airlock at 1756, having been out

---

[3] The experiment was reassigned on STS-98.

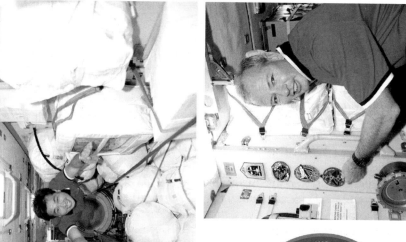

Zarya's main compartment had been converted into a warehouse with barely enough room for Koichi Wakata to pass through (upper left), and Michael Lopez-Alegria, is barely visible at the far end (right). Brian Duffy adds STS-92's patch to the growing collection in Zarya.

for 6 hours 56 minutes and taken the EVA total for this flight to 27 hours 19 minutes. Duffy and Melroy followed the EVA by a third and final series of thruster firings to raise the ISS's orbit.

Flight Day 9 was the last full day docked to the ISS. With the EVAs behind them, the crew spent the day working inside the station. Having opened the hatches between Discovery and Unity, they made their way right through to Zvezda. Melroy and Wisoff collected samples from a number of surfaces inside Zarya in an attempt to identify any bacterial growth. They also wiped down some surfaces and stowage bags with a weak fungicide to inhibit any future growth. They then reopened the pressurised dome between Unity and the Z-1 Truss Structure to complete the connections. At the same time, McArthur and Chiao used a laptop to command heaters to warm the four CMGs in the Z-1, then spun them up to 100 revolutions per minute to test their speed calibration and power consumption. The CMGs would be made operational once the Destiny Laboratory Module was in place. Meanwhile, Wakata used the RMS cameras to make a photographic survey of the 10 tonnes of hardware that they had installed on the station's exterior. All seven crew members combined to transfer equipment from Discovery into Unity and Zarya. The Protein Crystal Growth experiment that the STS-106 crew had delivered was retrieved. It was the first microgravity experiment to be performed on the ISS. During the time that the station had been untended, a number of crystals had grown. Melroy and Wisoff spent the evening unblocking the solid-waste pipe on the Shuttle's toilet. The day's activities had taken somewhat longer than planned so the retreat from the station was later than planned, but by the time Duffy was able to close the final hatch they had spent 27 hours 4 minutes on the station.

During the day, Energiya announced that the Expedition 1 launch had been put back by one day to 31 October. This postponement was because Discovery's reboost manoeuvres had placed the ISS beyond the Soyuz-TM spacecraft's reach on the original date for the rendezvous.

After breakfast on Flight Day 10, Discovery's crew prepared to undock. This occurred at 1108 on 20 October (one orbit later than in the flight plan due to the previous day's late finish) after having been docked for 6 days 21 hours 23 minutes. Melroy withdrew to a safe distance but, in contrast to earlier flights, did not make a photo-documentation fly-around, she simply made the separation burn 45 minutes later. Once clear, the crew was given four hours of free time to relax after the hectic pace of the previous week. On the flight plan, Flight Day 11 was to be the final full day in space. While most of the mission specialists spent the morning stowing equipment, Duffy, Melroy and McArthur tested Discovery's systems and prepared for re-entry. Flight Day 12 was to see the return to Earth, but crosswinds at the Cape's Shuttle Landing Facility threatened to delay re-entry and bad weather was also expected at Edwards Air Force Base in the next few days, so a 24-hour wave-off was issued. The astronauts then spent the day relaxing, and took the opportunity to 'chat' with their families via computer. The Cape crosswinds were no better the following day, 23 October, so it was decided to prepare for a landing at Edwards at 1758. At 1630 this had to be put back two orbits due to rain, but conditions did not improve and Discovery was obliged to remain in orbit for another day. 24 October

The state of the International Space Station after the STS-92 visit. PMA-3 had been mated on Unity's nadir CBM, the Z-1 Truss Structure had been mounted on its zenith CBM and the Space to Ground Antenna had been deployed.

offered only one opportunity to land in Florida, and this was only if the weather improved, which it did not, so planning switched to Edwards. In the event, the first opportunity had to be waved-off because the weather was still below the minimum requirement, but conditions later improved and Discovery was able to sneak down onto the dry lake at 1700. Although not exactly home, STS-92 was at least back on Earth after an unusually long flight of 12 days 21 hours 43 minutes.

EXPEDITION CREWS CONFIRMED

On 19 October 2000 Rosaviakosmos confirmed the prime and backup crews for the first seven long-duration ISS crews.

Table 13.3 Phase Two expedition crews

Flight	Prime crew	Backup crew
Expedition 1	William Shepherd Yuri Gidzenko Sergei Krikalev	Kenneth Bowersox Vladimir Dezhurov Mikhail Tyurin
Expedition 2	Yuri Usachev James Voss Susan Helms	Yuri Onufriyenko Carl Walz Daniel Bursch
Expedition 3	Frank Culbertson Vladimir Dezhurov Mikhail Tyurin	Valeri Korzun Sergei Treshchov Peggy Whitson
Expedition 4	Yuri Onufriyenko Carl Walz Daniel Bursch	Gennadi Padalka Steven Robinson Edward Michael Fincke
Expedition 5	Valeri Korzun Sergei Treshchov Peggy Whitson	Alexander Kaleri Dimitri Kondratyev Heidemarie Stefanyshyn Piper
Expedition 6	Kenneth Bowersox Donald Thomas Nikolai Budarin	Carlos Noriega Donald Pettit Oleg Kotov
Expedition 7	Yuri Malenchenko Sergei Moshchenko Ed Lu	Sergei Krikalev Maksim Surayev Paul Richards

(Original list compiled by Igor Marinin and Sergey Shamsutdinov of OOO IID Novosti Kosmonavitki)

EXPEDITION 1

SOYUZ-TMA 31 – SSAF-2R

Soyuz Commander:	Yuri Gidzenko
ISS Commander:	William Shepherd
Flight Engineer:	Sergei Krikalev

When Discovery departed, the ISS was in an attitude in which PMA-3 on Unity's nadir CBM was pointed towards Earth to satisfy thermal requirements, but in the run up to the launch of the Expedition 1 crew the flight controllers in Korolev and Houston began to prepare the ISS for occupation. On 29 October, the ISS was rolled through 180 degrees, so that PMA-3 faced open space, and then the station was put through a 'dress rehearsal' for the docking with Soyuz-TMA 31 in which all of the commands that would be transmitted to the ISS during the docking were sent to verify the station's functionality. Afterwards, the remaining propellant contained in Progress-M1 3's tanks was pumped aboard the ISS. In the final week, one of the ISS's three controlling computers automatically tripped off-line, but the two remaining computers continued to function, and controllers in Korolev set about determining what had prompted the third to shut down.

With more than three years of training behind them, the three members of the Expedition 1 crew were probably the most highly trained crew ever to launch. Their prolonged training was the result of the two-year delay in fabricating Zvezda.

At 0253 on 31 October 2000 Soyuz-TMA 31 set off from the same Baikonur pad that had been used to launch Sputnik, the first satellite, and Yuri Gagarin, the first human in space. The staging was performed perfectly, and 9 minutes after leaving the pad the spacecraft was in orbit. The launch was observed by the backup crew of Vladimir Dezhurov, Mikhail Tyurin and Ken Bowersox. Once in orbit, the prime crew extended the spacecraft's solar panels and antennae and set it up for the two-day rendezvous. After two phasing burns on orbits 3 and 4, the trio retired. Upon being awakened at 1930, they were given trajectory data to refine the rendezvous. Meanwhile, Korolev commanded Progress-M1 3 to undock from Zvezda at 2302 to clear the station's wake port. Upon being de-orbited two orbits later and burning up in the atmosphere, it disposed of the trash that the Shuttle crews had loaded on board. After manoeuvres at 0348 and 0839 on 1 November resulting in a 243 × 275-kilometre orbit, Soyuz-TMA 31's rendezvous was going well. During a communication session with Korolev, Shepherd was able to talk to his wife, who was a physical therapist for NASA. At 1900, after their second sleep period, the crew prepared for the automated sequence. The first manoeuvre took place at 0225 on 2 November. Starting at 0357, the spacecraft performed a fly-around of the ISS to take up a position 152 metres from Zvezda's now empty wake port. Once over Russia and properly aligned, the spacecraft began its straight-in approach at 0415, and docked on the first attempt 6 minutes later. After powering down their spacecraft, pressure tests were made and the hatches were opened, at which time Shepherd assumed

Soyuz-TMA 31 is prepared and launched with the International Space Station's first Expedition crew.

command and requested the flight controllers to use the name 'Space Station Alpha'.[5] At 0523 he suggested that Gidzenko and Krikalev enter Zvezda ahead of him. Having already visited the ISS on STS-88, Krikalev became the first person to enter it twice. Their first activities included checking communications, activating food warmers, charging the batteries for power tools, starting the water processors and activating the toilet systems. The food warmer was supposed to take 30 minutes to set up, but it actually took several hours.

Until the Elektron electrolysis system could be activated, three VIKA oxygen-producing canisters (similar to those used on board Mir) would have to be 'burned' each day. On 3 November Gidzenko and Krikalev set out to install the Vozdukh regenerative air-scrubbing system in Zvezda. The installation quickly fell behind schedule, as the procedure took longer in weightlessness than it had in training. Once operational, the carbon dioxide that the Vozdukh drew from the atmosphere would be vented overboard. The system that used a series of canisters of lithium hydroxide would henceforth serve in a backup role.

Krikalev also experienced difficulties with cable connections in installing Zvezda's central computer, which delayed his completion of the task. Gidzenko and Krikalev then set about troubleshooting the battery in Zvezda that had failed following its installation by the STS-106 crew. They found a bent, or broken pin in one of the battery's connectors. It was decided that the engineers would carry out further investigations of the fault before a repair of the connector was attempted. With the crew falling behind schedule, the Korolev controllers urged them to reestablish the flight plan, fraying the crews tempers somewhat until it was realised that the crew would have to be allowed to work at their own pace, at least until the start-up activities were complete, when the planning based on experience on board Mir would probably be able to be applied. They ended the day well behind schedule. The slippage was due to the fact that the ISS had only intermittent communication with Korolev – each orbit provided at most 15 minutes of communications through the network of Russian ground stations, so questions concerning the location of equipment had to be saved up until a communications session. The situation would persist until the systems for the TDRS relay were installed to provide continuous communication. In an effort to keep track of where everything was, they started assembling a database on a laptop.

A routine was soon established. The two sleeping quarters in Zvezda were assigned to Shepherd and Krikalev. Gidzenko set up his sleeping bag in a corner of Zvezda's main compartment. The day began with a wake-up call at 0100. After 40 minutes to wash and dress, breakfast was followed by a day's work. In the early days each member of the crew undertook an hour's exercise prior to lunch, with a second hour before dinner.

Normally, the ISS Expedition crews were to pursue a standard American Civil Service week involving an 8-hour day from Monday to Friday and then Saturday and Sunday off. However, the crew elected to work through Saturday 4 November to try to catch up. After they applied electrical power to the Vozdukh and Korolev

---

[5] For the sake of continuity, this text will continue to refer to it as the International Space Station.

verified that it was removing carbon dioxide from the air, they were authorised to cease using the lithium hydroxide canisters. They then set up the Elektron, which electrolysed water donated by visiting Shuttles as a by-product of their fuel cells. The next task was to install the compressor on Zvezda's air-conditioning unit. At that time, Zvezda was operating at 75 °F air temperature and 40 to 50 per cent humidity. The air-conditioning unit was brought on-line the next day. The moisture that it removed from the air would be stored in a condensate collection system until it was emptied. They rounded out the day by finishing the installation of the interface which would enable laptops to access Zvezda's central computer system.

Although Sunday 5 November was a day of rest for the crew in space, the ground controllers spent it tracking a piece of space debris whose trajectory would take it within about 2 kilometres of the ISS at 1343. Because this was deemed to be a safe 'miss distance', no object avoidance manoeuvres were scheduled. The next day, 6 November, they resumed setting up the Elektron. The fact that one of two batteries in Zvezda that were refusing to hold their charge had been repaired allowed the crew to bring the Elektron on-line as the primary source of oxygen late on 8 November. The eighth battery in Zvezda remained off-line. They also continued to set up the computer network that would allow them to e-mail the control centres in Korolev and Houston. Shepherd revealed that they were having difficulty finding the proper cables for the different Russian and American laptops. Interestingly, the first use of English was heard at this time. Korolev's attempt to help was frustrated by the limited communications. The crew also experienced difficulty in starting some of their various computers. Gidzenko and Krikalev installed the monitor and hand controls for the TORU backup rendezvous and docking system in Zvezda. They completed the installation the following day. This would be used if a Progress failed to dock automatically using its own Kurs system. The second Progress was to be launched on 16 November, and to dock at Zarya's nadir port two days later. The plan called for it to be unloaded in time to depart before STS-97 made its approach to PMA-3 in the first week of December. All three men exercised on the bicycle ergometer for the first time. This was the beginning of a daily routine of exercise to ameliorate the physiological effects of prolonged microgravity. Even when in perfect condition, the station required continuing maintenance. When the air-conditioning unit's condensate tank filled up on 8 November the system tripped off-line, so Krikalev emptied the tank and reactivated the unit. The retrieved water would be used for personal hygiene.

On 9 November, with increased solar activity expected over the next 48 hours, the ISS crew were told to set up a radiation monitor inside Zvezda to monitor their exposure. Although the increase in activity was not expected to pose a serious threat, they were instructed to spend their next two sleep periods in Zvezda's aft section, close to the Transfer Compartment where the module had been fitted with additional radiation shielding for just such occasions. The crew went to sleep two hours later than usual in order to move their working day by that amount in future. The next day, they configured cables for Zarya's nadir port in anticipation of the Progress's arrival. Finally, they resumed configuring Zvezda's computer network to interface to laptops.

Following the hectic first week on the ISS, the crew were given Saturday, Sunday and Monday off. The only task assigned to 11 November was to commence wiring up Zvezda's 'ham' radio station. During the ISS's life in orbit, many astronauts and cosmonauts would use this radio equipment to talk to schools and enthusiasts as the planet rotated beneath them. Rested after their extended weekend off, the trio started their new working week with a dress rehearsal for Progress-M1 4's docking, during which Gidzenko and Krikalev were particularly busy as they manoeuvred the ISS to the correct attitude to support the spacecraft's final approach and tested the TORU system. On Tuesday 14 November, each man assessed how well his hearing was standing up in the noisy environment. They also measured their body masses and exercised on the stationary bicycle. While Gidzenko and Krikalev concentrated on the docking rehearsal, Shepherd carried out a routine inventory of equipment on the station. In a related task, Krikalev started work to update the software of the computer which read barcodes to maintain an inventory of where items were at any given time. The following day, the inventory became the primary task for all three men in an effort to record everything already on board before the Progress delivered another 1,815 kilograms of stores. During the day, Krikalev assisted Gidzenko as he performed another TORU simulation. At Korolev's request, Krikalev unplugged the cable from the battery in Zvezda that remained off-line and confirmed that one of the pins in its plug was bent, which suggested that it would be straightforward to repair.

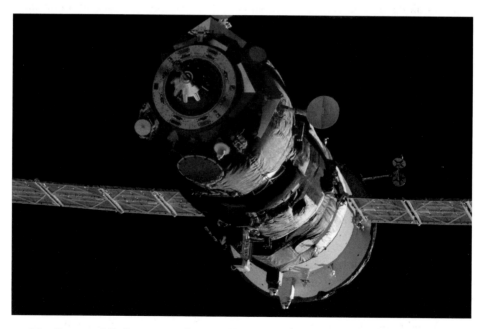

After Progress-M1 4's automated Kurs system became confused, Yuri Gidzenko steered it towards Zarya's nadir port using the TORU remote-control system.

PROGRESS-M1 4

While the crew slept on board the ISS, Progress-M1 4 was counted down and launched from the Baikonur Cosmodrome at 2033 on 15 November. Ten minutes later, upon achieving orbit, it deployed its photovoltaic arrays and antennae and settled into the standard two-day rendezvous. It approached the ISS and flew around the station to align itself with Zarya's nadir for a projected docking at 2208 on 17 November, but then its Kurs automatic system failed to lock onto Zarya's radar transponder. Progress's docking system camera showed the ISS moving in and out of the field of view as the spacecraft manoeuvred in search of its target. Deciding to intervene, Gidzenko took the TORU hand-controllers, stabilised the spacecraft facing the station, and began to command its thrusters to fire to slowly manoeuvre it in. As the vehicle approached, the view on Gidzenko's screen progressively degraded because the Sun was shining into the lens. The view was further impaired by the fact that the lens was fogged. Once the spacecraft had approached to within 10 metres, Gidzenko fired its thrusters to slow its approach and it drew to a halt 5 metres out. Nevertheless, the docking target was still not visible due to the unfavourable lighting conditions. A hasty visual check out of the station's porthole showed that the vehicle was misaligned. Korolev told Gidzenko to hold the ferry in this position until sunset, at about 2245. Once in the Earth's shadow, the view cleared up and Gidzenko commanded the Progress to back out to a distance of 35 metres and then made a second approach, leading to a docking at 2248, some 40 minutes behind the flight plan. Looking at the vehicle's television camera through a pair of binoculars, Gidzenko confirmed that there was a thin coating of ice on the lens. While this incident underlined the wisdom of installing the TORU system, it also highlighted the impracticability of having a ferry approach the station when it was silhouetted by the Sun; it was a significant procedural error on the part of the flight planners. The crew spent several hours powering down the Progress's systems and then retired. After their second weekend off, the crew began unloading cargo on 20 November. Each item was unbolted from its rack, logged into the ISS's computerised inventory, and then transferred to the appropriate position on board the station. By Wednesday, 70 per cent of the cargo had been transferred. The unloading was due to be finished on Friday. After their third weekend off, the crew performed routine maintenance on the ISS's air-conditioning system, toilet and treadmill, set up new apparatus, and replaced a faulty blade on a ventilation fan in Zvezda. As oxygen was pumped into storage tanks on board the ISS, the crew loaded unwanted packaging into the cargo ferry. By the mid-point in the second week of unloading cargo, the engineers at Korolev had determined that the ferry's initial approach had failed because its Kurs system had been unable to distinguish between Zarya and Zvezda. Such a situation did not arise when approaching from the rear. This was the first time that a docking from the side had been attempted. The engineers had developed a software patch to overcome the problem. The engineers called for the Progress to undock and redock to verify the new software. Gidzenko would redock it manually if the Kurs experienced difficulty. But with STS-97's launch only a few days away, Korolev did not make an immediate request to have the docking test added to the flight programme. The plan would remain under discussion by programme managers throughout the first half of December.

STS-97 ADDS POWER

STS-97 – Endeavour – SSAF-4A

Commander:	Brent Jett
Pilot:	Michael Bloomfield
Mission Specialists:	Joseph Tanner; Carlos Noriega; Marc Garneau (Canada)

STS-97 lifted off from Launch Complex 39's Pad B at 2206 on 30 November. Due to the late hour, the exhaust plume lit up the dark night sky. When the SRBs burned out 2 minutes 30 seconds later, telemetry indicated that one of two explosive charges used to separate the base of the left-hand SRB from the strut on the ET had failed to fire, but a redundant detonator had done the job. Once in orbit, Endeavour's five-man crew quickly configured the vehicle for the standard two days of solo flight leading up to a rendezvous with the ISS. The principal cargo in the payload bay was the P-6 photovoltaic power system, a 13.7-metre-tall truss segment with a pair of solar 'wings'.

As Endeavour entered orbit, the ISS was travelling over the southeastern Indian Ocean, its crew asleep. After the wake-up call at 0006 on 1 December, their primary task for the day was to finish loading Progress-M1 4, close the hatches, and dispatch it so that it would not impede the Shuttle's access to PMA-3, which was on Unity's nadir close alongside Zarya's docking unit. After undocking at 1023, the spacecraft manoeuvred into a parking orbit, some 2,500 kilometres in trail of the ISS, where it was to remain throughout STS-97's visit, so that it could later return and dock in order to verify the new software.

Flight Day 2 for the STS-97 crew was spent preparing for the construction tasks ahead. Garneau and Bloomfield put the RMS through a series of standard tests, used its cameras to conduct a video survey of the payload bay, and completed tests of the Space Vision System. Noriega and Tanner prepared the EMUs for their spacewalks. Electrical power was applied to the photovoltaic unit to verify that it had not been damaged during the launch. Jett and Garneau performed manoeuvres, at 1341 and 2224 on 1 December to refine the rendezvous. The diurnal cycles for the two crews were out of phase. As Endeavour's crew retired at 0100 on 2 December, Shepherd's crew was awakened. When Shepherd had enquired on 17 November whether they were to adjust their cycle in order to be in phase with their visitors, he was informed that there would be no change. This meant that the two crews would work independently and remain in their own vehicles most of the time, combining only for short periods to unload Endeavour's cargo.

Flight Day 3 saw the fourth rendezvous manoeuvre at 1100 on 2 December. The

TI burn followed at 1233. Upon arriving directly below the ISS, Jett and Bloomfield verified that everything was as it should be and Jett then made a manual approach to the station, while his crew provided range and range-rate data using handheld laser rangefinders. At a separation of 152 metres, he reoriented Endeavour 180 degrees so that its tail was aligned along the velocity vector, and then closed in to 9 metres and halted. On the ISS, both Zarya and Zvezda's photovoltaic arrays had been locked in a position that would minimise contamination of their transducers by efflux from Endeavour's thrusters. It was a standard safety precaution that was used each time a Shuttle docked with the station, but this was the first time that a Shuttle had docked on the nadir-facing PMA. Once everything was just right, Jett nudged Endeavour upwards to soft-dock with PMA-3. Both the Shuttle's and the ISS's attitude control systems were deactivated so that the Orbiter Docking System's mechanism could be retracted to draw the vehicles together, which occurred over Kazakhstan at 1500. Zarya and Zvezda's arrays were released to resume tracking the Sun. Endeavour was to assume responsibility for attitude control for the docked combination.

Once Noriega and Tanner had checked the air pressure between Endeavour and PMA-3, the hatches were opened. On the far side of PMA-3, they opened the hatch to Unity's vestibule and left supplies for the ISS crew to retrieve when they awoke. These supplies included a laptop, a hard drive for a Russian laptop, headsets for the station's two-way video-teleconferencing system, a variety of tools, fresh food, bags of water from the Shuttle's fuel cells and a package of personal items. Noriega and Tanner then retreated and closed the hatches. At 1717, Garneau activated the RMS and used its end-effector to engage the P-6's grapple pin, and having released the

On mission STS-97, Endeavour approaches the International Space Station with the P-6 photovoltaic array unit filling its bay.

hold-down attachments that had secured it in the payload bay, he lifted it and 'hung' it out at 30-degrees to the payload bay to achieve thermal stability prior to mounting it on the Z-1 Truss Structure. At 0438 on 3 December, four hours after Jett's crew had retired, Shepherd's awoke for their 33rd day of occupation. Their first task was to enter Unity through PMA-1 – which was the first time that they had done so, because it had previously been off limits due to insufficient power – to retrieve the items left there by Noriega and Tanner. After these supplies had been transferred to Zvezda, the hatches to Unity were closed once again.

Flight Day 4 on board Endeavour saw Noriega and Tanner open the airlock at 1335 on 3 December for their first EVA. After preparing the tools and aids that they would use, they made their way to the top of the Z-1, on the far side of the station, to provide verbal instructions to help Garneau to manoeuvre the RMS to place the P-6 unit into a position beyond his line of sight. With a little physical assistance from Noriega, the P-6 was in position at 1432, and Noriega and Tanner secured the bolts at its four corners. Once Garneau had released and removed the RMS, Noriega put a work station on the end-effector and mounted its foot restraint. Bloomfield assumed control of the RMS and moved Noriega around the Z-1/P-6 so that he could connect nine power, command and data cables. Noriega extended the two photovoltaic array attachment joints and deployed and locked into place the pivoting bars on each unit. Meanwhile, Tanner had uncovered the photovoltaic array blanket boxes on the P-6 unit. However, when Jett sent the command to release the pins holding the boxes closed, nothing happened. After repeated attempts, the pins on the starboard box released, but one pin on the port side still failed to unlatch. Tanner and Noriega then released the solar array wing launch-restraints before stowing their equipment and returning to Endeavour's airlock. The EVA ended at 2108, after 7 hours 33 minutes. At 2120, Jett sent commands to close the securing pins on both of the blanket boxes, then repeated the command to open them. All of the pins on both boxes disengaged, leaving the crew free to proceed with the deployment of the photovoltaic arrays. In Houston, it was decided to defer deploying the port side unit until the failure of the single pin to release had been investigated. The photovoltaic array blankets on the starboard side were commanded to deploy at 2123, more than two hours later than scheduled, and allowed to do so in one movement. A motor extended the telescopic mast, which drew out the photovoltaic arrays on either side. The deployment took approximately 13 minutes, but resulted in a 'wave' which oscillated back and forth along the length of the blankets. At the end of the deployment, it was observed that one set of tension lines were much looser than they should have been. Two hours later, the first of a trio of radiators was deployed. As Jett's crew retired in the early hours of 4 December, Shepherd's crew woke up and set about installing a new dust collector fan, collecting condensate water samples, replacing a microprocessor and photographing glaciers in Patagonia, South America.

Flight Day 5 on board Endeavour was a relatively quiet day of internal activities. Interviews took place with three American television networks. Garneau completed a photographic survey of the P-6 tower and its starboard photovoltaic arrays, using the cameras on the RMS. The highlight of the day was the deployment of the port side. The command to extend the telescopic mast was sent at 1952 on 4 December,

Having deployed the P-6 photovoltaic arrays, Carlos Noriega waves to the camera.

and the deployment was made in a series of stop–start moves designed to damp out travelling 'waves' in the blankets. As a result, the process was not completed until 2146. The rest of the day was spent on housekeeping chores. Overnight, the dozen storage batteries in the P-6 tower (six for each pair of arrays) were charged up, ready to supply fully four times the power from the arrays on Zarya and Zvezda.

On Flight Day 6, Noriega and Tanner left the airlock at 1221 on 5 December for their second EVA, and made a visual inspection of the starboard photovoltaic arrays. In Houston, planning was underway to have the two astronauts attempt to improve the tension on the slack array during their final EVA. Noriega moved to the starboard side of the Z-1, where he moved cables from one connector to another to allow the transfer of power from the P-6's batteries. He then removed a thermal cover from a power conditioner, and Tanner removed the cover from a signal processor. The main task of the day was to install the S-Band Antenna Subassembly. After retrieving it from the temporary site on the Z-1 where it had been placed by the STS-92 crew, they used a leapfrogging movement to pass it between them to the top of the P-6 tower, where they installed it on the Integrated Equipment Assembly. While at that location, they took the time to study the take-up reels on the starboard photovoltaic arrays to determine if the tension lines had jumped off during the one-stage deployment used for the blankets on that side. Meanwhile, Shepherd's crew entered Unity for the second time, to reconfigure cables to allow the module to accept power produced by the P-6 arrays. This one-hour activity kept them up past their intended 1636 bedtime. Outside, Noriega and Tanner released the bolts holding the second of the trio of radiators to the side of the P-6 tower, which was to be deployed after their return to Endeavour. This radiator would dissipate heat produced by the electrical equipment in the American Laboratory Module, once it was in place and operational. They also unplugged a number of electrical umbilicals from PMA-2, plugged them into a nearby dummy connector, and prepared it for its transfer during the Laboratory's attachment process. This done, Noriega and Tanner returned to the airlock at 1858, having been out for 6 hours 37 minutes.

After a long day, Jett's crew retired at 2330. Half an hour later, Shepherd's crew were awakened. They replaced the air-conditioning unit that had failed earlier in the week with a similar unit that had been ferried up by Progress-M1 4, and replaced a malfunctioning fan in the Vozdukh carbon dioxide removal unit and brought that unit back on-line. At 0530 on 6 December, Shepherd re-entered Unity to install electrical outlets and air ducts, and to separate the power feeds to the Early Communications System and the newly installed S-Band system. With P-6 supplying power, and the environmental control systems repaired, Unity could be opened. The engineers in Houston and Korolev worked with Shepherd, Gidzenko and Krikalev to route the P-6 power through the relevant interfaces to the various modules. American–Russian power converters allowed the P-6 batteries to provide an additional 3 kW of power for use in Zarya and Zvezda, to bring their power levels up to 7 kW each.

Flight Day 7 on board Endeavour was quiet, so that Noriega and Tanner could rest before their final EVA, and they prepared the Floating Potential Probe (FPP) that they would install on top of the P-6 tower. When Houston relayed the decision that they should try to increase the slack tension on the starboard array, they ran through the relevant procedures then took the rest of the day off. Shepherd's crew awoke at 0130 on 7 December, spent most of the day packing up the items that were to be transferred to Endeavour, then set up the wireless instrumentation system that was to be used in the future in an attempt to model the structural integrity of the station while the Shuttle's thrusters were firing.

Noriega and Tanner left Endeavour's airlock for the third and final time at 1113 on 7 December, on Flight Day 8, and promptly set about retensioning the starboard array. During the previous day, they had watched video transmissions showing how astronaut David Wolf had done so on the simulated equipment in Houston. Having discussed the procedures, they had expressed their confidence that they would be able to do it. The *ad hoc* procedure called for Jett to partially retract the starboard telescopic arm so as to release the tension on the deployment cables. This allowed Noriega to place the slack cables back in their spring-loaded take-up wheels. Tanner then manually rewound the take-up wheel until the lines were taut. As the reels were then allowed to unwind again, Noriega ensured that the cables remained within the grooves. The mast was re-extended and the blanket was extended with the correct tension. Noriega and Tanner then fitted a print of an evergreen tree (taken from the side of a cargo bag) on top of the P-6 tower in a ceremony mimicking the topping-off ceremony used by terrestrial builders when their structures reach their highest point. The next task was to install a cable on the exterior of Unity for the centreline camera that would assist in the laboratory module's berthing. Finally, they installed the Floating Potential Probe which would measure the electrical potential in the plasma surrounding the station; plasma contractor units mounted on the P-6 tower emitted xenon atoms which, by completing an electrical circuit, warded off electrical arcing as the station flew through the ionosphere. With time on their hands, they undertook a photographic survey of the station's new additions, before making their way back to the airlock at 1623, having been out for 5 hours 10 minutes.

At 0936 on 8 December, on Flight Day 9, the two vehicle crews finally met when they opened their respective hatches on either side of PMA-3. Until that time, at least one hatch had always been closed between the vehicles because Endeavour's cabin had been kept at a reduced pressure in order to support the EVA programme. After a brief welcoming ceremony and safety briefing for the residents' first visitors, the six astronauts and two cosmonauts started their joint programme, which included a structural test which established that the docked vehicles behaved almost precisely as the computer simulations had predicted. As new equipment was transferred from Endeavour to the ISS, rubbish and other items were offloaded. Both crews held press conferences in the early evening. Shepherd voiced a complaint over the heavy workload that Korolev's controllers had placed on his crew during their first five weeks, and explained that the issue was that many activities took longer in reality than had been allowed for in the flight plan drawn up on the basis of trials on the ground. After monitoring telemetry from the Floating Potential Probe, Houston announced that there was little or no risk of arcing taking place.

On Flight Day 10, following the farewell ceremony, Jett's crew returned to

Endeavour and PMA-3's hatch closed at 1051 on 9 December. After undocking at 1413, Bloomfield withdrew to a distance of 137 metres and initiated a tail-first 360-degree loop around the station's X-axis. This had been scheduled to enable the IMAX camera mounted in the payload bay to film the station with its new solar power unit, but the camera had malfunctioned earlier in the flight. Instead, a photographic survey was undertaken using still and video cameras. With the fly-around complete, Bloomfield manoeuvred Endeavour clear. Late in the day, Jett's crew spoke with journalists and schoolchildren in Canada, then took the rest of the day to wind down, then retired. Two hours later, Shepherd's crew awoke for a day off, which they spent housekeeping and talking to their families by radio. Shepherd gave Houston a video tour of Unity, which was full of the items which had just been delivered, asked that an entire day be set aside for housekeeping in order to stow it all away, and announced that as a temporary measure PMA-3 would be used as a closet. The Shuttle's crew spent most of Flight Day 11 preparing Endeavour to return to Earth. The next day, retrofire was on schedule, and Jett made a perfect touchdown on the Shuttle Landing Facility at 1803 on 11 December.

BUSINESS AS USUAL

Alone once again, the ISS crew spent several days checking the station's systems and conducting some of the small number of experiments on board, spent two days storing the items delivered by Progress-M1 4 and STS-97 and logging them into their computerised inventory, and continued routine maintenance of Zvezda's Vozdukh, the carbon dioxide scrubber, one of whose fans had once again malfunctioned. When the Press suggested that with only one spare fan in store, the crew was one failure away from having to evacuate the ISS until the Vozdukh could be repaired, Shepherd countered that he was sure they would be able to fix it, and in any case they had the lithium hydroxide canisters as a backup. Spare parts for the Vozdukh were added to the manifest for Progress-M1 5, which was due in early February 2001.

THE HOLIDAY PERIOD

The Expedition 1 crew had a relaxing holiday period, performing only sufficient light duties as necessary to maintain the ISS in good working condition. On 25 December, they spent some time admiring the view of the Earth passing below. All three men had private conversations with their families. Shepherd lamented that there was no Christmas tree on board, but he voiced the opinion that he was sure there would be one for Christmas 2001.

On 19 December, NASA had approved the plan for Progress-M1 4 to return to dock at Zarya's nadir port on 26 December, on the understanding that the attempt would be aborted at the first sign of trouble. Once the Kurs system had manoeuvred the ferry to the stationkeeping point 200 metres out, Gidzenko activated the TORU and flew it in to dock at 0630. Once the hatches had been opened, the electronics for the Kurs system

The STS-97 and ISS crews in Zvezda: Sergei Krikalev (left front), Brent Jett, William Shepherd and Joseph Tanner, Marc Garneau (left rear), Carlos Noriega, Yuri Gidzenko and Michael Bloomfield.

The state of the International Space Station after the STS-97 visit. The scale of the P-6 photovoltaic array is evident.

was removed so that it could be returned to Earth for analysis. Over the next few days, the crew performed biomedical experiments and reviewed the plan for the installation of the American Laboratory Module, Destiny, in January. Shepherd observed New Year 2001 in keeping with the traditions of the US Navy, with the person on duty at midnight on 31 December writing something appropriate in the ship's log in prose; he wrote a nine-verse poem which compared the ISS to an old-style sailing ship.

COMMENCING 2001

The Expedition 1 crew spent the first week of the year, their tenth week in space, conducting experiments and performing routine maintenance. The ongoing technical problem was Battery 3 in Zvezda, which engineers in Korolev took off-line to allow Krikalev to check the pins on the plug of a cable between the battery and its current converter. With the Sun shining on Zvezda's photovoltaic arrays at an oblique angle, Korolev decided to conserve electricity by restricting the use of the Elektron system, so additional VIKA canisters were used to supplement the oxygen supply. As soon as the solar illumination improved, the Elektron was returned to full power. The next week was spent preparing for the arrival of Destiny, which was due for launch on 19 January on STS-98. The passageways that would be used during the Shuttle's visit were cleared, and the locations of the items inventoried. In preparation for Destiny's installation, the CBM on Unity's ram end, to which PMA-2 was currently mated, was depressurised and Houston attempted to cycle its latches. The first latch cycled correctly, but the second was fouled by a piece of air duct. Once the PMA had been repressurised, Shepherd cleared the obstruction and the test was completed. On 14 January, the Shuttle's launch was slipped to February. In space, Shepherd's crew replaced the malfunctioning current converter on Zvezda's battery and continued to inventory items as they tidied up. The week ending 25 January was spent rehearsing emergency and evacuation procedures after a simulated air leak. Such simulations would become a routine part of life on board the ISS. They also performed routine maintenance tasks and continued to update and expand the Inventory Management System. At one point, Shepherd said that although they were now acclimatised to working in weightlessness, and wanted to do everything on time, many tasks still took much longer to achieve than had been allowed in the flight plan.

PAYLOAD OPERATIONS CENTER OPENED

The ISS Payload Operations Center (POC) at NASA's Marshall Space Flight Center in Huntsville was opened on 2 February. Once the Destiny laboratory was installed, and the Expedition 2 crew were on board, Houston would handle operational issues and the POC would be deal with all activities pertaining to the science programme, including scientific payload safety, activity planning, execution and troubleshooting malfunctions. It would operate on a 24-hour basis with a staff of up to 19 controllers on each 8-hour shift, many of whom had supported Shuttle Spacelab missions, and call on technical support from universities around the world. Once it was fully operational, the ISS was expected to sustain four times the scientific capacity of the Spacelab module.

STS-98 WITH DESTINY

STS-98 – Atlantis – SSAF-5A

Commander:	Kenneth Cockrell
Pilot:	Mark Polansky
Mission Specialists:	Robert Curbeam; Marsha Ivins; Thomas Jones

---

This important crew was initially assigned experienced Shuttle astronaut Mark Lee, who had flown on four previous Shuttle flights, as a mission specialist, but he was withdrawn in September 1999 as a result of "an internal astronaut office matter". In fact, the removal order was issued by George Abbey, the director of the Johnson Space Center, as a disciplinary action "to maintain the high level of professionalism necessary to insure the integrity of space flight". It was made clear, however, that Lee "remains eligible for a future assignment". Similar disciplinary action had previously been taken against astronauts David Walker and Robert Gibson, both of whom had later commanded missions. Although STS-98 commander Ken Cockrell complained at Lee's removal, citing the amount of training that his crew had already completed and the short time that remained to integrate Robert Curbeam, Lee's replacement, Abbey was insistent. That same month, the Russians withdrew Talget Musabayev from STS-98, and Nikolai Budarin was removed from STS-92. No reason for these removals was given at the time, but Musabayev was reassigned to command the Soyuz flight laid on to take Mircorp's Dennis Tito to Mir, and when it was decided to de-orbit Mir this 'tourist' mission, undoubtedly to Musabayev's delight, was reassigned to fly to the ISS.

Following the problem with the separation charges on STS-97's left-hand SRB, all of STS-98's SRB separation charges were X-rayed and crumbling was noted on the shielding on one cable in the left-hand SRB's forward skirt and three other cables. As a result, on 19 December Atlantis's roll out was postponed until 3 January 2001. The engineers were instructed to test the thousands of re-usable electrical cables in the two SRBs. That testing identified a number of data cables running the length of the SRBs that transmitted test signals only intermittently. Some of the cables were as much as 17 years old, and had not been inspected during the earlier tests. As 'single point' failures that could threaten a mission, they had to be repaired immediately. On 14 January, NASA rescheduled the launch for "no sooner than 6 February". After being rolled out to the pad on 26 January, STS-98 Atlantis lifted off at 1813 on 7 February. At that time, the ISS was east of Newfoundland and a video of the launch was relayed to its crew as their day ended.

When the ISS crew awoke just after 0500 on 8 February, they started their 100th

On mission STS-98, Atlantis approaches the International Space Station with the Destiny laboratory in its bay.

day in space. Korolev ordered Progress-M1 4 to undock from Zarya's nadir port at 0626. A few hours later, the rubbish-filled spacecraft was de-orbited and destroyed during re-entry. Flight Day 2 for Cockrell's crew began at 0713, so for once the two space crews were on essentially the same cycle. Curbeam and Jones checked out the EMUs. When they tested the spare unit they thought it had a leak, but upon further inspection this proved not to be. Meanwhile, Ivins powered up the RMS and used it to perform a video inspection of the payload bay and the Space Vision System that she would use to transfer the 8.5-metre-long and 4.3-metre-diameter Destiny module to the ISS. Shepherd's crew divided the day between their regular exercise regime and reviewing plans for the up-coming joint activities.

The TI burn at 0924 on 9 February, on Flight Day 3, marked the final phase of the rendezvous. One orbit later, Cockrell took control and manually steered Atlantis to a docking with PMA-3, which was on Unity's nadir, at 1151. The hatch to Unity was opened at 1431. After hearty greetings, Cockrell's crew set about transferring cargo to the ISS. The supplies included three bags of water produced by Atlantis's fuel cells, a backup laptop computer for Zvezda, and several internal cables which would have to be fitted before Destiny could be activated. The residents also received gifts from their families, fresh food and movies. At 1803, Cockrell's crew retreated to Atlantis, closed the hatch, and reduced the cabin's atmospheric pressure to assist Curbeam and Jones in their pre-breathing. However, it was soon discovered that three external cables that they would require on their first EVA had inadvertently

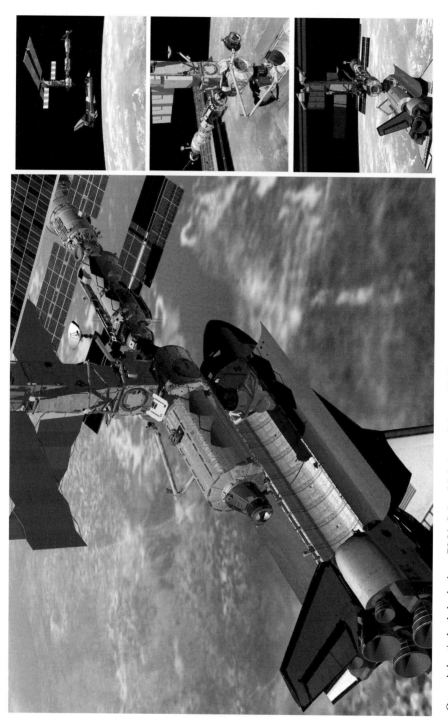

Once Atlantis had docked on PMA-3 (on Unity's nadir), its RMS moved PMA-2 to a temporary berth on the Z-1 Truss Structure, installed Destiny and then relocated PMA-2 on its end.

been transferred, so PMA-3 was used as an airlock to pass them back without adjusting the Shuttle's internal pressure.

Flight Day 4 was to be the highlight of the mission. While Curbeam and Jones donned their EMUs, Ivins powered up the RMS and used it to remove PMA-2 from Unity's ram CBM and store it temporarily on an attachment on the Z-1 Truss Structure. As soon as they left the airlock at 1050 on 10 February, the spacewalkers set about their individual tasks. Curbeam disconnected the power cables and cooling umbilicals which had sustained Destiny in Atlantis's payload bay, and removed the launch covers from its CBMs. Meanwhile, Jones made his way on to the exterior of the ISS, to verbally guide Ivins as she relocated PMA-2. Ivins attached the RMS to Destiny at 1223, lifted it clear of the bay, rotated it through 180 degrees, and then positioned it in front of the now-vacant end of Unity. Destiny had a CBM on either end. One would be used to mate with Unity, the other would be used as a mounting point for PMA-2. With verbal cues from Jones, Ivins mated the two modules at the first attempt, at 1357. Curbeam and Jones connected a number of electrical, data and cooling lines from one to the other. A small ammonia leak occurred as Curbeam was connecting one cooling line. It posed no risk to the astronauts, but led to the use of a decontamination protocol at the conclusion of the EVA to prevent any contaminants entering Atlantis's cabin. Curbeam remained in direct sunlight for 30 minutes to vaporise any ammonia crystals which may have settled on his EMU. Jones brushed off his partner's suit and equipment. Once in the airlock, they partially-pressurised, depressurised and then fully repressurised the airlock in an attempt to remove any lingering ammonia crystals. Cockrell, Polansky and Ivins wore face masks for 20 minutes to allow any ammonia crystals which did enter the cabin's atmosphere to be filtered by the air-conditioning system. The decontamination protocol lengthened the activity by over an hour, giving an EVA of 7 hours 34 minutes.

Destiny could accommodate 24 'standard' racks of equipment. Of the 11 racks assigned to systems to run the laboratory, only three were installed for launch. The 13 science racks (all of which had been left vacant due to the Shuttle's launch constraints) would be ferried up a few at a time by later missions in Multi-Purpose Logistics Modules. With the laboratory in place, the ISS had more pressurised volume than any previous space station, including Skylab and Mir.

With Atlantis's cabin pressure again matching that inside the ISS, the hatches were reopened at 2050 and Shepherd came on board to work with Cockrell at a laptop on the aft flight deck to commence powering up Destiny's systems and initiate the flow of cooling fluids to its avionics. When the two crews finally retired, two hours later than planned, Houston continued the initialisation work. Flight Day 5 got off to a late start because the crews were allowed to sleep longer. Everyone gathered in Unity at 0938 on 11 February when Shepherd opened Destiny's hatch. As the commander of the Shuttle that had delivered it to the ISS, Cockrell was the first person to enter the module. Most of the day was devoted to activating the laboratory's air conditioners, ventilation systems, electrical outlets, computers, internal communications system and fire extinguishers. When they installed and activated a rack of equipment designed to purify the module's air, they not only augmented the Vozdukh system in Zvezda but also provided a welcome degree of

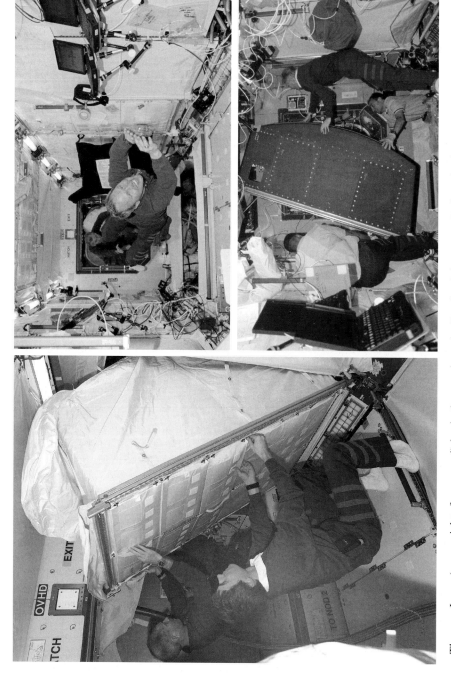

Thomas Jones (upper right, foreground) leads the way into the Destiny laboratory. Robert Curbeam, Mark Polansky and Kenneth Cockrell move a rack into position (lower right), and Kenneth Cockrell and Sergei Krikalev (left) set up hardware.

redundancy. In the afternoon, Atlantis made a set of thruster firings to boost the ISS's orbit to counter decay, raising the perigee by 29 kilometres. Cockrell took his crew home and the hatches were sealed at 1740 so that Atlantis's cabin pressure could be lowered in preparation for the second EVA.

On Flight Day 6, Ivins powered up the RMS and grasped PMA-2 at 1020 on 12 February. Curbeam and Jones left the airlock 20 minutes later. Their first task was to provide Ivins with verbal cues as she moved PMA-2 from its temporary location on Z-1 to its permanent position on the end of Destiny, where it would act as the docking system for the majority of future Shuttles visiting the ISS. Curbeam and Jones also added an electronic Power and Data Grapple Fixture and a video signal converter unit for the Canadian-built Space Station Remote Manipulator System that was to be mounted on Destiny while the Integrated Truss Structure was being assembled. Then they installed thermal covers over the pins which had held the module in place inside the payload bay, fitted a vent to its air system and installed a number of EVA handrails around its exterior. Finding themselves ahead of the flight plan, they then made a start on tasks assigned to EVA-3, including connecting the cables between Destiny and PMA-2 to complete its reinstallation, uncovering the module's high-fidelity porthole and installing an external shutter. After relocating a movable foot restraint that they had taken out at the end of the first excursion they finally retreated to the airlock at 1749, having been out for 6 hours 50 minutes. As the EVA took place, Shepherd's crew continued to activate Destiny's systems. It held the electronics for the four CMGs in the Z-1 Truss Structure, so Houston began the process of spinning them up at 1643. When Cockrell's crew retired, the hatches were closed to conserve the Shuttle's air. Overnight, Houston completed the activation of the Atmosphere Revitalisation Rack within Destiny and commanded the bolts holding PMA-2 in place to fully tighten. Cockrell's crew had half of Flight Day 7 off in order to relax. At 0700 on 13 February, Houston activated Destiny's computers and made a series of tests to verify their readiness to take over attitude control from the computers in Zvezda. Over the next few days, control was switched back and forth between the two modules. The CMGs worked as expected throughout these 'control authority' tests, so they were commanded on-line on 13 February, thereby significantly reducing the rate at which the station would consume propellant. When he awoke, Shepherd gave Houston a status report, telling them that some loose wires had been found in Zvezda's treadmill and a washer had been lost inside Destiny, but neither issue was serious. When the residents returned to Destiny to resume methodically testing its systems, and powered up the carbon dioxide removal system, its pump failed, so they turned it off and left Houston to troubleshoot the fault. Meanwhile, Cockrell and Polansky used Atlantis's thrusters to make the second series of manoeuvres to raise the ISS's orbit. Later still, Cockrell and Ivins activated the RMS to inspect an area of bubbling paint on one of the ISS's cooling radiators, but the radiator was working normally so there was no immediate concern. The day concluded with Atlantis's crew taking part in the traditional press conference.

The highlight of Flight Day 8 was the third EVA for Curbeam and Jones. They left the airlock at 0948 on 14 February. In fact, this was NASA's 100th EVA and the event was celebrated by the two astronauts holding up a plaque in Atlantis's payload bay. "This achievement, this golden anniversary so to speak," Curbeam told the controllers in Houston, "is a tribute to all the people who have done spacewalks, all the people who have designed the Gemini, Apollo, Skylab and now Shuttle suits. And we salute all of you and appreciate your hard work and thank you so much." Then it was down to work. Polansky, reading from the flight plan, guided them through the installation of a spare communication antenna on the ISS's exterior. After checking the electrical cable connections between PMA-2 and Destiny, they inspected the condition of the P-6 photovoltaic arrays, then released a radiator that would be deployed later. Their final activity, inherited from STS-92, was an attempt to demonstrate a method by which a spacewalker could assist an incapacitated colleague. The EVA ended at 1513, after 5 hours 25 minutes. At one point, the ISS's crew, working in Destiny, filmed their colleagues outside using an IMAX camera. The hatches between Atlantis and Unity were reopened at 1814 and the two crews resumed the transfer of equipment over to the station. Atlantis also made another series of orbit-boosting manoeuvres.

On Flight Day 9, the final day of docked operations, the two crews finished the transfer of the supplies to the station. One of the items was the EMU that had been used by Jones. The one worn by Curbeam and the spare had to be retained on board the Shuttle in case an EVA had to be made to close the payload bay doors. Once the Joint Airlock Module was delivered by STS-104, the EMU would facilitate EVAs from the station. Once 'expired', the EMU would be returned to Earth for servicing, and replaced by another. A large amount of trash and equipment that was no longer required was then transferred to Atlantis. Without a Spacehab module in the bay in which to store this to-Earth payload, it had to be accommodated on the mid-deck. During the day, the station's computer automatically took one of the four CMGs off-line, and shut it down. Houston spun it back up to 6,600 revolutions per minute and then, without making it available, proceeded to monitor its performance. Only two units were required to maintain the ISS's attitude, so the remaining three were more than adequate. CMGs are notoriously quirky pieces of equipment. When the fourth unit displayed no further problems, Houston brought it back on-line early on the following morning. Later, Atlantis made a further series of thrusters firings to raise the ISS's orbit. By now, whatever cargo it delivered, each Shuttle undertook a routine of essential but unremarkable tasks. At 1337 on 15 February, the two crews gathered for a 40-minute press conference with journalists in Houston and Korolev. On Flight Day 10, the residents bade their visitors farewell and then Cockrell led his people back to Atlantis. The hatches were closed at 0718 on 16 February. Atlantis undocked at 0906. Shepherd thanked them for their effort and wished them a safe journey home. Polansky flew Atlantis in a half loop from directly below to directly above the ISS so that his crew could photograph Destiny and the relocated PMA-2. As Atlantis performed a separation burn, at 2145, Shepherd's crew were exercising. Alone once again, they were to resume their standard routine of sleeping at 1630 and awakening at 0100. Cockrell's crew spent Flight Day 11 tidying up and reviewing re-entry procedures. Landing was scheduled for just after noon on 18 February, but it was not to be. Out-of-tolerance winds over the Shuttle Landing Facility prompted a wave-off. Bad weather ruled out both the Cape and Edwards Air Force Base the next

A view of the Z-1 Truss Assembly and the base of the P-6 unit showing the extent of external activity.

The state of the International Space Station after the STS-98 visit.

day. The winds ruled against the Cape on 20 February too, but Edwards was open and Cockrell brought his spacecraft to a safe landing on the dry lake at 1533, after a mission which had lasted 12 days 21 hours 20 minutes.

The addition of the American Laboratory Module marked the completion of Phase Two. Its installation was to initiate the flow of American experimental results as soon as possible in the programme. Nevertheless, it would be some time before all the racks of scientific equipment were in place. In addition to delivering these racks, some Shuttles would linger, to 'utilise' the module's facilities.

At this time, France's Space Agency announced that in addition to its payments towards the ISS as a result of its ESA membership, it would pay Russia $500 million to have its astronauts fly to the ISS during routine Soyuz-TM exchanges. This was a logical extension of France's long-standing participation in the Russian space station programme. The first such mission would provide a week-long visit in October 2001.

In a separate agreement, Russia agreed to fly ESA astronauts to the ISS as Flight Engineers on Soyuz flights once per year during the period 2002 to 2006. The third couch on a Soyuz crew would be offered to ESA and would be paid for by the national space agency of whichever astronaut was selected to fly. If ESA refused the offer of a flight, then Russia would be free to assign one of its own cosmonauts. The first astronaut scheduled to fly under this arrangement was Italy's Roberto Vittro.

Table 13.4 Phase Two flight log

Zarya
SSAF-1A/R
Launched	2040	20 Nov 1998 by Proton
Objective	The 'Control Module' was the first element of the ISS to be launched.

STS-88
Endeavour
93rd mission
SSAF-2A
Launched	0335	4 Dec	1998 from Launch Complex 39 Pad A
Docked	1948	5 Dec	1998 at Zarya's ram
Undocked	1525	13 Dec	1998
Recovered	2254	15 Dec	1998 at KSC
Objective	Formed the Zarya–PMA-1–Unity–PMA-2 stack.

STS-96
Discovery
94th mission
SSAF-2A.1
Launched	0649	27 May	1999 from Launch Complex 39 Pad B
Docked	0024	29 May	1999 at PMA-2 on Unity's ram
Undocked	1839	3 Jun	1999
Recovered	0203	6 Jun	1999 at KSC
Objective	Double Spacehab logistics.

STS-101
Atlantis
98th mission
SSAF-2A.2a

Launched	0611	19 May 2000 from Launch Complex 39 Pad A
Docked	0032	20 May 2000 at PMA-2 on Unity's ram
Undocked	1903	26 May 2000
Recovered	0220	29 May 2000 at KSC
Objective	Double Spacehab logistics.	

Zvezda
SSAF-1R

Launched	0056	12 Jul 2000 by Proton
Docked	1945	25 Jul 2000 at Zarya's wake
Objective	The 'Service Module' provided a habitat for a crew of three, and additional power, and assumed responsibility for attitude control and orbital reboost.	

Progress-M1 3
SSAF-1P

Launched	1426	6 Aug 2000 by Soyuz launch vehicle
Docked	1613	8 Aug 2000 at Zvezda's wake
Undocked	2302	13 Oct 2000
De-orbited	–	1 Nov 2000
Objective	Deliver logistics and propellant and take away trash.	

STS-106
Atlantis
99th mission
SSAF-2A.2b

Launched	0845	8 Sep 2000 from Launch Complex 39 Pad B
Docked	0152	10 Sep 2000 at PMA-2 on Unity's ram
Undocked	2344	17 Sep 2000
Recovered	0356	20 Sep 2000 at KSC
Objective	Double Spacehab logistics and in-orbit Zvezda upgrade.	

STS-92
Discovery
100th mission
SSAF-3A

Launched	1917	11 Oct 2000 from Launch Complex 39 Pad A
Docked	1345	13 Oct 2000 at PMA-2 on Unity's ram
Undocked	1108	20 Oct 2000
Recovered	1700	24 Oct 2000 at Edwards AFB
Objective	Put the Z-1 Truss Structure on Unity's zenith and PMA-3 on Unity's nadir.	

Soyuz-TMA 31
SSAF-2R (1S)

Launched	0253	31 Oct 2000
Docked	0421	2 Nov 2000 at Zvezda's wake
Undocked	2221	5 May 2001

| Recovered | 0142 | 6 May 2001 |
| Objective | Delivered the Expedition 1 crew to the ISS. It was undocked at 0506 on 24 Feb 2001 and then redocked on Zarya's nadir at 0537 to make Zvezda's wake available for Progress-M 44. At 0740 on 18 April it was undocked from Zarya's nadir and at 0801 redocked at the wake to create clearance for STS-100 to swing a MPLM onto Unity's nadir. It was ultimately taken away by the 'taxi crew' that delivered Soyuz-TMA 32. | |

Progress-M1 4
SSAF-2P

Launched	2033	15 Nov	2000 by Soyuz launch vehicle
Docked	2248	17 Nov	2000 at Zarya's nadir
Undocked	1123	2 Dec	2000
Redocked	0600	26 Dec	2000 at Zarya's nadir
Undocked	0626	8 Feb	2001
De-orbited	0850	8 Feb	2001
Objective	Deliver logistics and propellant and take away trash.		

STS-97
Endeavour
101st mission
SSAF-4A

Launched	2206	30 Nov	2000 from Launch Complex 39 Pad B
Docked	1500	2 Dec	2000 at PMA-3 on Unity's nadir
Undocked	1413	9 Dec	2000
Recovered	1803	11 Dec	2000 at KSC
Objective	Mounted the P-6 photovoltaic power module on the Z-1 Truss Structure.		

STS-98
Atlantis
102nd mission
SSAF-5A

Launched	1813	7 Feb	2001 from Launch Complex 39 Pad A
Docked	1151	9 Feb	2001 at PMA-3 on Unity's nadir
Undocked	0906	16 Feb	2001
Recovered	1533	20 Feb	2001 at Edwards AFB
Objective	Removed PMA-2 from Unity's ram and temporarily mounted it on a fixture on the Z-1 Truss Structure, then mounted the American Laboratory Destiny on Unity's ram at 1357 on 10 Feb; PMA-2 was later retrieved and mounted on Destiny's ram.		

(All times US Eastern)

Table 13.5 Phase Two EVA Log

	Start time	Finish time	Duration
STS-88			
Ross and Newman	1710 7 Dec 1998	0031 8 Dec	7h 21m
Ross and Newman	1533 9 Dec 1998	2235 9 Dec	7h 02m
Ross and Newman	1533 12 Dec 1998	2232 12 Dec	6h 59m
STS-96			
Jernigan and Barry	2000 29 May 1999	0651 30 May	7h 55m
STS-101			
Williams and Voss	2148 21 May 2000	0432 22 May	6h 44m
STS-106			
Lu and Malenchenko	0045 11 Sep 2000	0701 11 Sep	6h 14m
STS-92			
Chiao and McArthur	1027 15 Oct 2000	1655 15 Oct	6h 28m
Wisoff and Lopez-Alegria	1030 16 Oct 2000	1722 16 Oct	7h 07m
Chiao and McArthur	1030 17 Oct 2000	1718 17 Oct	6h 48m
Wisoff and Lopez-Alegria	1100 18 Oct 2000	1756 18 Oct	6h 56m
STS-97			
Noriega and Tanner	1335 3 Dec 2000	2108 3 Dec	7h 33m
Noriega and Tanner	1221 5 Dec 2000	1858 5 Dec	6h 37m
Noriega and Tanner	1113 7 Dec 2000	1623 7 Dec	5h 10m
STS-98			
Curbeam and Jones	1050 10 Feb 2001	1824 10 Feb	7h 34m
Curbeam and Jones	1040 12 Feb 2001	1749 13 Feb	6h 50m
Curbeam and Jones	0948 14 Feb 2001	1513 14 Feb	5h 25m

(All times US Eastern)

PROGRESS-M 44

As the International Space Station programme advanced into Phase Three, the crew had to undertake an operation that had been essential in operating the replenishable Salyuts but had been rare in the case of Mir, namely flying their Soyuz around from one port to another to accommodate a tanker. After spending a week shutting down most of the ISS's systems just in case they were unable to redock and had to return to Earth, they secured themselves in Soyuz-TMA 31, Shepherd passed command to Gidzenko, the Soyuz Commander, who undocked from Zvezda's wake port at 0506 on 24 February 2001, drew back 100 metres, flew around to line up on Zarya's nadir port, and then closed in and redocked at 0537. This was an unfortunate overhead on the station's operations. If they had been obliged to return to Earth, the ISS would have had to have been left vacant until the Expedition 2 crew could be launched. As soon as the tunnel to Zarya had been verified airtight, Gidzenko powered down the Soyuz. As they re-entered the ISS, Shepherd resumed command. After restarting the station's basic systems, they were assigned an extended sleep period. One problem that arose concerned the Floating Potential Probe. It had been deactivated prior to the manoeuvre, and when it was reactivated it failed to downlink data. Progress-M 44 was launched from Baikonur at 0318 on 26 February, carrying propellant, water, supplies, personal effects for the crew and equipment for Zvezda. The Kurs system automatically steered it into Zvezda's wake port at 0450 on 28 February without incident. The hatches were opened two hours later. The ISS crew spent the next week unloading its cargo.

HARDWARE DELAYS

As the Expedition 1 flight drew to a close, much of the ISS's Phase Three schedule was derailed. The US Habitation Module, which was due to be added in September 2005, was late and overbudget. The Russian Docking Compartment, due for launch in June 2001, was behind schedule and suffering technical and financial difficulties. The module was shipped to Baikonur in March 2001, but by that time the launch was in doubt because the Progress-based propulsion unit which would deliver it was incomplete and finishing it would take an injection of $1 million. The Russian programme managers were of the opinion that it would not be able to be launched in June. At the same time, the Russian Universal Docking Module had not advanced beyond a paper study. And work on Russia's Science Power Platform (the Russian acronym for which was NEP) had practically ceased. Two prototypes and some flight hardware had been developed, but substantial additional funding would be required before the flight unit could be built. NASA had it scheduled for October 2002, on SSAF-9A, but this was unrealistic. Energiya planned to develop a scaled-down version which would supply a smaller amount of electrical power to the Russian portion of the ISS than initially planned. At the same time, Spacehab and Energiya announced that the Enterprise Module that they were to develop jointly would replace the Docking and Stowage Module. The agreement to do so had been reached on 16 February 2001. As the month ended, Rosaviakosmos asked NASA to review the planned replacement. The new Proton-launched module would be large enough to include living quarters for three additional crew members, bringing the total to six. Enterprise would be leased in a fee package that would include a Soyuz ferry which would dock with the module. The plan was to introduce the Enterprise Module in 2003, at least two years before the US Habitation Module and X-38 combination on the current schedule, so it would significantly increase the station's utility. However, this plan caused a conflict with the Commercial Space Module which Boeing and Khrunichev had announced in July 2000 that they would build using the Zarya backup vehicle. The issue was that both consortia wished to use Zarya's nadir docking port. It was strange to have companies fighting over access to the station.

NASA's submission for Fiscal Year 2002 requested $14.5 billion for science, space and technology, which represented a 2 per cent increase over the current level. The figure included an increase of expenditure on the ISS, which was expected to experience a cost overrun of $1 billion in Fiscal Years 2001 and 2002, and $4 billion for each of the subsequent five years. With total spending on the ISS capped, NASA was obliged to halt work on the Habitation Module, X-38 and Propulsion Module and reassign the money to cover the overruns. George W. Bush's administration went further, warning NASA to expect these three items to be cancelled. With the loss of the Habitation Module, the ISS's crew capability was reduced from a proposed seven to just three people when no Shuttle or visiting Soyuz was in attendance. The X-38's cancellation placed the onus on Russia to supply Soyuz ferries in a timely manner. When several of the materials science experiment racks planned for the Destiny laboratory were cancelled, NASA estimated that only 20 per cent of the crew's time on board the ISS would be devoted to science. The only glimmer of hope for increasing the science yield in the medium term was the commercial modules.

STS-102: LEONARDO'S DEBUT

STS-102 – Discovery – SSAF-5A.1

Commander:	James Wetherbee
Pilot:	James Kelly
Mission Specialists:	Andrew Thomas; Paul Richards
Expedition 2 (up):	James Voss; Yuri Usachev (Russia); Susan Helms
Expedition 1 (down):	William Shepherd; Yuri Gidzenko; Sergei Krikalev

The delay in launching STS-98 had a knock-on effect, slipping STS-102, so the first ISS expedition crew were obliged to extend their programme. Discovery was finally launched from Launch Complex 39's Pad B at 0642 on 8 March 2001, and initiated the standard two-day rendezvous.

In addition to delivering the Expedition 2 crew to the ISS and retrieving the Expedition 1 crew, STS-102 marked the first flight for the Multi-Purpose Logistics Module (MPLM) 'Leonardo', which carried six systems racks and two storage racks for Destiny. Another pallet carried the tightly folded seven-jointed arm of the Space Station Remote Manipulator System (SSRMS), a 17-metre-long strengthened form of the Shuttle's RMS. The arm's Mobile Base System would be delivered by SSUF-2 and the Special Purpose Dexterous Manipulator would be ferried up by SSUF-4. Once it was fully installed, the SSRMS would be able to self-relocate by 'walking' across the station's exterior, attaching itself to a set of Power and Data Grapple Fixtures using an end-effector at each end of the arm. The system was built by Canada as its contribution to the station.

On Flight Day 2, Voss and Helms inspected the EMUs that they would use on their two assigned EVAs. Meanwhile, on the ISS, Shepherd's crew prepared to hand the ISS over to their successors.

With the launch of the Expedition 2 crew, the Payload Operations Center at the Marshall Space Flight Center in Huntsville came on-line. The Shuttle Operations

Coordinator (SOC) would oversee the transfer of scientific experiments to the ISS, and would act as the principal interface between the POC and the Johnson Space Center in Houston. On board Discovery on Flight Day 3, as Wetherbee and Kelly prepared for the rendezvous, Usachev, Helms and Voss, the Expedition 2 crew, made a start on the science programme by undertaking the first test of the Hoffman Reflex (H-Reflex) neurological experiment to measure the response of the spinal cord to a stimulus applied to a muscle in the leg. This was done early in the flight in order to monitor the rate at which the body adapted to microgravity. After the test, they had a conference with the experiment's principal investigator, who was at the Payload Telescience Center of the Canadian Space Agency in Montreal.

Discovery approached the ISS from below. When 200 metres from the station, Wetherbee began a 90-degree arc, halting in front of the ISS with the payload bay facing PMA-2. At that point, he had to wait because Houston could not confirm that one pair of photovoltaic arrays on the P-6 unit had locked in the correct edge-on attitude to reduce contamination from Discovery's thruster efflux. When the signal was finally received, Discovery docked with PMA-2 at 0138 on 10 March, just over an hour late. A short time later, and for about half an hour, Discovery's downlink was unable to be relayed by the TDRS facility at White Sands in New Mexico to the Johnson Space Center's Shuttle Control Room, and had to be routed through the ISS Control Room. Meanwhile, once the hatches had been opened at 0351, Usachev, the ISS's new commander, was the first to enter the station, and after the welcoming ceremony and safety briefing, installed his couch liner in Soyuz-TMA 31, swapping with Gidzenko, the ferry's commander, who transferred his to Discovery and, in so doing, formally joined Wetherbee's crew. Two hours later, when the hatches were closed again, Usachev remained onboard to be briefed by Shepherd and Krikalev on the state of the station. On Discovery, meanwhile, the cabin pressure was reduced in preparation for the EVA on Flight Day 4.

Voss and Helms left Discovery's airlock at 0012 on 11 March. Thomas operated the RMS and Richards directed their activities. During the early preparations, a foot restraint became untethered and drifted away. Voss made his way to a storage locker mounted on Unity's exterior to fetch a replacement. The first scheduled task was to disconnect eight cables so that PMA-3 could be removed from Unity's nadir CBM. They then made their way round to the port CBM and removed an antenna for the Early Communications System. Their main task was to collect the Cradle Assembly from the payload bay and install it on Destiny's exterior so that the central segment of the Integrated Truss Structure could be anchored to the module during a later mission. Connecting the Cradle's power cables was slipped to the second EVA. After a cable tray for the SSRMS had been affixed to Destiny, they retreated to the airlock, out of the way. Then Thomas grasped PMA-3 with the RMS, Wetherbee released its bolts, it was swung around to Unity's port CBM, and Wetherbee re-engaged the bolts. If Thomas had encountered difficulties, Voss and Helm would have gone back out to render assistance, but as soon as the transfer was successfully completed they pressurised the airlock, ending the EVA at 0908 after 8 hours 56 minutes, which was a record for an EVA from a Shuttle.

The hatches were reopened on Flight Day 5, at 2215 on 11 March, and Krikalev

STS-102 ferried up the Leonardo Multi-Purpose Logistics Module (left), which was mated with Unity for unloading. Yuri Gidzenko is dwarfed within it (lower right). Andrew Thomas and Paul Richards transfer cargo (upper right).

promptly exchanged couch liners with Voss. During the day, the Expedition 2 crew re-ran the H-Reflex experiments. At 2237, Thomas grasped Leonardo with the RMS. At 2310 Wetherbee released the restraining bolts, and Thomas hoisted the 11-tonne module from the bay. The berthing operation on Unity's nadir was delayed while Shepherd routed the output from the CBM's centreline camera to Discovery so that Thomas would be able to line up Leonardo. Once it was in place, Wetherbee engaged the bolts, and at 0102 on 12 March, with the seals confirmed, Thomas retrieved the RMS. Shepherd opened the cargo module at 0651. On Discovery, Richards, Thomas and Helms serviced the EMUs. After having been open for 8 hours 24 minutes, the Shuttle was sealed at 0639 so that its pressure could be lowered in preparation for the second EVA. After breakfast on Flight Day 6, Richards and Thomas moved into the bay at 0023 on 13 March. Their first assignment was to install the electrical cables between Destiny and the Cradle Assembly that had been deferred from the previous EVA. The cables would supply power to the SSRMS once it was on the Integrated Truss Structure. Meanwhile, on the ISS, Shepherd deactivated and then reactivated Leonardo's DC–DC power converters and verified that the Cradle Assembly was functioning correctly. Richards and Thomas made their way up to the top of the P-6 unit. Three of four struts had latched correctly when the P-6 had been installed on the Z-1 Truss Stucture but one had failed, so they tapped it into position. While up there, they also inspected the Floating Potential Probe, which had fallen silent after the Soyuz-TMA 31 transfer, and reported that its status lights were not illuminated. Their next task was to retrieve the External Stowage Platform from Discovery's payload bay, mount it on the exterior of Destiny and connect the cables to power its heaters. Then, after inspecting the connector of a cable on Unity's exterior which supplied power to one of the node's heaters, they photographed vents on Destiny and Zvezda. Throughout the EVA, Kelly used the RMS to manoeuvre the spacewalkers around the exterior of the ISS, and Helms supervised. Richards and Thomas retreated to the airlock at 0644, after having been out for 6 hours 21 minutes. Meanwhile, on the ISS, Shepherd, Usachev and Voss began to unload Leonardo. One by one, four equipment racks were unbolted, floated through Unity into Destiny, and installed. One rack contained the first medical experiments for the Human Research Facility. This EXPRESS rack was comparable in size to a stand-alone phone booth, and had its own computer workstation and laptop interface. Data on how the human body adapted to spaceflight would be stored in the computer, and dumped to Earth when convenient. It would also handle the data from the suite of radiation monitors which would characterise the radiation inside and around the station. While the EVA was in progress, Gidzenko and Krikalev, on Discovery, were exercising in an effort to condition their space-adapted muscles for the stress of re-entry which would conclude their 4.5-month-long mission.

The hatches between Discovery and Destiny were opened again on Flight Day 7, and the two crews combined to unload Leonardo's cargo, which had to be done while Discovery was present because, although the MPLM was designed to be left between missions, on this occasion it was to be returned to Earth. They were soon ahead of schedule. During the day, Shepherd and Helms exchanged couch liners in the Soyuz and she took her place on the Expedition 2 crew. However, the official

Having been ferried up by STS-102, ISS Expedition 2 crew members Yuri Usachev (right) and James Voss inspect the Destiny laboratory.

handover of command from Shepherd to Usachev would not occur until Discovery was about to depart. After assessing the ability of the Shuttle to control the docked combination's attitude, Wetherbee set up Discovery's autopilot to make the reboost manoeuvres during the night. Voss and Helms spent most of Flight Day 8 installing the SSRMS's workstation inside Destiny, and then tested the system that was to route television pictures from docked Shuttles to the SSRMS operator. Meanwhile, Thomas acted as loadmaster, directing Usachev, Shepherd and Richards in loading rubbish and other items into Leonardo for return to Earth, taking care to ensure that the placements did not disturb Discovery's centre of mass. When the data from the first two series of H-Reflex experiments were transmitted to Earth, the principal investigator reported that it confirmed that the astronauts' spinal chords were adapting to microgravity in the expected manner.

At breakfast on Flight Day 9, the welcome news was relayed that Discovery's mission had been extended by one day to provide time to ensure that Leonardo was properly loaded. Later, the two crews took time off for a joint press conference with journalists across America and in Korolev. Shepherd summed up the Expedition 1 tour, saying: "We basically put the station in commission. We have taken something that was an uninhabited outpost and we now have a fully functional station where the next crew can do research. I think that is the substance of our mission." The last of Leonardo's cargo was loaded on Flight Day 10, and the module was closed. Early on 17 March, Wetherbee, Richards and Thomas answered questions from American journalists, and then an hour later Usachev, Gidzenko and Krikalev talked to journalists in Korolev. Flight Day 11 was the last day of docked operations. Discovery's systems had become cold by this time and ice had formed in some of the water lines. In an attempt to overcome the problem, Wetherbee was instructed to power up two of the four computers to produce additional heat to thaw the ice. The final task was to retrieve Leonardo. Thomas grasped Leonardo using the RMS. When the vestibule was depressurised, a pressure leak prevented it being removed. After tracking down a loose connector on an air hose, Voss tightened it, and when the depressurisation was repeated the seal was verified. The module was unberthed at 0540 on 18 March, and Thomas put it back on its mount in the bay at 0708.

Captain Shepherd USN had run the ISS employing a series of naval traditions, including installing and ringing a ship's bell at appropriate times. He made the formal handover of command to Usachev on Flight Day 12, lining up the two ISS crews facing each other. After a short speech, he announced: "Now I am ready to be relieved." Usachev replied: "And I relieve you." The two crews said their farewells, the visitors and the retiring crew departed, and the hatch was sealed at 2037 on 18 March. Kelly undocked Discovery at 2332, backed away, made a fly-around, then executed the separation manoeuvre at 0048 on 19 March.

Discovery's new crew members spent the remainder of the day exercising, talking to their families and enjoying some free time. The following day they took part in a press conference. On Flight Day 14 the first landing opportunity was waved-off due to clouds and wind as a low-pressure area crossed over Florida, but the conditions had improved sufficiently on the next orbit for retrofire at 0126 on 21 March. As had

The state of the International Space Station after the STS-102 visit, with PMA-3 on Unity's port CBM. Soyuz-TMA 31 relocated to Zarya's nadir and Progress-M 44 on Zvezda's rear.

the astronauts retrieved from Mir, the retiring ISS crewmen strapped themselves into recumbent seats for the return to Earth. At 0231, Wetherbee set Discovery down onto the Shuttle Landing Facility. Having spent 136 days on the ISS, Shepherd, Krikalev and Gidzenko were now to be subjected to 45 days of monitoring to study how their bodies readapted to gravity.

EXPEDITION 2

Having bidden farewell to Discovery, Usachev, Voss and Helms settled down to life on the ISS. Their first day was highlighted by the activation of a fire alarm, which led to the shutdown of the ventilation fans in Destiny, a programmed response to delay the spreading of smoke through the station. A brief investigation showed that it was a false alarm, and Houston assumed responsibility for restarting the fans. However, several computers had also shut down and an investigation was launched to discover why they had crashed. Meanwhile, over the weekend, Huntsville uploaded new files to Payload Computers 1 and 2 to prepare them for the Expedition 2 crew's science programme. Light duties were scheduled for the first few days, in order to enable the astronauts to acclimatise to their new environment. On 21 March, using equipment delivered by STS-102, they activated the Ku-Band antenna which, because it offered a bandwidth 250 times that of the S-Band system, was a prerequisite for the science programme. Although the POC had used the Ku-Band to send the signal to start the two payload computers, the antenna suffered a pointing error and the system had to be deactivated while a software patch was written, tested and uploaded. This meant that the transmission of experimental data from Destiny's Human Research Facility had to be delayed but the data would meanwhile be accumulated on computer disks. A condensate venting system in Destiny was also misbehaving, so the temperatures in the thermal loops were raised in order to prevent the build-up of moisture in the module. To restore Destiny's carbon dioxide removal system, Usachev replaced the malfunctioning pump for one delivered by STS-102. Zvezda's Vozdukh system was working well, however. Meanwhile, because oxygen was being drawn from the tanks on Progress-M 44, Zvezda's Elektron system had been switched off.

In addition to maintenance, Usachev's crew set up new apparatus, including the Cycle Ergometer with a Vibration Isolation System (CEVIS) which would enable them to exercise with minimal disturbance to sensitive microgravity experiments. The first experiment in Destiny was set up by Voss on 23 March. The Bonner Ball Neutron Detector developed by Japan's National Space Development Agency was the first of two Expedition 2 radiation studies. The measurement of the radiation environment in and around the station was one of this mission's key scientific tasks. The apparatus measured the fluxes of neutrons in six different energy ranges and stored its data on its own disk. It was to be returned to Earth by STS-105, which was to retrieve the Expedition 2 crew. While setting up the experiment, Voss held the first conversation with Alan Johnson, who was the Payload Communications Manager (PayCom) in Huntsville. The other radiation study was ESA's Dosimetric Mapping (DOSMAP), whose two Dosimetric Telescopes (DOSTELS) Voss installed in

In a scene strongly reminiscent of the Mir 'base block', Yuri Usachev works at the table in Zvezda.

Destiny three days later. Its data would be collected fortnightly. On 29 March, umbilicals were installed to supply air, coolant, electricity, fluids and communications to the Human Research Facility, in preparation for activating the Ultrasound Imaging System that would provide three-dimensional imagery of internal organs, muscles and blood vessels, and the Gas Analyser System that would monitor the heart and lungs for the Metabolic Analysis Physiology (GASMAP) Experiment. Each member of the crew was to use the Human Research Facility's computer to fill out a questionnaire on a regular basis for the Interaction Experiment, to record how they were interacting. After the flight, behaviourists would analyse the results to investigate personal and cultural interactions. On 2 April, Voss made a start on the Crew Earth Observation (CEO) Experiment by photographing the Parana River Basin in Paraguay, a centre of new development, to help scientists to monitor the changes in the local environment. Every two weeks, the astronauts were to photograph preselected sites on Earth's surface using handheld 35- and 70-millimetre film cameras and electronic digital cameras. These observations were to be made through the high-fidelity window in Destiny. The digital imagery was able to be downloaded as part of the ISS's scientific data stream, but the film would be returned to Earth on visiting Shuttles. All CEO images were catalogued and added to the Earth image database in Houston. On 3 April, Helms began experiments with the Mid-deck Active Control Experiment 2 (MACE). This consisted of a 1.52-metre-long platform, four struts and five nodes. A hand-controller sent instructions to the experiment's computer to activate gimbals and wheels on one side of the platform to set it vibrating, and a support module detected the vibrations and attempted to damp them using a similar mechanism on the other side of the platform. It was a USAF–MIT joint experiment to investigate the effects of vibrations on moving structures in space. Later, the ability of Zvezda's computer to assume responsibility for the station's attitude control in an emergency was tested when a failure of the CMGs in the Z-1 Truss Structure was simulated.

With STS-100 due to visit the ISS in late April, preparations were made for the undocking of Progress-M 44. On 4 April, the spacecraft's thrusters were fired under the control of Zvezda's computers. This was the first time that a docked spacecraft had been commanded from the ground through Zvezda; previously commands had been transmitted directly to the spacecraft. On 6 April, Helms resumed the MACE experiments, and a few days later downloaded 610 megabytes of data to Huntsville. Meanwhile, Voss replaced the Bonner Ball radiation experiment's disk, which was full. The crew spent several days loading rubbish into Progress-M 44, as it pumped the last of its propellants into Zvezda. After activating the Ku-Band system, they televised their activities. Later in the week, they checked out two command and control stations in Destiny, in preparation for the SSRMS that STS-100 was to deliver.[6] With Progress loaded, the hatches were sealed and at 0430 on 16 April the spacecraft undocked and withdrew. The next task was to relocate Soyuz-TMA 31. Having powered down the station, they sealed themselves in their spacecraft and at

---

[6] To differentiate it from the 'Canadarm' Remote Manipulator System carried by Shuttles, the larger and more powerful Space Station Remote Manipulator System had been named 'Canadarm 2'.

0740 on 18 April undocked from Zarya's nadir. After the 21-minute fly-around, Usachev redocked at Zvezda's wake docking system. Following pressure checks, they re-entered the ISS and began the long task of reactivating the equipment that they had placed in automatic mode, or had shut down. The Earth Knowledge Acquired by Middle School (EarthKAM) Experiment allowed American school pupils to control a digital camera mounted on Zarya's nadir-facing observation window to photograph selected features on Earth's surface and see their results on the Internet. The crew also continued to monitor the radiation experiments, and switched one of the two DOSMAP sensors to a higher sampling rate during a solar flare.

STS-100 ADDS THE MANIPULATOR

STS-100 – Endeavour – SSAF-6A

Commander:	Kent Rominger
Pilot:	Jeff Ashby
Mission Specialists:	Chris Hadfield (Canada); John Phillips; Scott Parazynski; Umberto Guidoni (ESA); Yuri Lonchankov (Russia)

STS-100 lifted-off from Launch Complex 39's Pad A at 1441 on 19 April 2001. The ISS was south of India at the time, but 20 minutes later Usachev, Voss and Helms took time off from their busy schedule to watch an up-linked video of the launch. In Endeavour's payload bay was the MPLM 'Raffaello' and a pallet with the SSRMS. On Flight Day 2, Hadfield and Parazynski checked their EMUs, then unpacked the apparatus to be used during the final phase of the rendezvous. Guidoni and Ashby tested the RMS and Guidoni made preparations for the transfer of cargo to the ISS. Lonchankov, meanwhile, filled two large flexible tanks with the by-product from the fuel cells to serve as potable water on board the ISS. Usachev's crew spent the day packing items to be returned to Earth. The Vozdukh carbon dioxide removal system in Zvezda was not working at full efficiency. The problem was thought to be due to a clogged filter screen. Later in the day, Endeavour's cabin pressure was reduced in preparation for the first EVA. The next 'morning' on board the ISS, Usachev reported that the Vozdukh had resumed functioning at full capacity without any intervention.

At 0400 on 21 April, on Flight Day 3, Rominger commenced the sequence of manoeuvres to approach the ISS from beneath and draw to a halt 100 metres in front of PMA-2, leading to docking at 0959. Because the pressure of the Shuttle's cabin had been lowered, PMA-2 was used as an airlock to transfer equipment. Usachev's crew opened their hatch and placed a battery-operated drill in the vestibule which would be needed on the first EVA and then, once the pressure had been reduced, Endeavour's crew retrieved it and put in four bags of water, computer equipment

On STS-100, Endeavour arrives with the Raffaello Multi-Purpose Logistics Module and a pallet with the SSRMS.

and film for transfer to the ISS. During the transfer, Voss video-taped the visitors through the small inspection port. At 0645 on 22 April, on Flight Day 4, Hadfield and Parazynski began the flight's first EVA. In so doing, Hadfield became the first Canadian astronaut to make a spacewalk. Phillips served as coordinator and Ashby and Guidoni operated the RMS. After installing a UHF antenna on Destiny's exterior, a task that took approximately two hours, they set about the transfer of the SSRMS. This was first lifted on its pallet by the Shuttle's RMS and mounted on a fixture on Destiny's exterior. Once a number of power and data cables had been connected, Hadfield and Parazynski extended the SSRMS's two booms. When they tried to engage the bolts to lock them into position, they experienced difficulty in achieving the correct torque using the pistol-grip tool and had to switch the tool from automatic to manual mode to complete the task. Meanwhile, in Destiny, Helms and Voss worked at the Robotic Workstation to power up the SSRMS and sent a series of commands to verify the cable connectors. At 1453, Helms ordered the new arm to make its first motion, after which Hadfield and Parazynski returned to the airlock at 1455 having been outside for 7 hours 10 minutes.

At 0525 on 23 April, on Flight Day 5, the hatches were opened and the crews met for the first time. As her colleagues transferred items from Endeavour to the ISS, Helms powered up the SSRMS and at 0713, after three hours of testing its joints, 'walked' it off its pallet. It was able to do this because it had an end-effector at each end. One end was attached to the pallet. Helms first commanded the other end to engage a Power and Data Grapple Fixture on the exterior of Destiny, and then she released the pallet. In future, the SSRMS will be able to move in this end-over-end manner across the exterior of the station, from one pre-installed Power and Data Grapple Fixture to another. As Helms completed her work with the SSRMS,

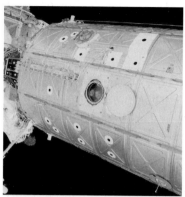

With the pallet holding the folded SSRMS prepositioned, Chris Hadfield assembled it (left). Susan Helms later 'walked' it to a Power and Data Grapple Fixture on the exterior of Destiny. Meanwhile, the Raffaello MPLM was placed on Unity's nadir (centre, bottom) and unloaded.

Parazynski, in Endeavour, powered up the Shuttle's RMS, grasped Raffaello and, at 1200, placed it on Unity's nadir CBM. Over the next few days, Guidoni, serving as loadmaster, was to oversee the unloading of the MPLM's 3 tonnes of cargo and then load it with items to be returned to Earth. Usachev, Helms and Voss started the unloading by transferring EXPRESS Racks 1 and 2 to Destiny. Rack 1 had to be transferred quickly because it contained experiments that could not be left without power for more than 30 minutes, and as soon as it was plugged in its systems were reactivated by a command from Huntsville. Three experiments – the Advanced Astroculture (ADVASC), the Microgravity Acceleration Measurement System (MAMS) and the Space Acceleration Measurement System (SAMS) – had been pre-installed to support plant growth, biological crystal growth, drug fermentation and vibration measurement experiments. The second rack was not to be activated until after STS-100 had departed, so its transfer was straightforward. The two crews having returned to their respective vehicles, the hatches were closed so that the Shuttle could reduce its pressure in preparation for the second EVA. Hadfield and Parazynski ventured out at 0834 on 24 April to connect power and data cables to the Power and Data Grapple Fixture currently holding the SSRMS. During this work the backup power supply failed, and the two men had to open a panel, disconnect cables and reconnect them to bring the supply back on-line. Next, they disconnected the power and data cables between the SSRMS and the pallet on which it had been delivered. Hadfield then made his way to Unity, where he removed an Early Communications System antenna that was no longer required. An electrical connector became loose and drifted behind the thermal cover on the starboard CBM, which was currently vacant. Although he tried to retrieve it, Houston told him to forget it. The airlock was sealed at 1615, having been open for 7 hours 40 minutes. Meanwhile, Helms used the SSRMS to lift the 1,360-kilogram pallet off Destiny and swing it around in a series of programmed manoeuvres; she then positioned it above Endeavour's bay, where it was left 'overnight'. The hatches were opened at 1815 so that the two crews could undertake a brief period of joint activities before retiring to their respective vehicles.

As they slept, Houston lost the signal from Command and Control Computer 1 (C&C-1), one of three identical computers operating in parallel. This resulted in an inability to transfer data between the ISS and the Space Station Flight Control Room in Houston. In an effort to regain the signal, Houston reconfigured C&C-1 overnight, switching it from 'primary' to 'standby'. When the crew awoke for Flight Day 7, the problem became the primary focus. Houston commanded C&C-1 to illuminate a light in Destiny and then switch it off again, which it did. But when Helms tried to transfer a file from C&C-1 to the Robotic Workstation the command failed. Houston spent the remainder of the day trying to correct the problem. Plans to work with the SSRMS were shelved. Meanwhile, both crews worked together in the ISS to unload Raffaello. The Commercial Generic Bioprocessing Apparatus (CGBA), Commercial Protein Crystal Growth-High Density (GPCG-H) and units 9 and 10 of the Protein Crystal Growth-Single Thermal Enclosure System (PCG-STES) were installed in EXPRESS Rack 1 and powered up. Having been supplied by the Centers for Commercial Development of Space that were jointly sponsored by

On STS-100, Scott Parazynski hooks up cables for the Power and Data Grapple Fixture on the exterior of Destiny.

NASA, academia and industry, ADVASC, CGBA and CPCG-H were the ISS's inaugural commercial payloads. ADVASC was to investigate growing plants from seed in microgravity, in the hope of producing and replanting seeds so as to follow successive generations. CGBA was to investigate why antibiotic production by microbes improves in microgravity. CPCG was to comprehensively investigate the detailed structure of proteins in more than 1,000 biological solutions. All three experiments would be returned to Earth at the end of Expedition 2. The Active Rack Isolation System (ARIS), the ARIS-ISS Characterisation Experiment (ARIS-ICE) and the Experiment on the Physics of Colloids (EXPPCS) were installed in Rack 2. To general consternation, C&C-3 then malfunctioned.

By the start of Flight Day 8, C&C-2 was therefore controlling the ISS without a level of redundancy because C&C-1 and C&C-3 were off-line with what appeared to be software faults. Overnight, two fault-protection computers in Unity, designed to prevent malfunctions in the other computers by automatically rebooting them, had also tripped. Helms worked with the flight controllers to try to correct the problems with the C&C computers. After working on C&C-1 using a laptop, at 0435 she was able to report that its command and data functions had been restored. In Houston, signals were sent to C&C-1 to dump the data relating to the problem so that it could be studied. Meanwhile, the other members of both crews loaded Raffaello with items to be returned to Earth. The computer problems meant that Raffaello's retrieval and the handover of the SSRMS pallet had to be postponed. Late in the day, the mission managers initiated plans to extend docked operations by two days to allow the two crews to overcome the computer issues, and NASA asked Russia to delay the launch of Soyuz-TMA 32 by one day in order to accommodate the extension because the Soyuz would have to pass uncomfortably close to the Shuttle's vertical stabiliser on its way to Zarya's nadir if Endeavour was still in place when the Soyuz arrived. While the astronauts slept, Houston continued to work on the malfunctioning computers in the hope of recovering at least the fault-protection computers. Upon awakening for Flight Day 9, the crews found that the engineers on the ground had solved many of the problems. A software patch had been uploaded to C&C-3, but the computer had not yet been rebooted. C&C-1 had been diagnosed as having a failed hard drive, so its problem had not been a software fault. During the day, the Expedition 2 crew removed C&C-1 for return to Earth and replaced it with Payload Computer 2. Meanwhile, the managers confirmed their wish for the Shuttle to remain docked for one extra day, with an option for a second day if this proved necessary. With the computers being brought back on-line, permission was given to retrieve Raffaello, which Parazynski did at 1520 on 27 April, stowing it back in the bay at 1658.

Meanwhile, the Russian State Commission, not having received NASA's request for a postponement, confirmed Soyuz-TMA 32's launch date as 28 April. However, the Russians agreed that if Endeavour was still present when the Soyuz arrived, the spacecraft would stand alongside and wait. Shortly after the wake-up call for Flight Day 10, Endeavour's crew was told that Soyuz-TMA 32 had launched successfully. They were also told that an attempt to upload new software to C&C-3 had failed, but a fault had been identified in the new software and a second attempt to upload it

The state of the International Space Station after the STS-100 visit, with the SSRMS temporarily mounted on Destiny.

would be made once it had been fixed. Endeavour's crew joined their colleagues in routine maintenance and installing new hardware. Helms put the SSRMS, which was still holding its pallet, through a series of planned manoeuvres and then positioned it over Endeavour's bay. At 1634 on 28 April, Hadfield raised the RMS and took hold of the pallet. At 1702 the SSRMS released its grip on the pallet to finish the 'hand over'. Hadfield stowed the pallet in the bay at 1651. This left only the computers on the list of tasks to be completed before Endeavour could undock. After a joint meal, the crews retired. Houston continued to work on the computers. On Flight Day 11, as Soyuz-TMA 32 continued its two-day rendezvous, Endeavour's crew, working on a strict schedule in an effort to avoid delaying the Soyuz's docking, prepared to depart. The hatches were closed at 1041 on 29 April and Ashby undocked at 1334, withdrew, made a three-quarter-circle fly-around to film the station using the IMAX camera in the bay, and then performed the separation burn at 1428. On board the ISS, Usachev, Helms and Voss had the afternoon off before preparing for the docking of Soyuz-TMA 32 at 0352 on 30 April. While they were asleep, Houston reformatted the hard drive in Payload Computer 2 (now serving as C&C-1) and copied the hard drive contents from C&C-2 to C&C-3, thereby giving the ISS two C&C computers operating in their primary mode. Bad weather obliged Rominger to return Endeavour to Edwards Air Force Base at 1211 on 1 May.

A NEW HABITATION MODULE?

On 19 April 2001 NASA announced that it would negotiate with the Italian Space Agency for a Habitation Module for four astronauts, to give the ISS a seven-person capability. The new module would replace the US Habitation Module that had been cancelled earlier in the year due to financial difficulties. NASA wanted a contract on a no-exchange-of-funds basis in which in return for constructing the module Italy would earn additional time on the ISS for Italian personnel and greater access to the existing research facilities. The module would almost certainly exploit the pressurised shell of the MPLM. The seven-person complement had come about because it was expected that three people would be needed to maintain the ISS once it was operational, and four people would be required to maximise the station's science programme. However, the development of the X-38 Crew Return Vehicle was still under financial threat. Installing a pair of Soyuz 'lifeboats' would accommodate six people, and although Russia was willing to supply the additional spacecraft, NASA's preference was for its European partners to develop a new vehicle – which ESA was willing to consider as long as it was a vehicle that it would be able to launch on the Ariane V and thereby – at long last – establish an independent human spaceflight capability.

THE FIRST 'SPACE TOURIST'

On 29 December 2000 Rosaviakosmos had announced that Talget Musabayev, Yuri

Baturin and Dennis Tito would fly the next Soyuz to the ISS. On 3 January 2001, Pytor Klimuk, head of the Yuri Gagarin Cosmonaut Training Centre, linked Musabayev's crew to Soyuz-TMA 32, the first Soyuz exchange flight. As a space engineer in the 1960s, Tito had worked on mission scenarios for Mars and Venus. In June 2000 he had paid Mircorp $20 million to become the first paying 'space tourist'.[7] The plan had been to fly to Mir, but when it was decided to de-orbit Mir the Russians offered to honour their contract by flying Tito to the ISS. NASA strongly disapproved of this plan, arguing that only professional astronauts and cosmonauts should fly to the ISS during its construction phase. However, the Russians argued that Tito had bought his seat on Soyuz privately and that NASA had no right of veto on whether he could fly on an all-Russian Soyuz flight. The issue was a contentious one because any delay in launching the Soyuz threatened the continuation of Expedition 2. Musabayev's crew was to retrieve Soyuz-TMA 31, which had delivered the Expedition 1 crew and was about to expire its 180-day operational life. Without a replacement, the Expedition 2 crew would be obliged to return to Earth, leaving the ISS vacant. The Soyuz crew had originally included Nadezhda Kuzhelnaya, a female Russian cosmonaut, but she had been withdrawn so that Tito could fly. Following negotiations, NASA and the other ISS partners agreed a series of rules. Tito was made to sign a document stating that he was flying by Russian invitation and none of the other ISS partners could be liable if he was seriously injured, or killed. He also had to agree to pay for anything that he broke or otherwise damaged while on the ISS. Finally, he was restricted to Zarya and Zvezda and could visit the non-Russian segment only if escorted by one of the Expedition 2 crew. On 11 April, after 900 hours of training, and having only a rudimentary knowledge of the Russian language, Tito was cleared for flight.

Soyuz-TMA 32

Commander:	Talgat Musabayev
Flight Engineer:	Yuri Baturin
Observer:	Dennis Tito

Soyuz-TMA 32 was rolled out to the pad on 26 April, as the issue of the ISS's computers was at its height. On that date, NASA requested that the launch be postponed a day to extend Endeavour's visit, but before the request arrived the State Commission confirmed the launch date. The next day, while Soyuz-TMA 32 was undergoing final preparations, Sergei Gorbunov, the senior press officer for the Russian Space Agency, claimed that NASA had withdrawn the postponement request, but NASA insisted that this was not the case. Clearly, communications between the two partners were less than they ought to have been. Soyuz-TMA 32 lifted-off at 0237 (1337 Baikonur time) on 28 April 2001. Upon entering

---

[7]In fact, a case can be made that Toehiro Akiyama was the first 'space tourist', because the Tokyo Broadcasting System paid a fee to the Soviet Union for his visit to Mir in December 1990.

microgravity, Tito was sick, but he recovered on the two-day rendezvous.[8] After docking on Zarya's nadir at 0358 on 30 April, the Soyuz crew were unable to open the hatch, so the ISS crew had to open it from their side. Somewhat behind schedule, a live telecast showed the visitors being welcomed on board the ISS. Musabayev was first to appear through the hatch. He was followed by Tito who, having pushed off from the wall too eagerly in weightlessness, had to be restrained as he drifted in. The two crews congregated in Zvezda, where Tito would be based. After the brief welcoming ceremony, Usachev gave the newcomers a formal safety briefing and pointed out the evacuation routes. Tito told controllers in Houston that Helms and Voss, both NASA astronauts, had shown him no animosity. Indeed, within an hour of arriving he had been given a tour of the entire station. Usachev, being a Russian cosmonaut, had always favoured such visits. By delivering the new spacecraft, the visitors had already achieved their main objective. Once the couch liners had been exchanged, the new spacecraft became the Expedition 2 crew's vehicle. They also transferred a crystal growth experiment from Soyuz-TMA 32 to the ISS.

As Tito spent his time admiring the Earth pass endlessly below, work on board the station continued unabated. On 1 May, the first commands were sent to the ADVASC to retrieve its performance data, which confirmed that it was functioning as intended. The seeds that it was cultivating were of the same family as radishes and cabbages. When they went to seed, the seeds were to be collected and returned to Earth with the Expedition 2 crew for examination. Unit 10 of the PCG-STES was activated on the same day. On 2 May, as part of the Human Torso Experiment, Voss installed the Tissue-Equivalent Proportional Counter. A sensor was positioned 0.33 metre from a full-sized model which matched the tissues and muscles of a male's head and torso; the torso contained 300 passive dosimeters to record radiation levels, and the head, neck, heart, stomach and colon each contained one active radiation detector. He also reinstalled the EarthKAM in Zarya for school pupils to control via the Internet. The MAMS and the six growth chambers in unit 9 of the PCG-STES were activated by Helms on 3 May. MAMS was one of two experiments designed to measure vibrations and accelerations imparted by crew activities and spacecraft dockings that might impair microgravity experiments – it was part of the effort to determine the quality of the station's microgravity environment. On this occasion it was to gather data continuously for a week. In an effort to make himself useful, Tito volunteered to make the meals for everyone, and performed other essential, but menial tasks so that they would have more time to pursue their work. He therefore not only did his best to keep out of the way, but he made himself useful. Despite NASA's concerns, his visit had minimal impact. "The American section", he explained during a telecast, "is at least 100 feet from here, from the place where we are sitting right now, where I spend most of my time. There is absolutely no way that my presence can interfere with their work." After a week, it was time to leave. MAMS monitored the low-frequency transients as Soyuz-TMA 31 undocked from Zvezda at 2221 on 5 May. Following retrofire at 0047 on 6 May it landed on the Kazakh steppe at 0142.

---

[8]A significant percentage of humans suffer a brief period of nausea upon entering weightlessness, and some actually vomit, but they all soon recover.

At a post-flight press conference, Musabayev opined the Expedition 2 crew had acted almost as if they had been told not to fraternise with Tito. He complained that upon arriving at the ISS, after two days cramped in Soyuz-TMA 32, his crew had had to wait three hours before receiving a meal. NASA defended the Expedition 2 crew, saying that they were exhausted after the visit of STS-100 and the computer problems. A NASA spokesman pointed out that the agency had "nothing negative" to say about Tito's visit. Nevertheless, NASA said that it would seek compensation from Russia for the time that the Expedition 2 crew lost through having an untrained visitor. With the flight over, the Russian Space Agency started a damage-limitation exercise, and agreed not to send any more 'tourists' to the ISS for at least two years. At his post-flight press conference, Tito summed up his experience simply, "I'd like people to see me as a serious man who had a dream, and pursued it in the face of great difficulty."

THE LONG HAUL

Having received two sets of visitors in almost as many weeks, Usachev, Helms and Voss were given one and a half days off from their work routine, although they had to continue their daily exercise regime in order to ameliorate the deterioration of their bodies in microgravity. On 7 May they returned to work. On 9 May, when Huntsville sent a command to the Human Research Facility's computer, the CGBA, and the Bonner Ball Neutron Detector and DOSMAP radiation monitors tripped off, and the POC set about troubleshooting the fault. The radiation experiments were reactivated the following day.

With C&C-2 performing as the primary computer, C&C-1 was a fully functional backup. C&C-3 was available, although without its disk, which had been removed. The crew assembled a new unit, and kept it available to replace C&C-3 if required. It was exchanged with C&C-3 on 11 May, and was subsequently loaded with software to enable it to serve the role of second backup to the C&C system. Spare computer parts were added to the cargo of Progress-M1 6, which was scheduled to be launched on 20 May.

In June STS-104 was to deliver the Joint Airlock Module, which the SSRMS was to install on Unity. In preparation, on 10 May the SSRMS was tested to verify the operation of the capture device in the end-effector by latching onto two fixtures on the exterior of the ISS, one on Destiny and one on the tunnel between Destiny and Unity. This test was to be performed on a weekly basis. On 15 May, the device that stored real-time scientific data whenever the ISS was unable to communicate by TDRS, shut down unexpectedly. Like the CGBA, it remained powered-off while Huntsville investigated the fault. Voss volunteered to undertake extra scientific tasks should the opportunity arise, and it was suggested that he document the experiments in Destiny's racks, so he downloaded a series of digital photographs. Meanwhile, in preparation for Progress-M1 6's arrival Usachev tested the TORU manual docking system. Maintenance was ongoing. A filter in a condensate pump line located behind a rack in Destiny was exchanged to enable the system's water to flow at full rate. On

17 May, when Voss and Helms made the weekly test of the SSRMS, this time using the backup software, one of the arm's shoulder pitch-joints suffered an intermittent software problem. Although the Canadian engineers who had built the arm set out to develop a software patch, NASA started planning for an EVA to replace one of the computers incorporated in the SSRMS's mechanism, to ensure that it would be fully functional for STS-104. Another problem developed as one of the motors that rotate the P-6 photovoltaic arrays started to draw an excessive current, so the motor was deactivated. Although this left its arrays locked in place, they continued to produce sufficient power to sustain operations. Nevertheless, NASA started to assess having a future Shuttle crew replace the motor. On 18 May, the EarthKAM was packed away. It would be reinstalled for the new school year. The Human Research Facility's GASMAP, which was to be used throughout the Expedition 3 tour, was checked out.

PROGRESS-M1 6

Progress-M1 6 was launched at 1833 on 20 May. The next day, Usachev tested the TORU system, and switched on the MAMS to measure the transients as it docked. While their commander was doing that, Helms and Voss put the SSRMS through a re-run of some of the manoeuvres which had shown up the problem, but there was no repeat of the intermittent pitch-joint fault. As they waited for the ferry, they started up some of the experiments in Destiny's EXPRESS Rack 2. Progress M1-6 docked on time at 1924 on 22 May. Almost half of its 2,750-kilogram payload was propellant. The crew worked through the following week to unload the dry cargo, which included a new computer disk to be installed in C&C-3, to restore this system to full capability. On 24 May, renewed problems with the SSRMS meant that Helms had to abandon a full dress rehearsal of the manoeuvres that it would employ to lift the Joint Airlock Module from STS-104's bay and emplace it on Unity. Routine maintenance and experiments continued. Helms resumed the MACE experiment, which had been begun by their predecessors. The third growth chamber in unit 10 of the PCG-STES was activated on 28 May, and the next day Voss set up the ARIS in EXPRESS Rack 2. Following the failure of several different software patches to fix the problem with the CGBA, it was decided simply to return the experiment to Earth. On 30 May, Helms tried to activate the EXPPCS in EXPRESS Rack 2. It was designed to study the microgravity behaviour of colloids such as in milk and paint, with a long-term view to developing new materials. In this case eight samples of three different colloids in its carousel were to be periodically rotated under a series of sensors so that their physical state could be recorded. However, it refused to start. Troubleshooting procedures were begun immediately. A patch was up-linked to the SSRMS's computer, but failed to rectify the problem with the shoulder pitch-joint. As a result, the launch of STS-104 was delayed from June until "no earlier than 12 July". Although, as a knock-on effect, STS-105 was slipped to August, one option if the SSRMS problem continued was to launch STS-105 in early August to deliver the Expedition 3 crew and retrieve the Expedition 2 crew and reschedule STS-104 for

September, by which time, one way or the other, perhaps by replacing the computer during an EVA, the arm would have been fixed. On 1 June, mission managers met to consider whether to add an EVA to Expedition 2 to replace the computer. While Helms continued the science programme, Usachev and Ross spent most of the first week of June preparing for an 'internal spacewalk'. On 7 June, a routine test of the Kurs automatic rendezvous and docking system to prepare for the arrival of the Russian Docking Compartment in September showed a fault. The next day, Usachev and Voss donned Russian Orlan suits and sealed themselves into Zvezda's spherical forward docking compartment. At 0921, with the compartment depressurised, they opened its nadir cover, secured the free-floating cover to a fixture on the inner wall, and replaced it with the conical docking drogue from the axial port, using a tool to lock the 12 latches on its circumference. At no time did either man venture out through the hatch. Within 20 minutes the repressurisation was underway. It was the first EVA by an ISS Expedition crew while there was no Shuttle in place. Helms remained in Zvezda to coordinate the activities.

Resuming the EXPRESS Rack 2 science programme, on 12 June Voss activated ARIS-ICE, which was designed to counteract vibrations on board the ISS that might interfere with sensitive microgravity experiments. Helms activated the fourth growth chamber in unit 10 of the PCG-STES for a four-day run. She also activated the Ultrasound Imaging System, a part of the Human Research Facility, and tested it by imaging her carotid artery. On 13 June, the ISS was turned so that Progress-M1 6's thrusters could be tested, but during the preparations a high-pressure reading in a manifold prompted an automatic switch to a backup system, so Korolev cancelled the test. On 14 June, with a software patch installed to monitor the behaviour of the SSRMS, Helms and Voss spent four hours putting the system through a dress rehearsal of the sequence of manoeuvres to install the Joint Airlock Module and verified that the backup software mode was capable of performing the task. The data would help the engineers to diagnose the intermittent fault with the shoulder pitch joint. Meanwhile, ground tests by the Canadian manufacturer had identified the most likely cause of the problem as lying in a faulty computer chip, rather than the mechanical shoulder joint. A new software patch would be uploaded to instruct the SSRMS to ignore errant commands from the chip in question. On 28 June Helms and Voss ran an additional rehearsal of the SSRMS manoeuvres, again without incident. On 15 June, the Expedition 2 crew marked their 100th day in space. Helms conducted MACE tests on 14, 15 and 18 June. The final pair of SAMS sensors were activated on 18 June, to bring that experiment up to full activation. On the same day, Voss deactivated the final growth chamber of unit 10 in the PCG-STES. Unit 9 of this experiment would remain active until mid-July.

On 19 June, the Shuttle managers decided to proceed with the launch of STS-104 rather than postpone it until after STS-105. Atlantis's planned roll out on 20 June was delayed due to overnight lightning around the Cape, but it was installed on the pad on 21 June. On board the ISS that day, Helms and Voss rehearsed the SSRMS manoeuvres to install the Joint Airlock Module, this time in the primary software mode. In preparation for the Shuttle's arrival, the MAMS and SAMS experiments were continuously monitoring the state of the microgravity environment to build up some context in which to interpret the results when the Shuttle was in place. In early July, the plants in the ADVASC experiment were prepared for return to Earth (it had produced seeds, which had reached maturity) and the SSRMS was subjected to a final test in its backup software mode. Tests of the ARIS-ICE over the weekend 7–8 July revealed that one of the eight actuators was sticking. The following day, Voss removed a possible obstruction, but could not eliminate the problem, but because the apparatus was designed to function properly with a minimum of six active actuators it was decided to leave it running throughout the STS-104 visit in its impaired state. Automated experiments continued to perform well and downlink data to Huntsville. In the POC, the control team for Expedition 3 had already taken over the day-shift. They would switch to 24-hour operations on 23 July, in preparation for the launch of the Expedition 3 crew on STS-105 in early August.

STS-104 WITH THE AIRLOCK

STS-104 – Atlantis – SSAF-7A

Commander:	Steven Lindsey
Pilot:	Charles Hobaugh
Mission Specialists:	Michael Gernhardt; Janet Kavandi; James Reilly

In the run-up to launch, STS-104 suffered several unusual problems. On 10 July, a group of illegal immigrants were landed on Cape Canaveral's restricted-access shore, and were rounded up by Kennedy Space Center security forces. The next day, the Moroccan TransAtlantic Abort site was closed by a threat of terrorist activity. With an alternative landing strip available at Zaragosa in Spain, however, it was decided to continue with the countdown. Unfortunately, the weather at Cape Canaveral threatened to preclude a launch, and predictions for the emergency landing sites also threatened the 24- and 48-hour recycle opportunities. Nevertheless, Atlantis lifted off on schedule, at 0504 on 12 July. This flight was testing the Block-II SSME, which incorporated a number of reliability-enhancing modifications. One of the new motors was clustered with two of the older type. The ascent was uneventful. After reconfiguring Atlantis, Lindsey's crew retired. On Flight Day 2, Gernhardt and Reilly checked the primary and backup EMUs. When a white deposit resulting from a potassium hydroxide leak was found close to the backup's battery, they removed the battery and sealed it in a bag, then replaced both the battery and the lithium hydroxide cartridge. It had been intended to leave the backup unit on the ISS (to make a pair with the one which had been left by STS-98) but Houston decided that the unit should be returned to Earth. Meanwhile, Kavandi powered up the RMS and used its cameras to survey the state of the payload bay.

The principal cargo on this flight was the Joint Airlock Module, which had been named 'Quest'. It was to allow EVAs by astronauts wearing American EMUs when no Shuttle was present, as the EMUs were too bulky to pass through the hatches on the Russian modules. It was a 'joint' airlock in that it could also accommodate the Russian Orlan pressure suits. Since it was to be installed on Unity's starboard side, and the Shuttle's RMS would not be able to reach this CBM, the transfer had to be made by the SSRMS which, fortunately, now seemed to be working properly. Quest was a composite of two co-axial cylinders, one short and fat and the other long and thin. Overall, it was 5.49 metres long and 3.96 metres across at its widest point, and had a mass of 5.8 tonnes. Its High Pressure Gas Assembly (HPGA) comprised two oxygen and two nitrogen tanks. In addition to repressurising the airlock following an EVA, these tanks would augment those in Zvezda. As the tanks would have to be replaced when they were empty, they had been designed to be attached to the periphery of the wide section of the airlock. Once the module had been installed, the SSRMS was to retrieve the tanks one by one from the pallet in the Shuttle's bay and mount them.

After the TI burn at 2033 on 13 July, on Flight Day 3, Atlantis drew up in front of the ISS. An IMAX camera in the bay recorded the docking with PMA-2 at 2308. The MAMS experiment registered the resulting vibrations. The results were to be used in planning future microgravity experiments. Once the hatches were opened at 0100 on 14 July, the two crews then spent an hour reviewing the plans for Quest's transfer, in particular the role that Gernhardt and Reilly would play on their first EVA. Then Helms powered up the SSRMS and rehearsed the manoeuvres. On Atlantis, Kavandi rehearsed how she would use the RMS to move the spacewalkers around. At 0640 the hatches were closed, so that Atlantis's pressure could be reduced in preparation for the first EVA.

Preparations took longer than scheduled, and Gernhardt and Reilly did not start until 2310 on 14 July. As Hobaugh supervised from the flight deck, they prepared Quest for removal from the bay. Gernhardt removed a cover known as the 'shower cap' from the berthing mechanism, then the covers from the module's various seals. Meanwhile, Reilly installed the securing points for the gas tanks. Finally, Gernhardt disconnected the cables that had powered the module's heaters while it was in the bay. The 'shower cap', seal covers and cables were stowed for return to Earth as rubbish. Helms, with assistance from Voss, swung the SSRMS down and grasped an attachment point on Quest's exterior, lifted it out of the bay and then manoeuvred it so that its broad end was close alongside Unity's starboard CBM. Kavandi used the RMS to manoeuvre Gernhardt and Reilly so that they could use handheld cameras to provide Helms with a variety of views of the module to supplement those from the SSRMS's own cameras. After a three-hour procedure, Helms finally berthed Quest at 0340 on 15 July. With the module in place, Gernhardt ran cables from Unity to power its heaters. Meanwhile, Reilly attached a number of foot restraints on the airlock's exterior to assist them during their second EVA. Once Quest's heaters were confirmed to be functioning properly, the two spacewalkers returned to Atlantis's airlock, closing it at 0509, having been outside for 5 hours 59 minutes.

On Flight Day 5, Lindsey led his crew back to the ISS to help the station's crew to prepare Quest. Having opened the hatches giving access between Unity and Quest, Usachev and Lindsey cut a white ribbon that had been put across the Crew Lock's hatch and so declared the new module open for business. They then set about mating

Having lifted the Joint Airlock Module Quest from Atlantis's bay on STS-104, the SSRMS is about to rotate it to be mated with Unity's starboard CBM.

umbilicals. When water was spilled from a coolant line designed to connect Quest to the ISS's Moderate Temperature Loop, there was a one-hour delay while the water was mopped up and the air bubbles that had prompted it were bled from the pipes. The other umbilicals would carry oxygen and nitrogen to pressurise the airlock at the end of an EVA. Since the module would be permanently attached, the bolt drivers were removed from the CBM to be returned to Earth. Although the water spillage led to the checkout of the EVA equipment being postponed, Helms and Voss verified the radios in Quest and in the two EMUs that were stowed in the Crew Lock. Kavandi, Gernhardt and Reilly spent much of the day transferring material from one vehicle to another. After breakfast on Flight Day 6, Voss and Kavandi worked on a leaky valve in Quest's Intermodule Ventilation Assembly (IMV). This comprised a series of fans and valves to circulate air between the ISS and Quest. They could not repair the fault, but managed to install a cap to stem the leak. This put them roughly half a day behind schedule. In the event that the valve needed to be replaced, there was one in Destiny. Lindsey and Helms tested the oxygen lines between the ISS and Quest that would be used to pressurise the airlock after their EVA. The relocation of the hatch that had been between Quest and Unity to its storage point between the Crew Lock and the Equipment Lock in Quest was delayed by the scheduling difficulties. At one point during the day, all outfitting work was halted so that a dry run of the airlock procedures could be made. The new module's atmospheric pressure was lowered for the first time, as it would be to help the astronauts in the Equipment Lock to pre-breathe oxygen prior to an EVA. After both crews went over the plans for EVA-2, Atlantis's crew retreated and the hatches were sealed at 0730 on 17 July so that the cabin pressure could be lowered. As the workday ended, mission managers were considering extending Atlantis's docked operations by one day to ensure that there was time to complete the flight plan. This decision was confirmed the following day, and EVA-3, which would be from Quest's airlock, was pushed back 24 hours to give the crews additional time to prepare the module's systems. Flight Day 7 began with Houston playing "Happy Birthday Darlin'" by Conway Twitty as the wake-up call to celebrate Kavandi's birthday. During preparations for EVA-2, the ISS's primary C&C computer crashed and had to be rebooted. The computer was vital to the EVA, because it would control the SSRMS. With the computer back on line at 2100 on 17 July, Helms and Voss were able to power up the SSRMS. Gernhardt and Reilly did not enter Atlantis's bay until 2340, somewhat later than planned. Gernhardt took up position on a workstation on the end of the RMS, and Kavandi moved him around. Once Helms and Voss had lifted the first gas tank from the bay and manoeuvred it alongside Quest, Gernhardt and Reilly bolted it into place. The two arms worked in a choreographed sequence. The procedure was repeated for the second tank. With the EVA running ahead of schedule, Houston decided to proceed with the third tank rather than leave it to the final EVA, as had been planned. Gernhardt and Reilly re-entered Atlantis at 0533 on 18 July, having been out for 6 hours 29 minutes.

The two crews spent Flight Day 8 working together on a number of items. Lindsey and Voss replaced the leaking valve in the IMV with one from Destiny. The valve in Destiny would not be needed until the second node was added, and that was

One by one, the SSRMS retrieved the gas tanks from a pallet in Atlantis's bay and affixed them to Quest's exterior. Later, James Reilly (facing the camera, top right) and Michael Gernhardt (obscured) ventured out from the new module to complete the installation.

not due for several years. After several hours of testing to ensure that the new valve was not leaking, Gernhardt, Reilly and Hobaugh checked the equipment inside Quest and then transferred from Atlantis the apparatus that they would need on the final EVA. Catching up on postponed items, Helms, Voss and Kavandi removed Quest's hatch and stowed it between the Crew and Equipment Locks, where it would be out of the way. The transfer was followed by a series of leak and pressure tests. Helms also changed the Payload Computer that had served as C&C-1 since that computer failed during STS-100's visit, installing a new one that had just been delivered. The Payload Computer was to be returned to Earth. Usachev devoted most of his day to routine maintenance of Russian equipment, but also joined Voss in finishing hooking up Quest to Unity. While the astronauts were asleep, Houston performed a pressure test and discovered a leak between the Equipment Lock to the Crew Lock. Following breakfast on Flight Day 9, everyone was involved in the preparations for the EVA from Quest. As Hobaugh and Voss operated Quest's controls, Gernhardt and Reilly ran a dress rehearsal of the suiting-up and transfer procedures. Later in the day, Lindsey gave a television tour of the new module, which was now certified ready for use on Flight Day 10. As the first astronauts to make an EVA from Quest, Gernhardt and Reilly employed a new regime of vigorous exercise to drive the nitrogen from their bloodstream. After donning their suits in the Equipment Lock, they moved into the Crew Lock and sealed the internal hatch. When everything was ready, a valve in the outer door was opened. It should have taken 7 minutes to evacuate the air but it took 40 minutes. An investigation was begun immediately to determine why. With the air finally vacated, the outer hatch in the wall of the narrow cylindrical compartment was opened at 0035 on 21 July. As Gernhardt and Reilly moved out, Voss used the SSRMS to retrieve the final gas tank from Atlantis's bay. Lindsey raised the RMS, which still had the workstation on it, so that Gernhardt and Reilly could be placed in position to secure the tank. Usachev and Hobaugh coordinated the activities of the arms from their respective vehicles. With their main task completed, Gernhardt and Reilly made their way to the top of the P-6 photovoltaic tower to inspect the motor that had been deactivated, but they saw no obvious cause for its behaviour. With the two astronauts back inside the Crew Lock, the EVA ended at 0437, after 4 hours 2 minutes. The Crew Lock is based on the 'internal airlock' which was originally incorporated into the mid-deck of the Shuttle, then removed when the Shuttles were converted to carry an Orbiter Docking System, which incorporates an airlock.

On Flight Day 11, the crews said their farewells, Hobaugh undocked at 0054 on 22 July, made the fly-around and then the separation burn at 0214. After a wave-off for weather, Lindsey set Atlantis down on Runway 15 at the Kennedy Space Center at 2339 on 24 July.

MORE BUDGET WOES

While STS-104 was in space, NASA announced a further $8 million increase in the ISS's cost overrun, as a result of underestimating the cost of the ISS hardware, the

The state of the International Space Station after the STS-104 visit, with Quest on Unity's starboard CBM.

need for additional ground support personnel and work on the life support system. This took the total predicted overrun to $4.8 billion. NASA also admitted that some items of hardware might have to be replaced earlier than planned. Meanwhile, the agency was criticised for cancelling 40 per cent of the scientific activity planned for the ISS as a result of its budget woes. Even the station's supporters argued that this decision seriously undermined the rationale for the programme. Another criticism was that it had failed to adhere to the appropriate procedures for the American Propulsion Module, which had just been deleted. The Office of the Inspector General accused the agency of failing to validate the module's requirements prior to a design review, resulting in $97 million being wasted. The report also argued that NASA had failed to justify awarding Boeing a sole-source contract. On the positive side, the House Appropriation Committee was considering adding $415 million to NASA's Fiscal Year 2002 budget to reinstate the X-38 Crew Return Vehicle, so that the crew complement could be increased if Italy supplied a Habitation Module.

THE FINAL STRETCH

After STS-104 departed, the Expedition 2 crew took two days off to slip their shift schedule back to what it had been, then settled into the familiar routine of tending experiments and maintaining the station's systems. On 2 August, Voss replaced the malfunctioning actuator on the ARIS apparatus. During the work, he noticed that a push rod operated by the actuator was bent so he replaced that too. The experiment was subjected to a week of tests to show that he had not impaired its functionality. Among the tests was the 'hammer test' in which he struck the EXPRESS Rack 2, in which the apparatus resided, using a small hammer to make sure that the experiment reacted correctly. The repaired ARIS would be available to record the vibrations as STS-105 docked.

As the Expedition 2 crew drew to the end of their programme, there was frantic activity on the ground. On 6 August, the Huntsville controllers who were to run the Expedition 3 science programme took over the operation of the POC. Preparations for STS-105, which would deliver the Expedition 3 crew, progressed. In Russia, the next Progress cargo ship was being prepared for launch in August and the Russian Docking Compartment was being prepared for launch in September. NASA was considering upgrading the Extended Duration Orbiter's cryogenic fluids package to enable Shuttles to remain in attendance for up to 30 days, thereby allowing up to seven astronauts to work alongside the residents to advance the science programme. The agency was also considering buying a second Soyuz ACRV in order to increase the resident crew to six. A plan was under discussion to "mothball" Columbia, the only vehicle in the fleet that had not been upgraded to dock with ISS. It would be placed in storage in a "ready to fly" condition. Columbia had been flying non-ISS missions. Grounding it would enable its upkeep costs to be reassigned. The Expedition 2 crew's final two weeks were spent preparing the ISS for the handover, but this was interrupted on 7 August when C&C-1 suffered a problem reading its hard drive. An attempt to reboot it from Houston failed. C&C-3 continued to serve in the primary role, and C&C-2 was upgraded from the standby mode to backup. A spare computer was due to be delivered on STS-105, so the problem was not serious.

STS-105 RESIDENT CREW HANDOVER

STS-105 – Discovery – SSAF-7A.1

Commander:	Scott Horowitz
Pilot:	Frederick Sturckow
Mission Specialists:	Daniel Barry; Patrick Forrester
Expedition 3 (up):	Frank Culbertson; Vladimir Dezhurov (Russia); Mikhail Tyurin (Russia)
Expedition 2 (down):	Yuri Usachev (Russia); James Voss; Susan Helms

STS-105 was originally scheduled for launch in July, but it was slipped to August when the difficulties with the SSRMS prompted a one-month delay to STS-104, and once this mission was safely off STS-105 was set for 9 August. Bad weather caused a 24-hour postponement. During the countdown, the launch was advanced by five minutes to 1710 on 10 August in order to avoid approaching weather – this was the first time that this had been done. After configuring Discovery for orbital operations, the crew retired.

On Flight Day 2, while Horowitz and Sturckow pursued the rendezvous, Barry and Forrester prepared their EMUs. The RMS was powered up and used to survey the bay. The main payload was the Leonardo MPLM, which was to be temporarily installed on Unity. As usual, the rendezvous was made on Flight Day 3. Horowitz docked with PMA-2 at 1442 on 12 August. Following pressure checks, the hatches were opened at 1641. Culbertson, as the Expedition 3 commander, was the first man through. On Flight Day 4, Culbertson, Tyurin and Dezhurov installed their couch liners in Soyuz-TMA 32, and Usachev, Voss and Helms put theirs in Discovery. In an emergency, the Expedition 3 crew now would leave in the Soyuz. They officially took over the ISS at 1415 on 13 August. The Expedition 2 crew had spent 148 days on the station. If Discovery left on schedule, they would have spent 163 days on the ISS and 167 days in space. Horowitz and Sturckow powered up the RMS, hoisted Leonardo out of the bay and mounted it on Unity's nadir at 1155 on 13 August. The hatch was opened at 1447, and the Shuttle crew commenced two days of unloading Leonardo, which included EXPRESS Racks 4 and 5 (Rack 3 was to be delivered on a later flight). One item was a third 'bedroom' to supplement the compartments built into Zvezda. It was to be set up in Destiny, where one member of the crew had been sleeping. Expedition 3 would continue 10 experiments and introduce 8 experiments. Usachev and Dezhurov installed new software into Zvezda's computers to improve Korolev's command functions to prepare for the Russian Docking Compartment's arrival. On Flight Day 6, with Leonardo empty, Horowitz's crew began offloading the items that had been carried on the mid-deck. Barry and Forrester spent most of

The STS-105 crew joined the International Space Station's second and third crews for supper in Zvezda.

the day checking out their EMUs. Usachev, Helms and Voss joined Horowitz's crew on the Shuttle and at 1752 on 15 August the hatches were closed so that the cabin pressure could be reduced.

At 0810 on 16 August, on Flight Day 7, Culbertson, Dezhurov and Tyurin made remarks to celebrate the 1,000th day since Zarya's launch. As soon as Discovery's crew had finished breakfast, they concentrated on preparations for the 25th EVA devoted to the ISS. Barry and Forrester egressed at 0958. Their primary task was to mount the Early Ammonia Servicer – which held spare ammonia to augment the ISS's cooling system – on the P-6's support frame. Horowitz operated the RMS while Sturckow directed activities. Their second task was to install the Materials International Space Station Experiment (MISSE) on Quest's exterior. This was the first experiment to be mounted outside the ISS. It held 750 samples of materials, ranging from lubricants to photovoltaic cells to be exposed to the space environment to determine how they degraded. After 18 months it would be recovered and returned to Earth. Barry and Forrester returned to Discovery at 1614, having been out for 6 hours 16 minutes. Meanwhile, Culbertson and his crew had continued to load Leonardo with items that were to be returned to Earth.

On 17 August, on Flight Day 8, everybody gathered for a news conference and the commanders of the two Expedition crews held a ceremony to mark the handover of command.

On Flight Day 9, while Expedition 3 loaded Leonardo, Barry and Forrester exited from Discovery's airlock at 0942 on 18 August. Horowitz used the RMS to move them around the exterior of the ISS so that they could install heater cables for the Starboard-0 (S-0) segment of the Integrated Truss Structure which was later to be attached to Destiny using the already installed Cradle Assembly. This done, they returned to Discovery at 1511, after 5 hours 29 minutes. To date, the ISS's assembly had accrued a grand total of 167 hours 24 minutes of EVA time. Despite concern expressed early on that it would be folly to rely upon extravehicular activities, the spacewalkers had excelled themselves. With the hatches open once more, the two crews retired. Meanwhile, Houston rescheduled the next day's activities to give the crews some free time in the afternoon. The morning of Flight Day 10 was spent finishing the loading of Leonardo. Forrester unberthed the module at 1414 on 19 August and secured it in the bay at 1515. Both crews enjoyed some free time before retiring. The farewells were said at breakfast on Flight Day 11, and during a TV broadcast, Culbertson, who had served as the Shuttle–Mir programme manager, produced two plaques which had flown on Mir. He said that he had brought them to the ISS in order to "tie the two stations together". Turning to the departing Expedition 2 crew, he told them, "Yuri, Jim, Susan, I know it's a tough day. I know it's hard to say goodbye. But we really, really are proud of all you've done. We will do our best to keep up the good tradition you have started and to maintain just as high a standard of excellence if we can. Have a happy landing." With that the Shuttle crew retreated.

The hatches were sealed at 0830 on 20 August. Sturckow undocked at 1052 and followed the fly-around with the separation burn at 1212. Discovery's crew spent

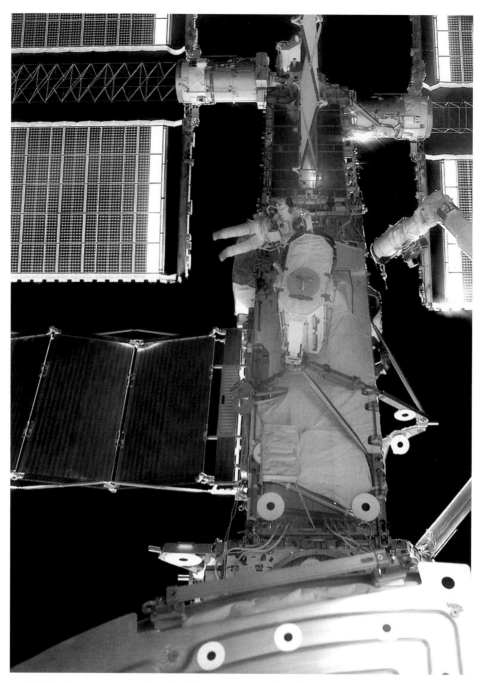

Patrick Forrester, installing the Early Ammonia Servicer on the P-6 unit during the STS-105 mission, is dwarfed by the external assemblies of the International Space Station.

Flight Day 12 configuring the Shuttle to return to Earth. Usachev, Helms and Voss installed their recumbent seats on the mid-deck. The first landing opportunity was waved-off due to cloud but the second was acceptable and Horowitz flew Discovery to a perfect touchdown on Runway 15 at 1423 on 22 August. Usachev, Helms and Voss were home after having spent 167 days in space, but outside of America – in fact some would argue outside of NASA – few people either knew or cared.

EXPEDITION 3

Although 21 August was Culbertson's crew's first day alone in space, they had to begin work immediately because Progress-M 45 had just been launched and they had to prepare Progress-M1 6 for departure. This done, they made a start on activating the experiments in EXPRESS Rack 4 in Destiny by hooking up coolant lines and applying electrical power to the rack's eight lockers and two drawers. The first new experiment was the Dynamically Controlled Protein Crystal Growth (DCPCG) that was to refine the methods for growing biological crystals by real-time control of the protein solution. Once started, it was to run unattended throughout September. For the first time, scientists in Huntsville would be able to observe the growing crystals and control their development by using nitrogen gas to regulate the evaporation rate of the solution surrounding the crystals. The Biotechnology Specimen Temperature Controller (BSTC) loaded with 32 stationary tissue culture modules was set up the next day as part of the Cellular Biotechnology Operation Support System (CBOSS) – an interim platform for cell-based research prior to the arrival of the Biotechnology Facility. This particular experiment was significant because it would mark the first time that a culture had been grown in microgravity without ever having experienced Earth's gravity. The BSTC was one of four experiments that were to be performed within the CBOSS.

Progress-M1 6 departed at 0205 on 22 August, and Progress-M 45 slipped into the vacated wake port at 0551 on 23 August. In addition to daily physical exercise, routine housekeeping and experiment maintenance, Culbertson's crew spent the next week unloading cargo. During that week, the EXPPCS experiment was run for two 12-hour periods, and the Volatile Organic Analyser was installed to sample the atmosphere within the ISS on a daily basis in order to detect and identify contaminants. It was to be operated by Huntsville. One maintenance assignment was to replace a malfunctioning voltage converter unit on one of Zvezda's batteries. On 27 August, the growth in eight of the BSTC cell cultures was arrested and they were stored in Destiny's Biotechnology Refrigerator.

After a three-day 'holiday weekend', Culbertson's crew began their fourth week on the ISS by tightening a connector in the station's air-conditioning unit to stem a freon leak, inspecting suspect wiring on a treadmill to verify it was safe to use, and replacing a faulty tape recorder in Destiny.

Meanwhile, Russia released details of the scaled-down Science Power Platform. It would have four photovoltaic arrays rather than eight, would have neither thermal radiators nor a pressurised section, and in its initial configuration it would have only one pair photovoltaic arrays; the second pair would be added at a later stage. In the revised design, each individual Russian Research Module would incorporate its own radiators. Furthermore, it had been decided that the Enterprise Module would be delivered by Shuttle, and the initial section of the NEP would be carried on the same mission.

On 5 September, because a series of EVAs were imminent, a routine check of the GASMAP was carried out in preparation for an experiment related to the effects of EVA on lung capacity. The first measurements by the Pulmonary Function Facility were taken to establish a baseline for this experiment. On the same day, Culbertson oriented the SSRMS to enable its end-effector camera to watch a waste water dump from a vent on Destiny. The water was dumped on 7 September over a 10-minute period during which the camera recorded its instant freezing in the vacuum and the behaviour of the resulting ice crystals as they were expelled. The station's exterior surfaces were documented before and after the water dump to verify that the ice crystals were vented cleanly. On 10 September, the astronauts completed another round of H-Reflex measurements, recording the spinal chord's ability to respond to stimuli in an effort to determine whether, as suspected, its response decreases during a long mission. If this were shown to be the case, astronauts would be required to undertake a more intensive exercise regime to ameliorate such degradation. The EXPPCS experiment continued to function despite the failure of one of the two photon-counting modules; the remaining module was sufficient to allow the experiment to continue unabated.

AMERICA ATTACKED

On 11 September 2001, terrorists hijacked four aircraft soon after departing from a variety of east coast airports on early morning internal flights across America. Two aircraft were flown into the Twin Towers of the World Trade Center in New York, felling them with considerable loss of life to the multinational workforce and a group of tourists on the rooftop vista deck. Another aircraft was flown into the Pentagon in Washington, and the fourth, thought to have been heading for either the Capitol or the White House, crashed in Pittsburgh. American airspace was promptly closed and military fighters began patrolling the eastern and western seaboards. At the Kennedy Space Center, all four Shuttle orbiters were secured and the workforce evacuated, as were their colleagues at the Johnson Space Center, although there a team of flight controllers remained on duty to monitor the ISS.

Later that day, when the ISS passed over New York, Culbertson, Dezhurov and Tyurin observed the plume of smoke and dust being blown south from Manhattan over the New York Bay. "As we went over Maine," Culbertson told Houston, "we could see New York City and the smoke from the fires. Our prayers and thoughts go out to all people there, and everywhere else. I hope that the people responsible are caught and bought to justice as soon as possible." NASA's various facilities were reopened the next day. As the true magnitude of the attack was realised, Culbertson penned a rare email:

When the Twin Towers of the World Trade Center in southern Manhattan were attacked on 11 September 2001, the ISS crew had a unique view of the smoke plumes.

Well, obviously the world changed today. What I say or do is very minor compared to the significance of what happened to our country today when it was attacked by . . . by whom? Terrorists is all we know, I guess. Hard to know at whom to direct our anger and fear. I had just finished a number of tasks this morning, the most time-consuming being the physical exams of all crew members. In a private conversation following that, the flight surgeon told me they were having a very bad day on the ground. I had no idea. He described the situation to me as best he knew it at about 0900 CDT. I was flabbergasted, then horrified. My first thought was that this wasn't a real conversation, that I was still listening to one of my Tom Clancy tapes. It just didn't seem possible on this scale in our country. I couldn't even imagine the particulars, even before the news of further destruction began coming in.

Vladimir came over pretty quickly, sensing that something very serious was being discussed. I waved Michael into the module as well. They were also amazed and stunned. After we signed off, I tried to explain to Vladimir and Michael as best I could the potential magnitude of this act of terror in downtown Manhattan and at the Pentagon. They clearly understood, and were very sympathetic. I glanced at the World Map on the computer to see where over the world we were and noticed that we were coming southeast out of Canada and would be passing over New England in a few minutes. I zipped around the station until I found a window that would give me a view of New York City and grabbed the nearest camera. It happened to be a video camera, and I was looking south from the window of Michael's cabin. The smoke seemed to have an odd bloom to it at the base of the column that was streaming south of the city. After reading one of the news articles we'd just received, I believe we were looking at NY around the time of, or shortly after, the collapse of the second tower. How horrible. I panned the camera all along the East Coast to the south to see if I could see any other smoke around Washington, or anywhere else, but nothing was visible.

It was pretty difficult to think about work after that, though we had some to do, but on the next orbit we crossed the US further south. All three of us were working one or two cameras to try to get views of New York or Washington. There was haze over Washington, but no specific source could be seen. It all looked incredible from two to three hundred miles away. I can't imagine the tragic scenes on the ground. Other than the emotional impact of our country being attacked and thousands of our citizens and maybe some friends being killed, the most overwhelming feeling being where I am is one of isolation. . . . It's difficult to describe how it feels to be the only American completely off the planet at a time such as this.

On hearing that an old classmate had been the captain of the airliner that hit the Pentagon, he observed: "Tears don't flow the same in space."

CONTINUING ISSUES

As the last quarter of 2001 began, some people in NASA considered that the ISS would never develop the capability to support a six-person permanent crew. It was also suggested that the European and Japanese laboratories might not be launched for at least a further three years. Others in the agency were evaluating the possibility of supplementing the crew by flying up mission specialists by Soyuz to spend up to a month working in Destiny as a stop-gap until the Extended Duration Orbiter could be enhanced to allow Shuttles to make similar visits. It was considered that, until the station could be expanded, longer Soyuz visits would impose too great a demand on its logistics.

Goldin set up a Budget and Management Task Force to review the management and budget control of the entire ISS programme – with a view to changing the former to bring the latter under control. He promised that the review would not compromise ISS safety. The report was to be submitted by 1 November 2001. The $4.8 billion overrun that had led George W. Bush's administration to cancel the US Habitation Module and the X-38 had been blamed on gross mismanagement of funds by NASA. Last-minute design changes to the early modules had incurred $2 billion in overruns. Some items drew particular scorn – thousands of dollars had been spent on Unity's external name plaques.

PIRS

The first Russian contribution to the ISS since Zvezda was launched at 1936 on 14 September. Passing overhead, Culbertson reported seeing the launch vehicle climbing through the atmosphere. The Progress-style propulsion unit was ferrying Docking Compartment Module 1, which had been named 'Pirs' (Pier), and flew the standard two-day rendezvous. The final manoeuvres began just after 1830 on 16 September. Twenty minutes later, the station's photovoltaic arrays were turned edge on in order to minimise any impingement by thruster efflux as the ferry braked. The automatic docking at 2108 was flawless. "We felt that!" Culbertson reported, as the probe entered the recently installed drogue on Zvezda's nadir port. The MAMS instrument recorded the high-frequency transients propagating through the ISS's structure. A few hours later, they entered the pressurised compartment to inspect its cargo, and spent three days unloading this. The Kurs system and the docking probe were removed for return to Earth and re-use. The software was updated to enable Zvezda's computers to control Pirs's systems, and the communication, ventilation, lighting, and caution/warning systems were then installed and verified.

The customised propulsion module was mated to the Docking Compartment by a docking system, so when this departed on 26 September it gave the ISS a third port with a drogue. Supplementing Zarya's nadir and Zvezda's wake would greatly ease the operational overhead of transferring Soyuz from one port to another in order to accommodate Progress ferries. It would no longer be necessary to discard one cargo ship in order to accommodate another.

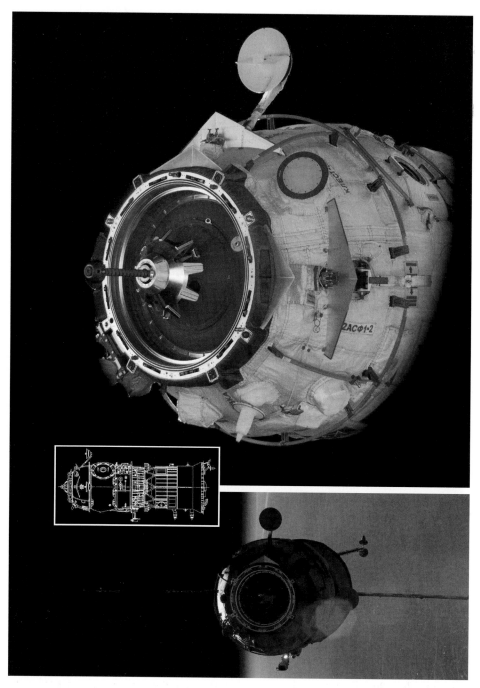

The Russian Docking Compartment, Pirs, approaches Zvezda's nadir port. Later, the Progress-style propulsion module was discarded.

In addition to providing another docking port, Pirs was equipped with a 1-metre-diameter side hatch to serve as an airlock for spacewalkers wearing Orlan pressure suits. Although the Quest airlock was already installed, its placement on Unity was designed to provide ready access to the Integrated Truss Structure, which was to be erected on the Cradle Assembly on Destiny. Situated on Zvezda, the Pirs airlock would provide easier access to the Russian part of the station.

On 8 October, Dezhurov and Tyurin ventured out from Pirs to lay power cables on its exterior to connect it to Zvezda. They also placed a cable to facilitate radio communications between the two sections of the ISS for spacewalkers using the airlock; an exterior ladder to assist with egress and ingress; and handrails to facilitate movement around the compartment's surface. With the preliminaries accomplished, they installed a Strela crane – which had just been delivered as part of the cargo – on Pirs's exterior. A similar crane had been assembled by Shuttle astronauts, but two were required to access the whole of the Russian part of the station. The package was so bulky that it had been necessary to employ Zvezda's forward compartment as an extension to the airlock, to make room to don their suits in the airlock itself. As this isolated Zvezda from the rest of the station, Culbertson had moved into the section that hosted Soyuz-TMA 32, just in case Zvezda's compartment could not be repressurised. In fact, it took five minutes and considerable wrestling with the mechanism to close the airlock's outer hatch.

Dezhurov and Tyurin ventured outside again on 15 October to mount packages on Pirs's exterior. The objective of this MPAC-SEEDS package, which contained a variety of materials in three briefcase-sized containers, was to investigate how they reacted to long-term exposure. The Russian national flag on Zvezda was retrieved for an exposure study, and replaced with a commercial logo. As previously, they found the outer hatch difficult to close.

MILESTONES

Soyuz-TMA 33

Commander:	Viktor Afanasayev
Flight Engineer:	Konstantin Kozeyev
Researcher:	Claudie Haignere

On 23 October Soyuz-TMA 33 docked at Zarya's nadir port. The 'taxi' crew was supplemented by Claudie Haignere, a Mir veteran and now an ESA astronaut who was to conduct experiments on behalf of the French Space Agency. On 2 November, two days after the visitors departed in the old Soyuz-TMA 32, the ISS passed the significant milestone of having been continuously inhabited for a year. NASA's ISS Program Manager, Tommy Holloway, was optimistic that the "unprecedented scale in orbital size and capability" of the station, would "be matched in the future by the scale of the benefits its research will bring to lives on Earth."

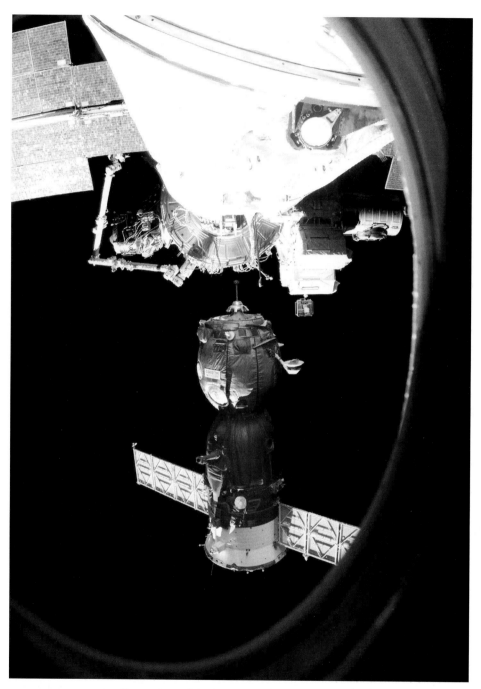

Soyuz-TMA 33 approaches Zarya's nadir port.

On 12 November Culbertson and Dezhurov made a five-hour spacewalk to make the final external connections for Pirs, to erect a Kurs antenna to facilitate dockings, to inspect and photograph a small panel of one of Zvezda's solar arrays which had not fully unfolded, and then tested the Strela by using a spacewalker as simulated cargo.

Meanwhile, having served longer than any previous NASA administrator, Dan Goldin announced his resignation, effective from 17 November 2001. In "the greatest job in the world", he had played a crucial role in involving the Russians in the programme, and had overseen its transformation from a dream to reality. He was superseded by Sean O'Keefe, the deputy director of the Office of Management and Budget.

As November wore on, the Expedition 3 crew prepared to hand over the ISS to their successors, who were to arrive on STS-108, but their routine was interrupted.

Progress-M 45 was discarded on 22 November. Progress-M1 7 was launched on 26 November. Upon arriving two days later, although the probe's latches were able to achieve a 'soft' docking, when the mechanism retracted the collars failed to establish a hermetic seal. A review of the docking camera video indicated that there was a strip of rubber trapped between the collars, so it was decided that Dezhurov and Tyurin should conduct a spacewalk to clear the debris from the docking system, which they did on 3 December, via the Pirs airlock. Upon retracting its probe a few minutes later, the ferry was able to achieve a hermetic seal. The O-ring had evidently become detached from Progress-M 45 as it departed.

NASA had postponed the launch of STS-108 in order to allow the Russians time to overcome the problem, because the transients propagated through the station by a Shuttle docking might have damaged the ferry craft swinging loose on its probe.

STS-108 – Endeavour – SSUF-1

Commander:	Dom Gorie
Pilot:	Mark Kelly
Mission Specialists:	Linda Godwin, Dan Tani
Expedition 4 (up):	Yuri Onufrienko (Russia), Carl Walz, Daniel Bursch
Expedition 3 (down):	Frank Culbertson, Vladimir Dezhurov (Russia), Mikhail Tyurin (Russia)

In addition to marking the significant milestone of the station's first 'utilisation' flight, STS-108 finally addressed the issue of the power spikes in the motors of the 'beta' gimbals, whose bearings were suffering thermally-induced distortions. Godwin and Tani exited Endeavour on 10 December and climbed to the top of the P-6 tower to wrap insulation blankets around the motors – so that was another problem fixed.

Half a century had passed since Wernher von Braun published his detailed design for a space station. The ISS was still at a work-in-progress stage but – contrary to early doubts by sceptics – the unprecedentedly complicated assembly task was

Soyuz-TMA 32 withdraws from the Pirs docking port.

unfolding to plan. It bore no resemblance to the classic 'wheel' that von Braun had envisaged, or the 'double wheel' that Stanley Kubrick had depicted in his visionary film *2001: A Space Odyssey*, but it is nevertheless a functional design and, crucially, it is capable of evolutionary growth.

Table 13.6 Phase Three flight log

Progress-M 44
SSAF-3P

Launched	0318	26 Feb 2001 by Soyuz launch vehicle
Docked	0450	28 Feb 2001 at Zvezda's wake
Undocked	0430	16 Apr 2001
De-orbited	0828	16 Apr 2001
Objective	Deliver logistics and propellant and take away trash.	

STS-102
Discovery
103rd mission
SSAF-5A.1

Launched	0642	8 Mar 2001 from Launch Complex 39 Pad B
Docked	0138	10 Mar 2001 at PMA-2 on Destiny's ram
Undocked	2332	18 Mar 2001
Recovered	0231	21 Mar 2001 at KSC
Objective	Perform the handover from the Expedition 1 and 2 crews, and deliver logistics in MPLM Leonardo – which was berthed on Unity's nadir at 0102 on 12 Mar and retrieved at 0540 on 18 Mar.	

STS-100
Endeavour
104th mission
SSAF-6A

Launched	1441	19 Apr 2001 from Launch Complex 39 Pad A
Docked	0959	21 Apr 2001 at PMA-2 on Destiny's ram
Undocked	1334	29 Apr 2001
Recovered	1211	1 May 2001 at Edwards AFB
Objective	Install the SSRMS on Destiny, and deliver logistics in MPLM Raffaello – which was berthed at 1200 on 21 Apr on Unity's nadir and retrieved at 1520 on 27 Apr.	

Soyuz-TMA 32
SSAF-2S

Launched	0237	28 Apr 2001
Docked	0358	30 Apr 2001 at Zarya's nadir
Undocked	2038	30 Oct 2001
Recovered	2359	30 Oct 2001
Objective	Lifeboat replacement. 'Space tourist' Dennis Tito visited the station and returned in Soyuz-TMA 31. The Expedition 3 crew had undocked Soyuz-TMA 32 from Zarya's nadir at 0548 on 19 Oct and redocked at Pirs at 0604 to vacate Zarya, and Soyuz-TMA 32 was later taken away by a 'taxi' crew.	

Progress-M1 6
SSAF-4P

Launched	1833	20 May 2001 by Soyuz launch vehicle
Docked	1924	22 May 2001 at Zvezda's wake
Undocked	0205	22 Aug 2001
De-orbited	0500	22 Aug 2001
Objective		Deliver logistics and propellant and take away trash.

STS-104
Atlantis
105th mission
SSAF-7A

Launched	0504	12 Jul 2001 from Launch Complex 39 Pad B
Docked	2308	13 Jul 2001 at PMA-2 on Destiny's ram
Undocked	0054	22 Jul 2001
Recovered	2339	31 Jul 2001 at KSC
Objective		Install the Quest airlock (it was berthed on Unity's starboard CBM at 0340 on 15 Aug) and fit its external gas tanks.

STS-105
Discovery
106th mission
SSAF-7A.1

Launched	1710	10 Aug 2001 from Launch Complex 39 Pad A
Docked	1442	12 Aug 2001 at PMA-2 on Destiny's ram
Undocked	1052	20 Aug 2001
Recovered	1423	22 Aug 2001 at KSC
Objective		Perform the handover from the Expedition 2 and 3 crews, and deliver logistics in MPLM Leonardo – which was berthed on Unity's nadir at 1155 on 13 Aug and retrieved at 1434 on 19 Aug.

Progress-M 45
SSAF-5P

Launched	0524	21 Aug 2001 by Soyuz launch vehicle
Docked	0551	23 Aug 2001 at Zvezda's wake
Undocked	1112	22 Nov 2001
De-orbited	1635	22 Nov 2001
Objective		Deliver logistics and propellant and take away trash.

Pirs
SSAF-SO-1

Launched	1936	14 Sep 2001 by Soyuz launch vehicle
Docked	2108	16 Sep 2001 at Zvezda's nadir
Objective		A Progress service module delivered the Docking Compartment Pirs, and then withdrew at 1136 on 26 Sep and was de-orbited later that day.

Soyuz-TMA 33
SSAF-3S

Launched	0359	21 Oct 2001
Docked	0544	23 Oct 2001 at Zarya's nadir
Undocked	–	

Recovered   –
Objective   Lifeboat replacement. French cosmonaut Claudie Haignere visited the
               station and returned in Soyuz-TMA 32.

Progress-M1 7
SSAF-6P
Launched   1324   26 Nov 2001 by Soyuz launch vehicle
Docked   0953   3 Dec 2001 at Zvezda's wake
Undocked   –
De-orbited   –
Objective   Deliver logistics and propellant and take away trash. The soft docking was
               achieved at 1543 on 28 November, but a 'foreign object' fouled the docking
               mechanism, and hard docking could not take place until the ISS crew had
               extracted this manually.

STS-108
Endeavour
107th mission
SSUF-1
Launched   1719   5 Dec 2001 from Launch Complex 39 Pad B
Docked   1503   7 Dec 2001 at PMA-2 on Destiny's ram
Undocked   1228   15 Dec 2001
Recovered   1255   17 Dec 2001 at KSC
Objective   The first 'utilisation' flight for the Destiny laboratory. The MPLM
               Raffaello was berthed on Unity's port CBM at 1255 on 8 Dec and retrieved
               at about 1330 on 14 Dec.

STS-110
Atlantis
109th
SSAF-8A
Objective   In Mar/Apr 2002 (after the STS-109 servicing of the Hubble Space
               Telescope) this mission will install the S-0 segment on Destiny, and thus
               start the assembly of the Integrated Truss Structure.

(All times US Eastern)

Table 13.7 Phase Three EVAs

	Start time	Finish time	Duration
STS-102			
Voss and Helms	0012 11 Mar 2001	0908 11 Mar	8h 56m
Kelly and Thomas	0023 13 Mar 2001	0644 13 Mar	6h 21m
STS-100			
Hadfield and Parazynski	0645 22 Apr 2001	1455 22 Apr	7h 10m
Hadfield and Parazynski	0834 24 Apr 2001	1615 24 Apr	7h 40m
Expedition 2			
Usachev and Voss	0921 8 Jun 2001	0940 8 Jun	0h 19m
STS-104			
Gernhardt and Reilly	2310 14 Jul 2001	0509 15 Jul	5h 59m
Gernhardt and Reilly	2204 17 Jul 2001	0533 18 Jul	6h 29m
Gernhardt and Reilly	0035 21 Jul 2001	0437 21 Jul	4h 02m
STS-105			
Barry and Forrester	0958 16 Aug 2001	1614 16 Aug	6h 16m
Barry and Forrester	0942 18 Aug 2001	1511 18 Aug	5h 29m
Expedition 3			
Dezhurov and Tyurin	0923 8 Oct 2001	1421 8 Aug	4h 58m
Dezhurov and Tyurin	0417 15 Oct 2001	1009 15 Oct	5h 52m
Culbertson and Dezhurov	1641 12 Nov 2001	2145 12 Nov	5h 04m
Dezhurov and Tyurin	0821 3 Dec 2001	1106 3 Dec	2h 45m
STS-108			
Godwin and Tani	1552 10 Dec 2001	1604 10 Dec	4h 12m

(All times US Eastern)
Note that when Orlan suits are used, Russian rules apply, and the duration of the spacewalk runs from hatch open to hatch closed, whereas NASA rules apply for an excursion from Quest and the duration starts when an EMU is placed on its battery for power to the time that the airlock is repressurised.

Stanley Kubrick's vision of a space station, as depicted in his film *2001: A Space Odyssey*.

Postscript

Despite the early criticism that the interlocking sequence of missions required to assemble the International Space Station would prove impracticable, the programme has unfolded remarkably smoothly. It is true that considerable time was lost when financial problems in Russia delayed the Service Module, Zvezda, but by mid-2001 the crew was in residence, NASA's Destiny Laboratory was in place and was being outfitted with microgravity experiments, and the first of a series of utilisation flights was imminent.

In contrast to serving as a 'way station' to deep space, as NASA had initially envisaged, the descoping of Freedom to make it affordable forced the agency to dedicate its laboratory to microgravity research, and the Europeans and Japanese followed suit. As a result, the programme will initially focus on life and materials sciences.

Critics have labelled microgravity research as 'ivory tower' science, with little commercial value, but much of the basic work has been undertaken by academics in concert with industry with a view to eventual commercial exploitation. It is true that the potential for commercialisation has not lived up to the initial hype, but much of the blame for this early over-optimistic expectation must be borne by NASA, for the way in which it promoted first the Shuttle and then, in the early 1980s, the role of a space station as a laboratory in space. Even if the agency did not originate the wilder ideas, it did little to correct such false promises. This initial mismatch of expectation versus practicality apart, however, when its Baseline Configuration is complete, the International Space Station will have three laboratories for microgravity studies, with other laboratories under construction. Research will be conducted on a 24-hour basis, not just because its crew will work in shifts but also because much of the apparatus will be able to be operated remotely, from Earth, enabling the principal investigators to conduct telescience.

One aspect of having a large crew – a benefit that will in all likelihood not be commented upon because it will be taken for granted – will be that while the commander and the flight engineer focus on station operations and take care of routine maintenance, the scientists will be able to work on their own programmes.

This point is worth noting formally because on Mir, with its two- or three-person crews, maintenance was performed at the expense of the science programme.

Although the expectation is that the station will eventually facilitate materials processing on a commercial basis, it will initially serve the Centers for Commercial Development of Space which NASA established with industrial and academic co-sponsorship, which completed their development work on Spacehab flights, and so are primed to exploit the opportunities presented by a permanent orbital facility.

The station's laboratories will be equipped with International Standard Payload Racks so that apparatus can be upgraded and modified as necessary. A modular approach is essential, too, because the laboratories will be required to support an integrated research programme, on a continuous basis, over a lifetime of several decades. This approach will enable the scientific focus to evolve. It will also enable the ISS to keep pace with state-of-the-art equipment. Furthermore, such plug-in compatibility will permit apparatus from different nations to be integrated in a pick-and-mix fashion to match specific requirements.

The station will transform the way in which experiments are undertaken. Scientists working in a terrestrial laboratory will repeat their experiment, fine-tuning its configuration in order to hone a crucial measurement, and then vary it in order to establish the sensitivity of the data to parameters, all of which takes time. However, on a Shuttle mission, *time* is a precious resource that must be shared between the *many* payloads that must be carried to justify the cost of a launch. On a station, the economics are different because the cost of the science is essentially *decoupled* from the Shuttle's running cost. Once an experiment is on board the station, it can be run for prolonged periods, and re-run at no further cost to the Shuttle. At long last, the Shuttle in *combination* with the station should slash the cost of operating *in* space, and so industrial materials-processing should become financially justifiable for those markets that can accommodate the premium imposed to meet the unique production costs.

To date, it has not just been the *overhead* that has deterred commercial processing in orbit, it has been the reliance of any such activity on the vagaries of the Shuttle's manifest – which has historically offered infrequent flight opportunities on an indeterminate schedule. It is difficult to justify commercial investment in a process that can be undertaken for at most a fortnight at a time, and at most twice a year. The ongoing nature of operations on board the station should relieve both of these constraints by making it possible to run a process on a semi-continuous basis. If raw material is ferried up in bulk, *mass production* will be practicable. The output will be collected whenever convenient. Production will have been *decoupled* from the vagaries of flight operations. If the process requires an extremely pure microgravity environment, it could be performed on a free-flyer. For little or no overhead, the station should be able to service a *flotilla* of free-flyers. The Space Vacuum Epitaxy Center at the University of Houston sought to develop industrial-scale manufacturing of crystals in space by epitaxy in which a crystal is built up layer by layer on a substrate in such a way that its lattice matches that of the substrate. All of the previous experiments in growing inorganic crystals had exploited the *microgravity* of space, but this would also rely upon the *vacuum* of space. It was designed as a free-flyer and because it would exploit the vacuum in the wake of its carrier it was called the 'Wake Shield Facility'. From NASA's point of view it was an ideal project, because the fact that its process *relied* upon the space environment meant that the product would not be able to be matched by terrestrial factories, so it would be a genuine *manufacturing-in-space* application. It would need to yield an output of just 200 kilograms per annum to provide a sufficient rate of return to recover the development costs over a financially acceptable period. It was not viable as a Shuttle application, but ongoing operations in concert with the station would make it so.

Research on Mir was hindered by a deficiency in the logistics infrastructure. It was *not* a problem of resupplying the station, because automated Progress craft docked every few months to replenish its fluids and dry cargo, and expired parts from the environmental system, food packages and other waste could be loaded on board a departing Progress, which then "ceased to exist", as the Russians put it, during re-entry. The problem which Mir faced was *returning* material. A Soyuz descent module can return three cosmonauts and up to 120 kilograms of compact cargo, but this is barely enough to return the film, crystals and log books that accumulate during a six-month tour of duty. To ameliorate this, a small capsule was developed for delivery as part of a Progress's cargo, and once it had been loaded with 120 kilograms of cargo it was mounted in the ferry's hatch so that it could be ejected as the main spacecraft de-orbited itself over the recovery zone. The root of the problem of returning a large cargo to Earth is that a descent capsule which uses an ablative shield to protect it during re-entry needs to be *small*. It was for this reason that, for post-Apollo, NASA opted for a transportation system based on a re-usable spacecraft having a large payload bay. With Shuttle–Mir, it became practicable to return large items for refurbishment and, indeed, the Russians took advantage of this to offload up to a *tonne* of apparatus each time a Shuttle docked. The significance of this is that the Shuttle will be similarly able to serve the International Space Station. In addition to delivering cargo it will return items that have expired their service life so that this life can be reassessed; worn out or broken objects will be returned for refurbishment and re-use; free-flyers will be returned for refitting; and accumulated product will be offloaded. If NASA had opted to keep the Apollo spacecraft and service a station assembled from Skylab-style modules, by now it would be facing the same constraint as hindered Mir operations.

As NASA's initial proposal to Nixon had argued, the Shuttle and the station form an integrated system; either without the other is deficient.

The International Space Station's primary asset, as on Skylab with its solar telescope, will be its crew. It is undoubtedly the case, as those critical of human spaceflight never miss an opportunity to point out, that developing an orbital facility capable of human habitation is an *extremely* expensive project. They have argued, rightly, that it would be cheaper to build automated satellites. But satellites have to be designed to fit specific mass and size constraints. They must be light enough to be lifted by their assigned rocket, and they must fit within its aerodynamic shroud. Once in orbit, they are required to operate autonomously. In the case of the larger and more valuable satellites, the development cost is usually offset by a long lifetime. Satellites usually have a suite of instruments, so if the power or communication systems fails the entire package generally has to be written off. Even if the satellite functions perfectly for 20 years, towards the end of its mission its instruments will, in all likelihood, be technologically obsolete, and in any case its sensors will have degraded. What is required is a platform which combines a long life with *both* reliability and flexibility. The most reliable and flexible 'component' that can be built into a system is a human operator. The International Space Station will, *in the fullness of time*, make astronomical and environmental research more cost-effective, although not necessarily when measured in terms of their individual disciplines and specific budgets, but together as a total package, alongside the ongoing microgravity work and the servicing of the free-flyers, because it offers an *infrastructure* for operating, servicing and upgrading their instruments. Not only will instruments be decoupled from their operational lifetime of their technologies and also from the pace of technological advance, but the experimenters will be able to design the instruments free of the usual constraints of mass and size, and of power and communications limitations. By being able to rely upon the Shuttle to provide transportation, and upon the station to provide utilities and ongoing technical support, the appallingly long time currently taken to conceive of, develop and fly an instrument will be reduced, which will, indirectly, benefit such endeavours. With the International Space Station operational, it would therefore be *negligent* not to exploit it to its fullest.

Despite the currently popular point of view that the presence of a human crew aboard a station effectively rules out any observational programme that requires the fine-pointing of instruments, Skylab demonstrated not only that solar observations were feasible but also that a telescope with a human operator was more flexible than an automated system. In the Shuttle era, some Spacelab missions carried clusters of telescopes on an Instrument Pointing System for solar and astronomical studies and, again, proved the value of a human presence. Plus, of course, the more massive the platform, the less significant will be the vibrations induced by the activities of its crew. It will be perfectly practicable to mount an IPS on the station's truss. Indeed, there is no reason why there should not be *several* mounts. As the Shuttle showed, telescopes can be operated by the crew, by the ground team, or automatically, in which case they need not impose an additional burden on the station's crew. The ongoing nature of operations will facilitate *comprehensive* studies, not just the snapshots that were so hectically gathered by the few Shuttles devoted to astronomy.

As for looking at the Earth, the instruments on the Upper Atmosphere Research Satellite, and the satellites of the Earth Observing System could also be mounted on the station's truss. By having a suite of sensors making simultaneous measurements, it will be possible to make *integrated* studies. With a platform that can be augmented and which provides a built-in capability for repair of attached payloads, it will make little sense to continue to build independent satellites unless they *require* a different orbit.

The International Space Station should therefore not be seen as a short-term venture. The Baseline Configuration is only the core of a complex that will be expanded. Indeed, there will be an *incentive* to expand it, because the overhead of adding a new facility will be small in comparison to the cost of developing the supporting infrastructure. However, even with telescience, as facilities are added the workload will exceed the capacity of the crew, so it will be necessary to add another habitat to increase the crew complement. If it is decided to keep free-flyers in space, and undertake maintenance *in situ* rather than back on Earth, it will become necessary to add a large airlock which can act as a berthing facility for free-flyers, so that their payloads can be reconfigured in a shirt-sleeved environment. This poses no problem, because the facility has been *designed* for evolutionary growth.

The Baseline Configuration represents only the completion of *Stage One* of the International Space Station – it will mark *the end of the beginning* of humankind's emergence from its cradle.

Acronyms

AASS	Alternative Access to Space Station (launch vehicle development programme)
ACRV	Assured Crew Return Vehicle
ADVASC	Advanced Astroculture experiment
AFB	Air Force Base
AM	Airlock Module
APM	American Propulsion Module
APU	Auxiliary Power Unit
ARIS	Active Rack Isolation System
ARIS ICE	Active Rack Isolation System ISS Characterisation Experiment
ATV	Autonomous Transfer Vehicle (ESA)
BSTC	Biotechnology Specimen Temperature Controller
CAM	Centrifuge Accommodation Module
CapCom	Capsule Communicator
CBM	Common Berthing Mechanism
CBOSS	Cellular Biotechnology Support System
CBOSS	Cellular Biotechnology Operations Support System
CEO	Crew Earth Observation experiment
CETA	Crew and Equipment Translation Aid
CEVIS	Cycle Ergometer with Vibration Isolation System
CGBA	Commercial Generic Bioprocessing Apparatus
CMG	Control Moment Gyro
CPCG-H	Commercial Protein Crystal Growth-High Density experiment
DCM	Docking Compartment Module
DCPCG	Dynamically Controlled Protein Crystal Growth experiment
DDCU	DC-DC Converter Unit
DoD	Department of Defence
DOSMAP	Dosimetric Mapping experiment
DOSTELS	Dosimetric Telescopes
EarthKAM	Earth Knowledge Acquired by Middle School experiment
EMU	Extravehicular Mobility Unit

EP	EXPRESS Pallet
ERA	European Robotic Arm
ESA	European Space Agency
EST	Eastern Standard Time
ET	External Tank
ETR	Eastern Test Range
EVA	Extravehicular Activity
EXPPCS	Experiment on Physics of Colloids in Space
EXPRESS	EXpidite PRocessing of Experiments to Space Station
FGB	Functional Cargo Block (Zarya)
FPP	Floating Potential Probe
GAO	Government Accounting Office
GASMAP	Gas Analyser System Metabolic Analysis Physiology experiment
HPGA	High Pressure Gas Assembly
HRF	Human Research Facility
ICC	Integrated Cargo Carrier (ATV)
ICM	Interim Control Module
IMV	Intermodule Ventilation Assembly
ISS	International Space Station
ITS	Integrated Truss Structure
JELC	Japanese Experiment Logistics Carrier
JEM	Japanese Experiment Module
JEM-EF	Japanese Experiment Module Exposed Facility
JEM-RMS	Japanese Experiment Module Remote Manipulator System
KhSC	Khrunichev State Research and Production Space Centre
KSC	Kennedy Space Center
LC	Launch Complex
LH2	Liquid Hydrogen
LOX	Liquid Oxygen
LSS	Life Support System
MACE	Mid-deck Active Control Experiment
MAMS	Microgravity Acceleration Measurement system
MBS	Mobile Base System
MCOR	Medium Rate Communication Outrage Recorder
MCS	Motion Control System
MIRTS	Russian acronym for Zarya's battery charge controllers
MISSE	Materials International Space Station Experiment
MIT	Massachusetts Institute of Technology
MPLM	Multi-purpose Logistics Module (originally: Mini-Pressurised Logistics Module)
MSS	Mobile Servicing System
MT	Mobile Transporter
NASDA	National Space Development Agency of Japan
NEP	Science and Power Platform (Russian acronym)
NRL	US Naval Research Laboratory

OAMS	Orbital Attitude Manoeuvring System
OCA	Orbiter Communications Adapter
ODS	Orbiter Docking System
OPF	Orbiter Processing Facility
ORCS	Orbiter Reaction Control System
OSVS	Orbiter Space Vision System
OTD	Orbital Transfer Device
P	Port
PA	Pressurised Adapter (Zarya)
PAS	Payload Attach Structure
PAYCOM	Payload Communications Manager
PCG-STES	Protein Crystal Growth-Single Thermal Enclosure System
PCU	Plasma Contractor Unit
PDU	Power Drive Unit
PHALCON	Power, Heating, Articulation, Lighting and Control Officer
PIC	Pyrotechnic Initiator Controller
PLSS	Portable Life Support System
PMA	Pressurised Mating Adapter
POC	Payload Operations Center
PTAB	(Russian acronym for) Storage Battery Current Regulator System
RACU	RussianAmerican Conversion Unit
RCS	Reaction Control System
RMS	Remote Manipulator System
RPCM	Remote Power Controller Module
RRM	Russian Research Module
RSA	Russian Space Agency
RTLS	Return To Launch Site (Shuttle abort mode)
S	Starboard
SAFER	Simplified Aid For Extravehicular activity Rescue
SAMS	Space Acceleration Measurement system.
SARJ	Solar Alpha Rotary Joint.
SASA	S-Band Antenna Sub-Assembly
SGANT	Space to Ground Antenna
SM	Service Module (Zvezda)
SMMOD	Service Module (Zvezda) Micrometeoroid and Orbital Debris Shield
SOC	Shuttle Operations Coordinator
SPDM	Special-Purpose Dexterous Manipulator
SRB	Solid Rocket Booster
SSM	Russian Docking/Stowage module (Russian Acronym)
SSME	Space Shuttle Main Engine
SSPF	Space Station Processing Facility
SSRMS	Space Station Remote Manipulator System
STS	Space Transportation System (the Space Shuttle)
TDRS	Tracking and Data Relay Satellite
TI	Terminal [Phase] Initiation (Shuttle rendezvous manoeuvre)

TLD	Titan Launch Dispenser
TVIS	Treadmill with Vibration Isolation System
UDM	Universal Docking Module
UDMH	Unsymmetrical Dimethyl Hydrazine
UF	Utility Flight
ULC	Unpressurised Logistics Carrier
USAF	United States Air Force
USHM	United States Habitat Module
VAB	Vehicle Assembly Building
VOA	Volatile Organic Analyser
X	Experimental Aircraft Designation (America)

Appendix 1: Launch Vehicles

Soyuz

The venerable Soyuz launch vehicle (known affectionately as the Semyorka[1]) is derived from the ballistic missile that was adapted to launch Sputnik. It comprises a central core (Block A[2]) and four 'strap-on' boosters. The 28.75-metre-long inversely tapered core is 2.95 metres in diameter at the top (its widest point). It is powered by a four-chambered RD-108 motor that burns kerosene in liquid oxygen (LOX) to deliver 96 tonnes of thrust for 314 seconds, and is steered by four vernier motors. It has a dry mass of 6.5 tonnes and a propellant load of 95.7 tonnes. Each booster is 19.8 metres in length and 2.68 metres in diameter, has a base that tapers to a point at the top, is powered by a four-chambered RD-107 motor that delivers 102 tonnes of thrust for 122 seconds, and has a pair of outboard vernier motors to assist with steering. Each strap-on block has a dry mass of 3.45 tonnes and a propellant load of 39.63 tonnes. Prior to launch, the boosters and the core are ignited to provide maximum thrust at lift-off. When their propellant has been consumed the boosters are jettisoned and the core continues to fire. The air-started upper stage is 8.1 metres long and 2.66 metres in diameter, and its four-chambered RD-461 motor delivers 30 tonnes of thrust for 240 seconds. It is connected to the core by a short framework truss, has a dry mass of 2.4 tonnes and a propellant load of 21.3 tonnes.

The payload is protected by an aerodynamic shroud during its ascent through the lower atmosphere. In the case of a Soyuz spacecraft, the shroud is topped by a tower containing the solid-propellant Launch Escape System to draw the crew re-entry compartment clear of any catastrophic failure. Four aerodynamic flaps on the exterior of the shroud deploy for stabilisation. The shroud for the Progress cargo spacecraft does not incorporate the escape system because, with no crew onboard, it is considered to be expendable.

As the only Soviet rocket 'rated' to launch a piloted spacecraft, the Semyorka will

---

[1] Semyorka means 'old number seven', and the rocket, designated the R-7, was the seventh design by the Korolev Bureau.

[2] The Russians refer to the separate 'stages' of their launch vehicles as 'blocks'. These are identified by the first few letters in the Cyrillic alphabet, the English equivalents of which run A, B, V, G and D. This scheme can become awkward, however. For example, the upper stage of the Proton is known as the Block D because it was derived from the fifth stage of the N-1 'moonrocket'.

serve the ISS by dispatching piloted Soyuz crew ferries, unpiloted Progress cargo vehicles and other specialised units such as the 'Pirs' Docking Compartment.

Proton

The three-stage Proton is used to launch 'large' Russian space station modules. The base of this vehicle looks as if it is a central core surrounded by six strap-on boosters, in a similar manner to the Soyuz launch vehicle, but it is actually a single 20.2-metre-long, 7.4-metre-diameter block with a ring of six RD-253 rocket motors that burn unsymmetrical dimethyl hydrazine (UDMH) in nitrogen tetroxide to deliver a thrust of 178 tonnes each for 130 seconds. It has a dry mass of 43.3 tonnes and a propellant load of 412.2 tonnes. The 13.7-metre-long, 4.15 metre-diameter second stage has a pair of motors (produced by the Kosberg Bureau) that deliver a total of 240 tonnes of thrust for 208 seconds. It has a dry mass of 13.2 tonnes and a propellant load of 152.4 tonnes. The 6.4-metre-long upper stage continues the 4.15-metre profile. Its single motor delivers 64 tonnes of thrust for 254 seconds, has a dry mass of 5.6 tonnes and a propellant load of 47.5 tonnes. Russian space station modules are self-propelled and after pursuing a rendezvous they dock automatically.

Space Shuttle

Since 1981, the Space Shuttle has served as the workhorse of the American piloted spaceflight programme. It is a combined launch vehicle and piloted spacecraft in a 'stack' comprising three major parts. The Orbiter – the piloted portion of the stack – has the superficial appearance of a delta-winged aircraft.

When a Shuttle lifts off from Launch Complex 39 of the Kennedy Space Center on Merritt Island on the Atlantic coast of Florida, a pair of 45.46-metre-long, 3.70-metre-diameter Solid Rocket Boosters (SRBs) each provide 1,316 tonnes of thrust, some 71 per cent of the total required for launch. They are connected to each side of the 14.63-metre-long, 8.53-metre-diameter External Tank (ET), which forms the stack's primary structural element and carries the Orbiter. The ET carries the liquid oxygen and liquid hydrogen propellant for the trio of Space Shuttle Main Engines (SSMEs) on the Orbiter's rear, each of which produces 170 tonnes of thrust at launch. The Orbiter has an Orbit Manoeuvring System (OMS) pod on each side of its vertical stabiliser, both of which contain a single 2,722-kilogram-thrust engine to assist with orbital insertion, make major orbital manoeuvres and the de-orbit burn at the end of the mission. Each pod also contains a dozen Orbiter Reaction Control System (ORCS) 395-kilogram-thrust motors. Together with 14 such motors in the Orbiter's nose, these thrusters effect roll, pitch and yaw attitude control motions. A pair of 11.3-kilogram-thrust engines in each pod, together with two such motors on the nose, facilitate minor orbital manoeuvres. These two systems are referred to as the Orbital Attitude Manoeuvring System (OAMS).

The Shuttle is launched vertically, as a rocket. The SSMEs are ignited first, and only when they have been verified to be functioning are the SRBs ignited to provide sufficient thrust for lift-off. The main task of the SRBs is to lift the Shuttle through the thickest portion of Earth's atmosphere. At burnout, 2 minutes after lift-off, the SRBs are jettisoned. They deploy parachutes and make a soft splashdown in the

Atlantic, from which they are recovered and towed to Cape Canaveral for return to the manufacturer for breakdown, cleaning, refurbishment and preparation for use on another mission. The rest of the stack continues to ascend, the SSMEs continuing to draw propellant from the ET. Once the requisite altitude and velocity have been attained, the three SSMEs are shut down, the ET is jettisoned, and the OMS motors are fired to ease the Orbiter into the desired orbit. The ET re-enters the atmosphere over the Pacific and burns up.

The main body of the Orbiter, between the wings, contains the 18.29-metre-long, 5.18-metre-wide, 3.96-metre-tall payload bay. At launch, the bay is enclosed by a set of longitudinal doors. Once in orbit, the payload bay doors are opened to expose the radiators mounted on their interior. If they fail to open the Shuttle must return to Earth before its electrical equipment overheats. With the doors open, the Orbiter is ready to perform the on-orbit portion of its mission. On ISS missions, the forward part of the bay is taken up by the Orbiter Docking System (ODS) and the rear accommodates ISS modules, Spacehab modules, Multi-Purpose Logistics Modules (MPLMs) and other cargoes. A Remote Manipulator System (RMS) is mounted on one side of the payload bay to manipulate the cargoes and assist in spacewalks. The forward part of the Orbiter contains the flight deck and, beneath it, the mid-deck. The lower deck contains spacecraft systems. The Commander and Pilot control the Orbiter from the flight deck. The RMS is operated by Mission Specialists from the Aft Flight Station on the flight deck, which has large upward- and rearward-facing windows. The mid-deck provides additional seating for the crew during launch and re-entry as well as sleeping quarters, galley and personal hygiene facilities. The mid-deck also contains the hatch giving access to the internal airlock, if it is carried, or to a tunnel to the pressurised modules in the payload bay.

The payload bay doors must be closed prior to the de-orbit burn. The Orbiter re-enters atmosphere in a nose-first, nose-high attitude. Once in the lower atmosphere, the aerodynamic surfaces on its wings and vertical stabiliser become operational. A microwave landing system guides the vehicle, now a glider, towards the landing strip and the three-point undercarriage is lowered just 30 seconds before landing. The twin main wheels touch down first, a drag parachute is deployed to slow the vehicle and line it up with the runway, the nose is lowered, the chute is released and, finally, mechanical braking is used to achieve wheel-stop. The vehicle is then towed off the runway and returned to the Orbiter Processing Facility for refurbishment.

Ariane V

The European Space Agency's Ariane V is a two-stage launch vehicle augmented by a pair of re-usable strap-on boosters for lift-off. Each booster is 31.6 metres in length and 3 metres in diameter and burns solid propellant to yield 536 tonnes of thrust at launch. After burning for 1 minute 12 seconds, they are jettisoned at an altitude of approximately 55 kilometres, deploy parachutes and are retrieved from the Atlantic for cleaning and re-use. The single Vulcain motor of the 30.7-metre-long, 5.4-metre-diameter first stage is powered by liquid hydrogen and liquid oxygen and yields 92 tonnes of thrust at launch. The thin walls of the propellant tanks in this stage means that the upper stage cannot be added until after the boosters have been mated to rigidise the first stage. Both the first stage and the boosters are ignited for lift-off and the first stage continues to fire through the booster separation sequence and for another 8 minutes 30 seconds. Aerospatiale is the prime contractor for the first stage and the boosters. The 3.3-metre-long, 3.9-metre-diameter DASA-built second-stage motor produces 2,800 kilograms of thrust. The restartable third-stage motor can be fired for a total of 18 minutes 30 seconds over three burns in order to place its payload into a specific orbit. The payload is protected during launch by one of two standard nose shrouds, depending on its size. Ariane launches take place from the equatorial site at Kourou, French Guiana. Ariane V was once assigned to launch the Columbus Orbital Facility to the ISS, but after its disastrous inaugural launch it was decided to transfer this crucial payload to the Shuttle. Nevertheless, it will launch the Ariane Transfer Vehicles which will replenish the ISS.

H-2

Japan's H-2 launch vehicle is a two-stage rocket with the first stage augmented by a pair of strap-on boosters for lift-off. The first stage is 28 metres long and 4 metres in diameter. Its LE-7 rocket motor burns liquid hydrogen and liquid oxygen and delivers 86 tonnes of thrust at launch. It burns for 5 minutes 46 seconds. The boosters burn solid propellants to give 160 tonnes of thrust for 1 minute 34 seconds, at which time they are jettisoned, and the first stage continues the climb on its own. The second stage is 10.6 metres long and 4 metres in diameter, and has a LE-5A motor that burns for 10 minutes 15 seconds (the first 6 minutes 43 seconds of which is to achieve low orbit, and the remaining propellant can be used later to place the payload into the required operating orbit). The main stages are manufactured by Mitsubishi Heavy Industries and the boosters are produced by the Nissan Motor Company. Depending on their size, payloads are protected by a range of different standard shrouds. The H-2 is launched from Japan's Tanegashima Space Centre. The initial H-2 has recently been superseded by a more cost-effective H-2A which will be used to launch unmanned cargo vehicles to the ISS where they will be used to resupply the station's Japanese facility.

Appendix 2: ISS Hardware

Nodes

Designed by the Boeing Company, the nodes will serve as connecting passageways between the ISS's various pressurised modules.

Node 1 (named 'Unity' prior to its launch on STS-88) is 5.4 metres long and 4.5 metres in diameter. Pressurised Mating Adapter 1 (PMA) is mounted on its wake Common Berthing Mechanism (CBM) to provide access to Zarya. It initially had PMA-2 on its ram end to accommodate a Shuttle. Its Early Communication System provided voice, data and low data-rate video communications with Mission Control in Houston to supplement the Russian communication systems during the earliest assembly flights. As the ISS grew, PMA-2 was removed, the American Laboratory 'Destiny' was mounted in its place, and PMA-2 was relocated on the Destiny's far end to accommodate Shuttles. Unity's zenith CBM became the mounting point for the Z-1 Truss Structure, a temporary, early mounting for an American photovoltaic array. Unity's starboard CBM has the American Joint Airlock Module mounted on it, and the port CBM, after temporarily hosting PMA-3 and a succession of Multi-Purpose Logistics Modules, will eventually hold a multi-windowed Cupola. The nadir CBM was to have accommodated the American Habitation Module, but this was cancelled. Unity contains four equipment racks, and resources such as life support systems, fluids, environmental control, electrical and data systems are routed through it for distribution to the attached modules.

Although Boeing manufactured Node 1 itself, and has a spare shell left over from manufacturing tests, Nodes 2 and 3 were built by Alenia Aerospazio in Italy for the European Space Agency (ESA), which agreed to construct them as partial payment for the launch of its 'Columbus' laboratory by a Shuttle. Each of these nodes has the standard arrangement of six CBMs, but because they are 6.4 metres long (one metre longer than Boeing's) they can host 8 (as opposed to 4) racks of equipment. Node 2 will be put on Destiny's ram end, and PMA-2 will go on its far end to accommodate Shuttles. Japan's Experiment Module and ESA's Columbus will be mounted on its port and starboard CBMs respectively, a Centrifuge Module has been assigned to its zenith CBM, but no module has currently been assigned to its nadir. Following the order in 1999 to Boeing to cease work on the Habitation Module while a comparison with an inflatable TransHab module was undertaken, it was decided to move the

Life Support System and Environmental Control System which had been assigned to the America Habitation Module to Node 3, and to mount this on Unity's nadir, with PMA-3 relocated to its far end. This was necessary because the TransHab does not have these systems built in, and either module could then be selected at a late date.

Pressurised Mating Adapters

The pseudo-conical Pressurised Mating Adapters (PMAs) can be mounted on any Common Berthing Mechanism to allow Shuttles and Russian ISS modules to dock. Multiplexer/demultiplexer units (computers) on the exterior of PMA-1 enabled early command and control of Unity's systems through Russia's Korolev Control Centre in Moscow via Zarya's systems. PMA-2 initially enabled Shuttles to dock at Unity. It was moved to allow Destiny to be mounted on Unity and was then relocated on the ram end, where it currently accommodates Shuttles. PMA-3 was originally to be stored on the Z-1 Truss Structure until it was required, then put on the exposed end of the US Habitation Module, but it has now been mounted on Unity.

Spacehab

Spacehab was privately developed in the 1980s to provide additional pressurised volume and experiment lockers for solo Shuttle flights. Spacehab modules are carried in the payload bay and are accessible through a pressurised tunnel. Both single and double modules are available. Spacehab's flat roof was specifically designed to allow Mission Specialists looking out of the Orbiter's aft flight deck windows to see over the top of the module to equipment in the rear of the bay – a problem identified by ESA's cylindrical Spacelab module. The original 'single' module was designed to be mounted at the front of the bay, in order to leave the rear of the bay free for other payloads. The 'double' module was manufactured by the simple expedient of mating a test article to the single module to produce a hybrid module in which only one half was wired for experiments, the extension was fitted for cargo. It was ordered to ferry logistics to Mir. Because the Orbiter Docking System had to be mounted at the front of the bay on those missions, the Spacehab module was relocated to the rear, in a position closer to the Orbiter's centre of mass, thereby enabling it's capacity to be increased to 4,536 kilograms. This configuration was later adopted for ferrying cargo to the ISS. The Italian-built Multi-Purpose Logistic Modules also serve this role, but they can be offloaded and left attached to the ISS, whereas Spacehab must remain in the bay.

Spacehab, the Multi-Purpose Logistic Modules and the Ariane Transfer Vehicle proposed by ESA (if it is developed) will provide regular cargo deliveries to the ISS and return to Earth experimental results and equipment that are no longer required in orbit.

Zenith-1 (Z-1) Truss Structure

Despite its name the Zenith-1 Truss Structure is *not* part of the Integrated Truss Structure. In the early stages of ISS construction, the growing station required more electrical power than the photovoltaic arrays on the Russian modules could provide. To overcome this problem, the Z-1 was mounted on Unity's zenith CBM to serve as a temporary mount for the P-6 section of the Integrated Truss Structure, with its photovoltaic arrays and cooling radiators. When the rest of main truss has been completed, the P-6 segment will be relocated to its final position on the far end of the truss. The Z-1's frame contains the Control Moment Gyroscope (CMG) attitude control system and a suite of communications equipment and associated antennae, so it will remain in place when the power module is moved. In the late 1990s there was a plan to mount propulsion units and propellant tanks on the Z-1 once it was free, but this plan was cancelled in favour of an American Propulsion Module incorporated into the left-over structural test article for Node 1.

American Laboratory Module 'Destiny'

The 8.5-metre-long, 4.26-metre-diameter American Laboratory Module 'Destiny' is constructed from three aluminium cylinders and a pair of end cones. A window is mounted in the central cylinder. The exterior of the cylinders is strengthened by a waffle pattern of aluminium, which is covered by a debris shield blanket made from material similar to the kevlar of bullet-proof vests, which, in its turn, is covered by a thin aluminium debris shield. The module holds 24 equipment racks, 11 of which are assigned systems necessary to support the laboratory environment and control the American segment of the ISS. Thirteen racks will host equipment for microgravity research and technology experiments. Destiny is mounted on Unity and has PMA-2 on its ram end.

Joint Airlock Module 'Quest'

The Joint Airlock Module, named 'Quest', has two sections which together support EVAs by astronauts and cosmonauts wearing either American EMUs or Russian Orlan pressure suits. It is 6 metres long and 4 metres in diameter at its widest point. This is the Equipment Lock, where spacewalkers can undergo 'pre-breathing' to remove the nitrogen from their blood streams, don and doff their suits and store their EVA equipment. The narrower cylinder of the Crew Lock will be used by the spacewalkers to egress and ingress. American EMUs do not fit through the external hatches of the Russian Zarya and Zvezda modules. The Airlock Module is mounted on Unity's starboard CBM.

Integrated Truss Structure

The Integrated Truss Structure will be constructed from preformed sections carried into orbit by a series of Shuttle flights. It will provide attachment points for external payloads, photovoltaic arrays, cooling radiators and systems equipment. It is based on the Integrated Truss Structure of Space Station Freedom. Its various segments are numbered Port (P) and Starboard (S) and their location. P-1 is the innermost segment and P-6 is the outermost segment. The exception to this sequence is the S-0 segment, which will be affixed by the Cradle Assembly to Destiny.

The starboard side of the truss incorporates four external attachment points for experiments, and the port side incorporates two more. These are located close to the pressurised modules in order to be within reach of the SSRMS. At each site, a series of Payload Attach Structures (PASs) will accommodate either a single large payload or a number of smaller payloads on an EXPRESS pallet. Such payloads will be able to draw electrical power from the ISS, but they have to provide their own thermal control system. At each site, the payload may be set to face any of four directions: zenith, nadir, ram, or wake.

Mobile Servicing System

Canada's Mobile Servicing System (MSS) comprises three principal components: (1) The Space Station Remote Manipulator System (SSRMS) is an advanced version of the robotic arm used on the Shuttle. It is 17 metres long and has seven motorised joints. The arm is not only capable of handling payloads but also of assisting with the docking of Shuttles. It is self relocatable, moving end-over-end using a Latching End-Effector at each end to 'walk' from one Power and Data Grapple Fixture on the exterior of the station to another. (2) The Mobile Base System (MBS) is a work platform able to move on rails along the length of the Integrated Truss Structure. It will provide lateral mobility for the SSRMS as it moves along the truss. (3) The Special Purpose Dexterous Manipulator (SPDM) is a small two-armed robot capable of performing delicate assembly tasks which otherwise would have to be undertaken by astronauts during extravehicular activities. This hardware replaces an earlier plan to develop a Telerobotic Servicer. It will be carried in the SSRMS's end-effector and operated as an adjunct to the main arm.

Cupola

The Cupola is to be placed on Unity's port CBM. It will provide a hemispheric exterior view to support extravehicular activity. The SSRMS's control station, initially set up in Destiny, will eventually be installed in the Cupola.

American Propulsion Module

During the two years of delays in launching the Zvezda module, Boeing commenced work on an American Propulsion Module to replace, (or at least to supplement) Zvezda's propulsion tasks. This development was assigned a budget of $540 million, but work was terminated when it reached $200 million over budget. Alternative plans included the installation of the Interim Control Module (ICM), or placing propulsion units on top of the Z-1 Truss Structure after the P-6 photovoltaic array had been repositioned to its permanent site. In September 2000, NASA extended Boeing's contract to include the conversion of Node 1's structural test article into an American Propulsion Module for launch in 2004. The module was to be mounted on Node 2's ram CBM, with the single propulsion unit in front of the Columbus module. One motor would be capable of satisfying one half of the ISS's propulsion requirements. If required, the module would be able to be retrofitted with a second motor. The propulsion module was not designed to be refuelled in orbit, so when its propellant supply was depleted it would have to be returned to Earth for servicing and replenishment. Some of the module's pressurised volume would be available for stowage, and the interior corridor would enable Shuttle crews to transfer through the module. When NASA's Fiscal Year 2002 budget was announced in March 2001, however, George W Bush's administration insisted that funding be directed away from the American Propulsion Module to meet ISS's cost overruns.

Interim Control Module

In 1996, at the height of the delays regarding the construction and testing of the Service Module, NASA began looking for a specialised vehicle to perform the orbital reboost portion of Zvezda's mission. The Titan Launch Dispenser (TLD) developed by the US Naval Research Laboratory (NRL) was an upper stage for the Lockheed Titan III and IV launch vehicles to deploy constellations of reconnaissance satellites into distinct orbits. In fact, it had originally been designed for carriage by the Shuttle, and had been adapted for the Titans following the loss of the Challenger on mission STS-51L in 1986. NASA hoped to be able to retrofit it for the Shuttle with only the minimum of design changes. However, the TLD was over-designed for the task that NASA had in mind. NASA and NRL collaborated on a second redesign to allow the TLD to serve as an Interim Control Module (ICM) to maintain the ISS's operating orbit. Work began in January 1997 with a budget of $120 million.

The ICM is 5.18 metres long and 4.5 metres in diameter. Its base is an octagon drawing the power for its systems from the photovoltaic cells on its slab-like sides. Four propellant tanks were to hold monomethyl hydrazine fuel and nitrogen tetroxide oxidiser. Two more tanks were for the propulsion system's pressurisation gas. Its 5,216-kilogram propellant load would be sufficient for a maximum of 18 months of reboost. As the original engine would have imposed an excessive dynamic load on the ISS when it was fired, it was replaced by a softer 489-newton engine. The stabilisation thrusters mounted on a pair of telescopic arms were relocated to accommodate the change from a spin-stabilised to three-axis stabilised mode. The thermal control system was similar to that used on NRL's Clementine spacecraft in the 1980s to overcome the thermal stresses that would result from the stabilised mode. The control system also had to be able to accommodate the shifting centre of mass as the ISS grew. Software was written to enable the ICM to perform its propulsion tasks as the station's configuration evolved between SSAF-2A and SSAF-7A. The ICM was to be carried into orbit by a Shuttle and transferred by the RMS to Zarya's wake docking system, where Zvezda was to dock, using a Russian-supplied probe/drogue system. It was to perform reboost tasks until such times as Zvezda was launched. At that time, the ICM would have been undocked to allow Zvezda to dock in its place. If Zvezda had been lost, the ICM would have been launched to serve its reboost function. However, the ICM could not be replenished in orbit, and if the ISS had ended up being reliant upon it, NASA would have had to alternate between two vehicles.

On 2 February 2000, Dan Goldin announced that if Zvezda was not in orbit by the end of 2000, NASA would launch the ICM. He said that NASA would have to make the decision as to whether to launch the ICM by the end of July 2000 if it was to meet the December launch date. He further stated that even if Zvezda was launched and docked to the ISS in 2000, the ICM would be launched in 2001 anyway because the combination of Zvezda and the ICM made the ISS "so much more robust". However, following the launch and successful docking of Zvezda, NASA decided to keep the ICM in storage as a contingency system.

EXPRESS Rack and Pallet
The EXPRESS (an acronym for **EX**pedite the **PR**ocessing of **E**xperiments to **S**pace **S**tation) programme consists of the EXPRESS rack for pressurised payloads and the EXPRESS pallet designed to fit to standard Payload Attach Structures for attached payloads on the Integrated Truss Structure and on the Exposed Facility outside the Japanese Experiment Module. Up to six individual payloads on each pallet can draw power, command and data handling and video from the ISS, but each must have its own thermal control system. The ISS crew will install, replace and control the payloads using the Special Purpose Dexterous Manipulator. The pallets will be returned to Earth periodically for servicing. The Brazilian Space Agency is providing EXPRESS pallets for the ISS.

US Habitation Module
Structurally, the US Habitation Module which was to provide sleeping quarters, messing facilities and leisure facilities was to be a clone of Destiny. It was to contain equipment for maintaining the astronauts' physical fitness to ameliorate exposure to weightlessness. Following completion of a test article, work was halted to await a comparison with the inflatable TransHab module. However, work on TransHab was curtailed in NASA's Fiscal Year 1999 budget. In the Fiscal Year 2001 budget George W. Bush's administration ordered that funding be directed away from the Habitation Module to meet the long-term cost overruns associated with the American portion of the ISS. In the same year, Italy expressed an interest in developing a Habitation Module using the same structure it had developed for the Multi-Purpose Logistics Modules.

TransHab
TransHab is intended to provide a large-volume Habitation Module for the ISS while simultaneously assessing its utility as a transit spacecraft for future flights into deep space and, ultimately, a mission to Mars. It is to be carried in a deflated condition, in the payload bay of a Shuttle. Its systems will be in the central core which will be accessible by the axial CBMs. After being transferred to Node 3 by the RMS, it will be inflated. The outer wall is a multi-layered shell to provide protection from micrometeorites and orbital debris. The inner layer is a web of composite materials to strengthen the outer layer as well as to provide structural supports for a series of water bladders on the inside. The 'floors' and internal partitions will be installed once the structure has been inflated. The upper level – that farthest from Node 3 – has exercise equipment, a hygiene facility and a storage area for 'soft' materials. The mid-level has six crew-quarter segments in the central core and life support equipment in the inflated areas. Half of this level forms an 'upper floor' in the crew's wardroom. The lower level is divided between a storage facility and the 'ground floor' of the wardroom, with the galley area in the central core.

NASA's Fiscal Year 2000 budget contained the stipulation that no funds were to be used to develop inflatable modules for use in space, so development of TransHab was passed to Alenia Aerospazia in Italy. When the 2001–2002 budget was passed in September 2000, this constraint was relaxed to permit NASA to lease an inflatable module if one was developed by a private company.

X-38 Crew Return Vehicle
The X-38, an American vehicle, was developed to serve as a Crew Return Vehicle for ISS. Its shape is based on the experimental lifting bodies tested in the 1960s. (A lifting body's fuselage creates enough lift as it passes through the air to support itself in stable flight, without the requirement for wings.) It was to employ off-the-shelf technology, rather than develop new items from scratch. The computer was to be a commercial model used by aircraft, as was its software. The vehicle would have a nitrogen-based attitude control system. It would have batteries for up to nine hours of operation, more than enough for a return to Earth in an emergency. The 9.1-metre-long and 4.4-metre-wide vehicle would be ferried up in a Shuttle's bay and transferred to the ISS by the RMS. In the event of an emergency evacuation of the station, the crew would seal themselves in the CRV and undock. Its capacity was up to seven astronauts. After separation, the CRV would perform a de-orbit burn and jettison the module containing its engine. Once in the lower atmosphere, the lifting body section would deploy a steerable parafoil and glide to an unpowered landing, using skids to absorb final landing loads. The de-orbit and landing procedures were to be automated, although the crew would have the option to use backup systems to control attitude, select a de-orbit site and steer the parafoil to a landing.

The X-38 programme started as an in-house study at NASA's Johnson Space Center in early 1995. Throughout that summer, the parafoil concept was tested by dropping pallets, with parafoils attached, over the Army's Yuma Proving Ground in Arizona. In early 1996, a contract was awarded to Scaled Composites Inc., Mojave, California, for three full-scale atmospheric demonstration vehicles. The first vehicle, designated Vehicle 131, was delivered in September, outfitted with avionics and other hardware, and then shipped to the Dryden Flight Research Center, California, on 4 June 1997 to be used in a series of captive flights, slung under the wing of NASA's B-52 aircraft throughout July and August 1997.

V-131 was successfully air-dropped on 12 March 1998. At release the vehicle rolled sharply to the right. When the parafoil deployed it was twisted and the test vehicle spun rapidly beneath it as it straightened out. The parafoil had two rips in it. Nevertheless, V-131 glided to a safe landing in the open brush area that had been chosen as the landing site after winter rains had left the planned dry lakebed too wet for use. In light of the rips, a series of 20 sub-scale drop-tests were conducted at Yuma in Arizona to measure the forces acting on the parafoil. As a result of those tests, the material at the bottom of the parafoil was strengthened. Flight-testing resumed on 3 February 1999, when V-131 made a captured flight slung under the wing of NASA's B-52. On 6 February 1999, V-131 was dropped from the B-52 on its second test. The test had been postponed from the previous day due to bad weather. On release from an altitude of 7,010 metres, the vehicle rolled right, but only half as much as on the previous test. The stabilisation parachute deployed from the rear of the vehicle and the parafoil deployment sequence began. With the parafoil deployed the vehicle was commanded to make a single turn before control was handed over to the X-38's computer. The vehicle turned left, turned into wind and then turned back to the left under computer guidance. The landing gear was deployed as planned. At 762 metres, the altitude data from V-131 ceased updating. Rather than reinstate manual control, it was decided to let the vehicle make an automatic landing. Soon thereafter, the altitude data resumed updating. It was later determined that condensation on the windshield had blinded the laser. When the condensation evaporated the laser could see the landing site once more, and new data was transmitted. The trajectory flared out and V-131 landed slightly 'long' on the Rogers dry lake after a 12-minute flight. After this flight, V-131 was returned to Scaled Composites to have its rear modified to reflect the revised plan to launch the CRV on Europe's Ariane V launch vehicle in a routine cargo-carrying role. Further changes to the shape of the fuselage would permit astronauts to be seated upright, rather than on reclined couches as was originally intended. The modified vehicle was redesignated V-131R, and used in tests to clear the way for Vehicle 201 to attempt a return from space in 2001.

Meanwhile, drop-tests continued using V-132, which was delivered to Dryden in September 1998. It incorporated full lifting body flight systems to allow it to fly independently after dropping from the B-52 aircraft and before the parafoil deployment. The programme's fourth free-flight was made on 9 July 1999 using V-132. The test vehicle was dropped from an altitude of 9,601 metres and permitted to free-fall for 31 seconds to allow the vehicle to build up sufficient velocity to act as a thorough test of the drogue parachute, which had been strengthened since the previous test. During the free-fall, V-132 made a slight roll to the right before stabilising in the correct attitude. The vehicle's aerodynamic surfaces were pulsed by the on-board control system as planned to verify their functionality. The drogue parachute was deployed from the rear of the vehicle, to reduce loads. It was de-reefed in three stages, then served to slow and stabilise the falling vehilcle. Prior to parafoil deployment, the drogue chute was repositioned from trailing behind the vehicle to above it. This caused the vehicle to pitch up, in a manoeuvre that was less dramatic than previously. The flaps were tested, then the parafoil was deployed in its reefed state. Unequal inflation caused the parafoil to twist during deployment, but it soon straightened out. The extention process was then initiated, with reefing lines being severed and the parafoil extending in length as areas of blue panels inflated, contrasting with the white panels of the reefed configuration. In fact, this was the parafoil's second flight, having been previously used on the 6 February drop test. A chase plane filming the deployment showed that the parafoil deployed without sustaining damage. The lifting body settled into a steady glide and executed two S-turns in the air to shed energy and reduce forward velocity. As it approached the dry lake, the landing gear was deployed. Initially, the left-hand landing skid only partially deployed, but it moved to its fully deployed position prior to landing. The vehicle made a stable landing after a flight lasting approximately 8 minutes.

In February and April 2000, a 697-square-metre parafoil was tested for use in the X-38 drop-test programme. The parafoil was claimed by NASA to be the largest ever manufactured. The test vehicle consisted of a pallet weighted to simulate a full-scale X-38. Once a parachute had drawn the pallet from the rear of a C-130 Hercules and been jettisoned, a 24-metre-diameter drogue was deployed to stabilise the pallet vertically prior to the parafoil commencing the 30-second-long 5-stage deployment sequence. After a descent lasting 11 minutes, the pallet touched down on the Yuma desert at 12 kilometres per hour.

The fifth drop-test was scheduled for 25 February 2000, but cancelled due to excessive winds. It was rescheduled for the following day. Despite early morning mist, the chase planes took off as planned, followed by the B-52/X-38 combination. At an altitude of 11,580 metres the ground team started receiving telemetry from the X-38 that they did not consider to be technically possible, so the drop was aborted. Two connectors were identified as likely causes of the spurious readings. Plans were drawn up for the B-52 to land and the connectors to be reinstalled with the X-38 still in place on the under-wing pylon. After the repair, the B-52 would resume its station so that the drop could occur as planned. Post-landing inspection revealed no loose connections, but as there was a fuel leak in the B-52's wing the second flight was cancelled. The unusual readings were finally traced to interference in the control system, and additional shielding was added to the cables within that system. With the B-52's fuel leak repaired, the drop-test was rescheduled for 30 March. It was the longest and fastest drop to date. V-132 was released from the B-52 mothercraft at an altitude of 11,850 metres and allowed to fall for 44 seconds to build up a velocity of 926 kilometres per hours, at which time an 18-metre drogue was deployed from the rear of the vehicle. Following de-reefing, the drogue was repositioned from behind to above the vehicle to slow it to 70.5 kilometres per hour, then the 510-square-metre parafoil deployed without incident. For the first time, flight control software steered the descent. On the final approach to landing on the dry lake, the landing skids were deployed. The plan to attempt a cross-wind landing was abandoned when one of the three landing skids failed to deploy, and the landing took place without incident after a 11.5-minute flight. This was the final test of the 510-square-metre parafoil. Future tests would be made with the modified V-131 (now redesignated V-131R) and the 697-square-metre parafoil. On 3 November 2000, V-131R was air-dropped for the first time in its final external configuration, in an attempt to simulate deployment dynamics of a return from orbit. It was dropped from the B-52's pylon at 11,125 metres, and promptly suffered a flight control problem which prompted a 360-degree roll manoeuvre. When the 24-metre-diameter drogue deployed 24 seconds later as planned, the vehicle recovered from the roll. Parafoil deployment commenced at an altitude of 5,791 metres, while the vehicle was in a nose-high vertical attitude. Even so, the huge parafoil deployed successfully and V-131R flew to a controlled landing on the dry lake.

The third vehicle, V-201, was due to be carried into orbit by a Shuttle in 2001 to make a full return to Earth from orbit. It was not to be. President George W. Bush's national budget for Fiscal Year 2002 drew funding from the X-38 programme to address the ISS's $4 billion cost overrun. However, work will probably be resumed in a few years, once the financial situation is relaxed.

Orbital Manoeuvring Vehicle 'Space Tug'
The Space Tug was conceived in the mid-1980s as an upper stage intended to collect satellites from a Shuttle in low orbit and deliver them to their operational orbits. It could also retrieve satellites and take them to the space station for repair or upgrade. The vehicle was to be a 1.5-metre-thick, 5-metre-diameter disk constructed in two parts. The inner core was the Short Range Vehicle (SRV) and the outer ring was the Propulsion Module (PM). As the name suggests, the SRV with the flight computers, avionics and the 24-thruster Reaction Control System, would have been able to be used independently of the PM on short-range flights, addressing perhaps 50 per cent of the envisaged missions. The SRV would remain in space and be refuelled by the station. The annular PM's four main rocket motors would have facilitated longer-range missions, ferrying satellite to and from their operational orbits. It was intended that this portion of the spacecraft could be replaced in orbit, and the depleted unit returned to Earth for servicing. It was intended that the Space Tug would be capable of making 100 flights over its 10-year operational life. Unfortunately, the Space Tug fell victim to the budget cuts which plagued the late 1980s and early 1990s.

Control Module 'Zarya'
The Zarya (Russian for Dawn) module was built for NASA by the Khrunichev State Research and Production Space Centre (KhSC) near Moscow, under subcontract to Boeing working in concert with Lockheed Martin, which had previously forged links with KhSC to commercialise the Proton launch vehicle. NASA bought the spacecraft outright to overcome American political doubts regarding the Russian government's commitment to the ISS. Although Lockheed Martin signed a development contract for the Zarya with KhSC on 8 February 1995, Zarya's construction had commenced several months earlier. The $200 million contract included the construction, launch by Proton and on-orbit delivery. Zarya was delivered to the Baikonur Cosmodrome in January 1998 for pre-launch testing. Its launch later that year began the assembly of the ISS. In some older documents, Zarya is referred to by its technical names, the Functional Energy Block, or the Functional Cargo Block (Russian acronym FGB).

Zarya shares its heritage with the Kvant 2, Kristall, Spektr and Priroda modules of the Mir space station, which were in turn developed from the TKS/VA vehicles designed in the 1960s by Vladimir Chelomei's design bureau for the Soviet Union's military space station programme. The modular design of the TKS meant that of the 30 basic systems in Zarya, 22 were developments of those in Kristall and the others were from Kvant 2. Like Kristall, Zarya has a spherical docking adapter at one end. On Mir this is called the 'node', but on the ISS that name had already been assigned to NASA's interconnecting modules. Zarya's spherical docking adapter has three docking ports. The one on the forward end was to link Zarya permanently to the American Unity node. The zenith and nadir parts have standard Russian drogues for Soyuz-TM and Progress-M spacecraft. The docking probe at the far end of the module was to link Zarya permanently to Zvezda, the Russian-built Service Module.

Once in orbit, the Zarya module activated its systems and deployed its antennae and photovoltaic arrays. The body of the module is 12.5 metres long and 4.1 metres in diameter at its widest point. Each of its two photovoltaic arrays is 10.6 metres long and 3.3 metres wide. They deliver power to six nickel–cadmium batteries, which in their turn supply an average of 3 kilowatts of power on demand to Zarya's systems. The module's photovoltaic arrays turn so that they remain pointing at the Sun, while Zarya remains in a stable attitude as it orbits the Earth. Its 16 tanks have a total capacity of 5,600 kilograms of propellant for the Attitude Control System's 24 large and 12 small thrusters. A pair of large rocket engines enabled the vehicle to climb to its operating orbit and to reboost the ISS's orbit during its early phase of assembly. Zarya's propellant tanks are routinely replenished by Progress-M tankers docked at Zvezda's far end. Although it is intended to have an orbital lifetime of 15 years, its functionality will be progressively superseded as the station expands. Its key role was in the early phase of assembly. Zarya provided orientation control, as well as communications and power for Unity prior to the Zvezda's launch, at which time it was largely powered down and many of its functions were superseded by Zvezda. Its photovoltaic arrays continued to supply power to the American elements of the station until flight 4A, however. It will eventually be reduced to a storage role, with particular attention on the capacity of its external propellant tanks.

In 2000, Boeing and Khrunichev suggested that they cooperate to develop the Zarya backup vehicle (FGB-2) into a Commercial Space Module (CSM), a rival to the Enterprise Module. That rivalry came to a head when it was realised that both projects were planning to utilise Zarya's nadir docking system.

Service Module 'Zvezda'
The Zvezda (Russian for Star) Service Module is similar to Mir's 'base block' and is the element which allowed the first Expedition crews to occupy the ISS on a permanent basis. Upon reaching orbit, it served as a passive target as the Zarya–Unity combination made a rendezvous and docked. It is 13.10 metres long and 4.15 metres in diameter at its widest point. A small spherical Transfer Compartment with three docking systems forms the forward end. The ram docking system will permanently link it to Zarya's wake unit. The two other docking units are mounted at the zenith and nadir locations. The zenith is assigned to the Science Power Platform and the nadir to the Universal Docking Module, but this may change. The Transfer Compartment can be isolated and used as an airlock for extravehicular activities using the Russian Orlan pressure suit. Like Mir, Zvezda was designed to house a crew of three on an ongoing basis and up to six people for short periods. The Work Compartment that forms the main body of the vehicle contains the living accommodation, individual sleeping quarters for two people, a toilet and hygiene facilities, and a galley containing a fridge-freezer and a folding table for preparing meals. The gymnasium includes a NASA-provided treadmill and a stationary bicycle. Waste-water and air condensation are recycled for use in the oxygen-generating devices on the module. Although recycled water will be used for personal hygiene, it is not considered to be potable. Zvezda's systems provide data, voice and video communications with the control centres in Moscow and Houston, initially via direct overflight of ground stations but later via NASA's geostationary relay satellites. There is a small cylindrical Transfer Chamber at the aft end surrounded by an annular unpressurised Assembly Compartment which contains the attitude

control and orbital reboost propulsion systems. The wake docking drogue at the far end of the Transfer Chamber allows Soyuz and Progress spacecraft to dock and the tunnel allows crew members to transfer from one vehicle to another. Plumbing incorporated into this docking system permits Progress tankers to transfer propellants from their own storage tanks to tanks on the ISS. Zvezda has 14 portholes, three in the forward Transfer Compartment, one in the Working Compartment, one in each individual crew compartment, and others placed for terrestrial, space and intermodule viewing. Zvezda's computers will control the ISS through to the completion of the Baseline Configuration. The computers were developed by ESA in return for Russian docking hardware for use on the Ariane Transfer Vehicle. Zvezda will serve as the ISS's main living quarters until the American Habitation Module (or some equivalent) is added, at which time Zvezda will serve as the core of the Russian portion of the station.

Progress-M1

The Progress-M1 logistical supply craft is a development of the Soyuz crew ferry, so it uses the same type of launch vehicle. Progress will perform four primary tasks in the context of the ISS: (1) orbital reboost, (2) replenishment of propellant for the Zvezda's attitude control system, (3) delivery of pressurised cargo and (4) removal of trash. They will dock automatically at Zvezda's wake port or Zarya's nadir port. In case of failure of the Kurs automatic system, the TORU remote-control docking system can be used by a crew member on the ISS to steer the ferry in. Propellant not needed for reboost operations can be transferred to tanks in Zvezda (or, via Zvezda to larger tanks on Zarya) for use by attitude control thrusters. Pressurised cargoes include oxygen, nitrogen, water, food, clothing and crew personal effects. Once its cargo has been removed, the vehicle is filled with rubbish. Following undocking, it is de-orbited and destroyed during re-entry over the Pacific. The Progress propulsion module was also used to deliver the 'Pirs' Docking Compartment to the ISS, then the propulsion module undocked, withdrew, and de-orbited itself.

Soyuz-TM Crew Transport

The Soyuz spacecraft has been the workhorse of Soviet-Russian human spaceflight since 1967. It was originally developed in three forms – Earth orbital, lunar orbital (Zond), and as the lunar orbital element of the human lunar landing programme (LOK). Upon the cancellation of the lunar effort, the Earth orbital version was simplified to serve as a '2-day ferry' to deliver crews to space stations and return them to Earth at the end of their flights. It has since evolved through several versions, expanding its capabilities. The Soyuz-TM, introduced to service Mir, will serve a similar role for the ISS. It will also serve as a Crew Return Vehicle (CRV) early in the occupation of the station. In this role, there will always be sufficient Soyuz spacecraft in place to evacuate the crew in an emergency. The limited duration of its propulsion system means that these vehicles will have to be replaced every six months. Short-duration crews will arrive in a 'fresh' spacecraft and return in the expiring vehicle, leaving the replacement to continue the CRV function.

The Soyuz spacecraft comprises three modules: (1) The Orbital Compartment, at the front, provided a pressurised 'living room' for the crew on early solo flights. In the LOK design it also served as an airlock for one cosmonaut to transfer externally to and from the lunar lander. The docking system is mounted at the nose and the probe assembly can be swung aside after docking to provide direct access to a space station. (2) The Re-entry Compartment behind it has the main flight controls and couches for a crew of three. It is the only part of the spacecraft capable of returning to Earth. Its bell-shape and offset centre of mass means that it can generate a small amount of aerodynamic lift during re-entry, to steer towards a specific landing point. Although it is capable of a water recovery, the preferred way is a dry landing. Final braking is provided by drogue and main parachute landing system. Immediately prior to touchdown, the aft heat shield is jettisoned and a series of retrograde rockets fire to cushion the impact. (3) The Instrument Compartment, at the rear, has the orbital manoeuvring and attitude control systems and the photovoltaic arrays which power the vehicle. Both the Instrument and Orbital Compartments are jettisoned following retrofire, leaving the Re-entry Compartment on track for re-entry.

When NASA astronauts proved either too tall or too short for the Soyuz-TM, the agency asked Russia to modify the Re-entry Compartment to accommodate such astronauts, so that they would not have to be excluded them from tours on the ISS. However, work on the Soyuz-TMA (A for anthropometric) was halted when the Russian government failed to fund its production and NASA, arguing that providing Soyuz spacecraft was part of the Russian contribution to the ISS, refused to allocate additional funding.

Russian Science Power Platform
The photovoltaic arrays of the Science Power Platform (Russian acronym NEP) was originally designed for Mir 2, the Russian equivalent of Space Station Freedom, but this was cancelled following the collapse of the Soviet Union. When Russia was invited to participate in the ISS, it offered the Mir 2 base block as the ISS's Service Module, and reintroduced the NEP to power the Russian portion of the station. The NEP has three major components. A 5.9-metre-tall, 2.2-metre-diameter pressurised module carries all the NEP's systems that are not capable of operating in the vacuum of space. It also contains a number of controls that cosmonauts can access internally from the ISS. The lower end of the pressurised module would be docked to Zvezda's zenith docking system. Spacewalkers would deploy a two-segment telescopic truss on top of the module. Two pairs of photovoltaic arrays will be fitted to the top of the truss to augment the power supply of the Russian portion of the ISS. The initial plan to launch the NEP using a Ukranian-built Zenit launch vehicle was cancelled when relations with the Ukraine deteriorated. The next plan was to deliver the NEP by Shuttle. Its eight photovoltaic arrays would be ferried up in two sets of four. The first set would accompany the pressurised module and truss; the second set would be delivered by a later Shuttle. The NEP was also to incorporate a thermal control system, a roll control system, and the power system for the European Robotic Arm, and host

standard mounting points for external experiment packages, which would be emplaced and serviced by this manipulator arm.

When the Russian Space Agency withdrew funding for the NEP in 1998, RKK Energiya constructed two prototypes and some flight hardware at its own expense, but it had to cease work in early 2000 so that funding could be redirected to other aspects of the Russian participation in the ISS programme. Despite the fact that the NEP was assigned to SSAF-9A.1 – scheduled to be launched in October 2002 – no new funding was forthcoming. In 2001, Energiya began planning towards a scaled down version of the NEP.

Russian Universal Docking Module

The Universal Docking Module (UDM) is a Russian element of the ISS with five probe-and-drogue docking ports which will serve a similar interconnecting role to the American nodes. After being released by its Proton launch vehicle, the UDM will complete an autonomous rendezvous and dock with Zvezda's nadir position. Once in place, the photovoltaic arrays which powered it during independent flight will be retracted to ensure that they do not obstruct the docking ports. As a conversion of a Zarya backup, it is a joint development by Khrunichev and RKK Energiya with the former supplying the docking ports, interface hardware and life support, and the latter undertaking the vehicle's integration. However, at the completion of Phase 2 of the ISS's assembly in January 2001, development had yet to start.

Russian Docking Compartment Module 'Pirs'

The Russian Docking Compartment Module 'Pirs' (Russian for Pier) built by RKK Energiya was delivered to the ISS by the propulsion module of a Progress-M cargo vehicle and left on Zvezda's nadir docking port. It will facilitate extravehicular activities employing Russian Orlan EVA suits during the station's assembly, and a third docking point for Soyuz and Progress vehicles (via the drogue on its exposed end, by which it had been mated with its carrier). It ferried up consumables and a second Strela telescopic boom, all of which were transferred to Zvezda following docking. It will be relocated during the build up of the Russian part of the station.

Russian Docking-Storage Module

The Russian Docking-Storage Module (DSM) was to use one of the backup hulls from Zarya's development. It was to deliver 5.6 tonnes of propellant, which would be pumped through the pipes in Zarya's nadir port. It would also accommodate six gyrodines for attitude control, and would serve as a storage facility in the Russian sector of the ISS. In 2000, however, it was superseded by the Enterprise Module.

Enterprise Module

On 10 December 1999, Spacehab Inc. announced that it would cooperate with RKK Energiya to supply the Enterprise pressurised module to expand the Russian part of the ISS. It would be the first orbital television station offering news and educational broadcasting from the ISS direct to the terrestrial networks. It would also serve as a laboratory for microgravity biotechnology and advanced materials processing experiments. Spacehab's partners supported the joint venture, which they described as "the first commercial real estate in orbit". The principals were "committed to making space commerce a reality". Enterprise was priced at $100 million, with Spacehab raising its $50 million share through private investments and a possible flotation of additional shares in Spacehab Inc. The original plan was to develop it as a 13-tonne payload for the Zenit-class launch vehicle which Kazakhstan intended to manufacture, but this plan was dropped when relations between Russia and Kazakhstan deteriorated. The revised plan was to revamp the Docking-Stowage Module that Energiya was to build and launch in 2003. On 16 February 2001, Enterprise officially replaced the DSM, which would enable it to become a 20-tonne payload for the Proton launch vehicle with accommodation for a crew of three, up to five windows and mounts for external payloads. Spacehab–Energiya offered the new module for rent to the ISS partners, with the fee including the Soyuz-TM which would deliver payload specialists and dock on the module's docking adapter. If the plan is fully funded, this package will provide the ISS with a six-person capability some two years earlier than would have been feasible using the American Habitation Module and the X-38 CRV. In March 2001 Rosavaikosmos asked NASA to consider this possibility.

Russian Research Modules
In the original concept, Russia was to supply three Spektr-style Research Modules. The first (also called the Russian Universal Module) will be a technology laboratory for both internal and external experiments. As a biotechnology facility, the second will have a controlled atmosphere for biological experiments. The Universal Russian Working Places for external payloads will be augmented by extravehicular activity as necessary. In 1998, budgetary constraints prompted the cancellation of the third, which was to have been devoted to Earth-resources studies. When the Russian Space Agency ceased using the Zenit launch vehicle manufactured in the Ukraine, it ordered the modules to be scaled down to enable them to be launched on Soyuz rockets, in which case they would be reduced to payloads comparable to the 'Pirs' Docking Compartment, which was delivered by a Progress-M propulsion module. However, when the Russian Space Agency and the Ukraine started to consider jointly developing a module, this opened the prospect of scaling this module to be launched on a Zenit after all. These modules, however, are still in the definition stage.

Multi-Purpose Logistics Modules
The Italian Space Agency has constructed three pressurised Multi-Purpose Logistics Modules (MPLMs) to ferry cargo to the ISS. The MPLM is carried into orbit as an isolated payload in the bay of a Shuttle, and transferred by RMS to the ISS and mounted on a CBM on an American node. While docked to the ISS, the MPLM's equipment racks can be removed and floated into the station. The MPLM includes systems for life support, fire detection and suppression, power distribution and computer functions. Once the MPLM is empty, equipment racks that are no longer required can be placed within the module for return to Earth, and the module

returned to the Shuttle's payload bay.

The acronym originally stood for Mini-Pressurised Logistics Module, the term 'mini' signifying that it was to be smaller than the 'full sized' logistics module that NASA had been obliged to abandon in the face of financial constraints. The MPLMs were built by Alenia Aerospazia in Turin, Italy, and presented to NASA in exchange for Italian access to American experiment time on the ISS. Each cylindrical module is 6.4 metres long, 4.5 metres in diameter, has a dry mass of 4,846 kilograms, and can accommodate 9,707 kilograms of cargo in 16 racks, five of which can be provided with continuous power, data and fluids for 'active payloads' and support a fridge-freezer. The three modules are named 'Leonardo', after inventor Leonardo da Vinci, 'Raffaello', after artist Raffaello Sanzio, and 'Donnatello', after sculptor Donato di Niccolo di Bardi.

In April 2001, NASA started negotiation with the Italian Space Agency for the construction of a four-berth habitat to replace the American Habitation module that had been cancelled earlier in the year due to financial considerations. If built, this will be based upon the MPLM's pressurised shell.

Japanese Experiment Module 'Kibo'

The Japanese Experiment Module (JEM) will be supplied by the Japanese National Space Development Agency (NASDA) with 10 racks for microgravity materials processing and life-sciences experiments. In return for a Shuttle delivering the JEM, the racks will be shared equally by Japanese and American experiments. The module has been named 'Kibo', the Japanese word for 'Hope'. It will mounted on Node 2, opposite ESA's Columbus. NASDA has constructed its own ground facility for Japanese astronaut operations and to process data directly from the module's experiments. In addition to the Pressurised Module (JEM-PM), there is an Exposed Facility (JEM-EF) for external packages. Access to the JEM-EF will be via a small airlock in the far end-cone of the module and packages will be emplaced, operated and retrieved by the twin Remote Manipulator Systems (JEM-RMS) controlled by astronauts observing through portholes in the module's end-cone each side of the airlock. These arms were supplied by the Canadian company that constructed the Shuttle RMS and SSRMS. The module has a berthing collar on its 'roof' for the Experiment Logistics Module (JEM-ELM). This will be delivered by Japan's H-2A launch vehicle. After making its rendezvous, it will be grasped by the SSRMS and transferred to the JEM's berthing mechanism.

Ariane Transfer Vehicle

The European Space Agency's Ariane Transfer Vehicle (originally referred to as the Autonomous Transfer Vehicle – ATV) is effectively an orbital stage for the Ariane V launch vehicle. In the ISS-support role it will dock at Zvezda's wake docking port to deliver cargo, replenish attitude control propellant and perform orbital reboost. Its modular design means that individual vehicles will be able to be optimised for pressurised and unpressurised cargoes as required. Alenia Spazio is responsible for developing the ATV's propellant tanks and pressurised cargo container. Resupply missions are expected to be made at 15-month intervals.

Columbus Orbital Facility
The European Space Agency's 6.7-metre-long, 4.5-metre-diameter Columbus Orbital Facility will have its own environmental and life support systems. It will undertake technology experiments and microgravity research in biological, life and physical sciences. In return for half of its 10 experiment racks being assigned to it, NASA will be delivered the module by Shuttle. It will be mounted on the starboard side of Node 2. Attachments on the two end-cones will enable external experiments to be aimed either towards the nadir or towards the zenith. It will be launched with its initial European payloads in-situ, the NASA experiments will be installed once the module is in place.

European Robotic Arm
The European Robotic Arm (ERA) is a European Space Agency system which, when launched, will be attached to the Russian Science Power Platform. The 10-metre-long arm will have twin end-effectors so that it can 'walk' end-over-end from one point to another. Its first task will be to position the photovoltaic arrays, thermal radiators and thruster systems on the NEP; thereafter, it will support cosmonauts working inside and outside the Russian part of the station. Its video system will enable it to make inspections of the station's external state without requiring an external human presence. It will have its own base platform and tool kit to enable it to support spacewalkers.

Index